ELECTRON AND COUPLED ENERGY
TRANSFER IN BIOLOGICAL SYSTEMS

Volume 1 | Part A

ELECTRON AND COUPLED ENERGY TRANSFER IN BIOLOGICAL SYSTEMS

Volume 1 / Part A

edited by TSOO E. KING

PROFESSOR AND CHAIRMAN
DEPARTMENT OF CHEMISTRY
STATE UNIVERSITY OF NEW YORK AT ALBANY
ALBANY, NEW YORK, U.S.A.

and MARTIN KLINGENBERG

DIRECTOR AND PROFESSOR
INSTITUTE OF PHYSIOLOGICAL CHEMISTRY AND PHYSICAL BIOCHEMISTRY
UNIVERSITY OF MUNICH
MUNICH, GERMANY

1971

MARCEL DEKKER, INC., New York

CHEMISTRY

MARCEL DEKKER, INC.
95 Madison Avenue, New York, New York 10016

LIBRARY OF CONGRESS CATALOG CARD NUMBER: 73-169184

ISBN : 0-8247-1375-3

PRINTED IN THE UNITED STATES OF AMERICA

PREFACE

About seven years ago the publisher approached us to write a treatise on electron transport and oxidative phosphorylation. We were immediately aware of the virtual impossibility for one or two people to accomplish such a task. Ironically, just because of this we felt the situation called for a comprehensive treatise. It was very difficult for the nonspecialist and, indeed, even for the specialist to follow all the concepts and important developments in the study of electron and coupled energy transfer in biological systems. Whereas in other areas of biochemistry, such as molecular genetics, enzymology, and metabolism, competent monographs are available; no comparable treatment exists so far for this field.

The present treatise was thus conceived with these considerations and should fill an important gap in biochemical literature. It is intended to present authoritative chapters which should serve both the specialist and nonspecialist who need advanced reviews. This book also invites promising graduate students and the fresh and imaginative mind to specialize in this fascinating field.

The authors were requested not to attempt to write in a stereotyped textbook style. Consequently, the many unsolved problems, contradictions, and speculations have not been avoided. The reader may also gain some individual points of view from the authors who are among the most active scientists in the area; these views usually cannot be found in current publications of conventional journals.

The organization of this treatise was not without difficulties. The problem was not so much to gain acceptance from the authors to write the chapters, but rather to obtain their finished manuscripts in time. For this reason, the originally planned organization of the two volumes (see the Appendix in Part B) could not be retained. It was not feasible for all those authors who had submitted their chapters earlier to rewrite them without repeating the cycle again. Therefore, various ways have been used to update the presentations. However, surprisingly enough the general validity of the material presented even without updating remains little affected. The prediction made in 1965 by a prominent scholar in the field that major problems in oxidative phosphorylation would be solved by 1969 was unfortunately too optimistic. At any rate, the second volume will contain all other chapters of the original plan and hopefully will have fewer problems than this present volume does.

Finally, we sincerely hope that the present treatise will serve its original purpose despite the usual shortcomings encountered in collecting contributions from active experimentalists. We would like to express our thanks to the authors, especially to those who sent in their contributions early and were ready to change their manuscripts again due to the long delay. We apologize for our faults to the authors and to those readers who have expected and inquired about this treatise for some time. Special acknowledgment must be given to the publisher for the patient cooperation and the skillful processing of the book in all stages.

Albany and Munich   Tsoo E. King
May, 1971   Martin Klingenberg

CONTRIBUTORS TO PART A

J. R. CRONIN, Department of Chemistry, Arizona State University, Tempe, Arizona

HENK DE KLERK,[1] University of California at San Diego, La Jolla, California, and Laboratoire de Photosynthèse, Centre National de la Recherche Scientifique, Gif-sur-Yvette, France

P. LESLIE DUTTON, Department of Biophysics and Physical Biochemistry, Johnson Research Foundation, University of Pennsylvania, Philadelphia, Pennsylvania

KARL M. DUS,* University of California at San Diego, La Jolla, California, and Laboratoire de Photosynthèse, Centre National de la Recherche Scientifique, Gif-sur-Yvette, France

TORGEIR FLATMARK,† University of California at San Diego, La Jolla, California, and Laboratoire de Photosynthèse, Centre National de la Recherche Scientifique, Gif-sur-Yvette, France

W. R. FRISELL, Department of Biochemistry, College of Medicine and Dentistry of New Jersey, Newark, New Jersey

PIH-KUEI C. HUANG, Institute of Biological and Medical Sciences, Retina Foundation, Boston, Massachusetts

F. M. HUENNEKENS, Department of Biochemistry, Scripps Clinic and Research Foundation, La Jolla, California

MARTIN D. KAMEN, University of California at San Diego, La Jolla, California, and Laboratoire de Photosynthèse, Centre National de la Recherche Scientifique, Gif-sur-Yvette, France

Present address:

[1] Shell Oil Company, Amsterdam, The Netherlands
* Department of Chemistry, University of Illinois, Urbana, Illinois
† Department of Biochemistry, University of Bergen, Bergen, Norway

RUFUS LUMRY, Laboratory for Biophysical Chemistry, Department of Chemistry, University of Minnesota, Minneapolis, Minnesota

B. MACKLER, Department of Pediatrics, University of Washington Medical School, Seattle, Washington

RICHARD L. PHARO, Institute of Biological and Medical Sciences, Retina Foundation, Boston, Massachusetts

D. R. SANADI, Institute of Biological and Medical Sciences, Retina Foundation, Boston, Massachusetts

ICHIRO SEKUZU, Department of Biology, Faculty of Science, Osaka University, Toyonaka, Osaka, Japan

JUI H. WANG, Kline Chemistry Laboratory, Yale University, New Haven, Connecticut

DAVID F. WILSON, Department of Biophysics and Physical Biochemistry, Johnson Research Foundation, University of Pennsylvania, Philadelphia, Pennsylvania

CONTENTS OF VOLUME 1, PART B

CONTENTS

End of Volume 1—Part A.

CHAPTER **I**

FUNDAMENTAL PROBLEMS IN THE PHYSICAL CHEMISTRY OF PROTEIN BEHAVIOR

Rufus Lumry

LABORATORY FOR BIOPHYSICAL CHEMISTRY
DEPARTMENT OF CHEMISTRY
UNIVERSITY OF MINNESOTA
MINNEAPOLIS, MINNESOTA

1

I. Introduction

The determination of structural descriptions of proteins by the X-ray-diffraction method has been the highlight of protein chemistry since the 6-Å structure of met-myoglobin was reported by Kendrew *et al.* in 1958 (*1*). The remarkable qualitative or semiquantitative models of proteins provided since that time have proved invaluable to the developing understanding of the mechanisms of the physiological functioning of proteins. However, they have not proved to be sufficient for this purpose, nor is it reasonable that they should. As is always the case, real understanding of chemical processes is a quantitative matter that rests on reasonably precise determination of entropy, volume, and enthalpy, or equivalent sets of potential-energy parameters at the several important intermediate points in a process. Thus for enzymic catalysis one must start with qualitative descriptions of molecular events and convert these into thermodynamic or equivalent quantitative descriptions of reactants, metastable intermediates, the important activated complexes, and the products. The molecular descriptions in themselves are a sufficient basis for the quantitative problem only if (a) the geometric parameters can be determined at the particular precision level required by the magnitude of the thermodynamic changes in the process under study, and (b) the geometric parameters can be related to the thermodynamic parameters through the use of small-molecule models for which these quantitative relationships have been established. To be more specific with respect to item (a), on the basis of the information now available the contributions to the thermodynamic description of specific protein functions from geometrical changes of the protein conformation system and its associated solvent are in the range of 5 to 25 kcal of enthalpy per mole of protein and from 20 to 75 entropy units per mole of protein. These upper bounds may be found to be overly conservative; but it is clear that the contributions are not large, although they are essential to function and indeed represent one of the unique features of protein reactions. Since the variations of the thermodynamic parameters of any molecule with variation in geometry can extend over a wide range of values determined by the nature of the forces among particular groups of atoms—van der Waal's attraction and repulsion, hydrogen bonding, simple Coulombic interactions, and interactions due to orbital overlap—there is no simple relationship between coordinate change and thermodynamic change. In fact, large atom movements in protein reactions must usually represent very weak dependencies of thermodynamic quantities on changes in these coordinates. Since the total thermodynamic changes in conformational mechanisms are not large, and the largest of these usually may be expected to be associated with small coordinate changes, it appears probable that many important coordinate changes will not be detectable in X-ray-diffraction experiments on proteins unless a very significant increase in resolution becomes possible.* Alternatively, as it now appears possible to demonstrate with myoglobin, quite large conformational rearrangements can occur in proteins without large thermodynamic changes.

As regards item (b) in the first paragraph, *a priori* applications to proteins of potential-energy functions of coordinates from small molecules is hazardous at best and is probably computationally impossible at any realistically useful level of attack. Even given adequately precise coordinate values this type of approach applied to conformational thermodynamics is quite impractical. Its alternative is to use a strictly phenomenological approach in the beginning and to supplement this in a progressively increasing way by the careful transfer of information from small molecules. We have called the

---

* It is not always clear to the noncrystallographer how the precision limits given in X-ray studies of protein structure can be converted into errors in distances between particular pairs of atoms. Of even more concern to the noncrystallographer protein chemist is the present real or apparent uncertainty about the quantitative reliability of statements made on the basis of Fourier-difference maps, whether in two or three dimensions, when based on an assumed set of phases. An important recent publication by Hoard (*138*) considers the latter problem in detail. According to Hoard the casual acceptance of such information as a support for chemical information may be unwise, and its use as a justification for the rejection of information from other sources is often unwarranted. Until such time as Hoard's criticisms are shown to be unjustified, it unfortunately appears to be necessary to severely limit our dependence on certain types of X-ray information about proteins.

latter method the *approach with models* (2) to contrast it with the *a priori approach from models*, and we will describe primitive approaches using this method in a subsequent section.

II. Magnitudes of Thermodynamic Change
in Specific Protein Reactions

Despite the low resolution enforced on X-ray investigations of hemoglobin by the poor quality of hemoglobin crystals, Muirhead and co-workers (3) have been able to demonstrate major rearrangements in the orientation of the subunits of hemoglobin in relation to each other when oxygen is bound. The atom rearrangements within the subunits upon which the reorientations depend, at least in part, have been undetectable in the X-ray work, but they now appear to be detected, though not described, in "spin-labeling" studies of McConnell and co-workers (4) and by recent immunological investigations (5). When hemoglobin binds oxygen, the thermodynamic changes that take place in the (protein + water) conformational subsystem, as distinguished from the chemical subsystem of ligands plus heme, have not been determined; but a study of the limited data available (6, 9) suggests that they cannot be greater than 60 entropy units and 18 kcal for all four oxygen-binding steps. These changes are relatively small when compared with the values that can accompany drastic changes in protein folding (2). They are nevertheless a measure of the thermodynamic characteristics of the important conformational processes responsible for the linkage among the chemical subsystems on a given molecule and the other unusual features of hemoglobin function.

Significant differences in the internal folding of myoglobin have been detected on comparison of X-ray data from metmyoglobin crystallized at slightly acid pH values with data obtained at basic pH values (7, 8). Presumably these differences in conformation are controlled by proton dissociation from the water molecule bound in the sixth or "distal" position of heme iron. Similar effects are produced in metmyoglobin when cyanide ion replaces this water molecule, but are not produced in the corresponding reaction with fluoride ion (7, 8) (A).* These three processes in solution can be compared using data collected by George (9) and shown in Table 1. There is clearly no pattern in the differences that distinguish hydroxyl and cyanide compounds from the fluoride compound and thus nothing that reveals the presence of the conformational rearrangement detected in X-ray studies. The standard

---

* The appearance of the symbol "(A)" in the text indicates that the material treated at that point requires revision on the basis of recent developments and is considered in the Addendum.

TABLE 1

DISPLACEMENT OF WATER BY IONIC LIGANDS IN METMYOGLOBIN AT 35°C

	Ligand		
	F$^-$	OH$^-$	CN$^-$
$\Delta F°$ (kcal/mole)	−2.0	−6.85	−11.5
$\Delta H°$ (kcal/mole)	−1.5	−7.65	−18.5
$\Delta S°$ (e.u./mole)	+1.8	−2.6	−24

thermodynamic changes for the reaction

$$(\text{Fe}^{III}\cdot\text{L}^-) + (\text{Mb}^{III}\cdot\text{H}_2\text{O}) \rightleftarrows (\text{Mb}^{III}\cdot\text{L}^-) + (\text{Fe}^{III}\cdot\text{H}_2\text{O}) \qquad (1)$$

in water, with $\text{L}^- = \text{F}^-$, OH^-, and CN^-, are given in Table 2. The large entropy changes do not contain significant contributions from the conformational rearrangements, since the entropy changes for the reaction

$$\text{H}_2\text{O} + (\text{Fe}^{III}\cdot\text{L}^-) \rightarrow (\text{Fe}^{III}\cdot\text{H}_2\text{O}) + \text{L}^-(\text{Aq}) \qquad (2)$$

are −49 e.u./mole for $\text{L}^- = \text{F}^-$ and −50 e.u./mole for $\text{L}^- = \text{OH}^-$ when the solvent is water (9). We can conclude that the entropy loss due to increased electrostriction about the ferric ion is the major part of the entropy change. A small negative entropy change in the protein balancing the entropy increase due to neutralization of the single positive charge on heme iron by the single negative charge on the ligand may be present, but it cannot be related to the conformational rearrangements under discussion, since these are produced by hydroxyl-ion binding but not by fluoride-ion binding. The situation is somewhat more complicated when $\text{L} = \text{CN}^-$ in Eq. (2), since $\Delta S° = -16$ e.u./mole for $\text{Fe}^{III}\cdot\text{CN}^-$ dissociation in water. However, even here there is no clear indication of a conformational rearrangement on binding the ligand. This situation provides a good object lesson, since the conformational rearrangements are associated with and perhaps triggered by

TABLE 2

THERMODYNAMIC CHANGES FOR REACTION (1)[a]

	Ligand		
	F$^-$	OH$^-$	CN$^-$
$\Delta F°$ (kcal/mole)	+4.9	+9.2	−2.0
$\Delta H°$ (kcal/mole)	−9.0	−6.4	−14.6
$\Delta S°$ (e.u./mole)	−51.0	−53.0	−40.0

[a] From George (9).

a change of the spin state of heme iron from the high-spin state found with water and fluoride ion to the low-spin state found in hydroxyl- and cyanide-ion binding. The binding of oxygen or carbon monoxide to hemoglobin in the Fe^{II} state is also accompanied by this high-spin to low-spin change and may produce similar conformational readjustments in the subunits of hemoglobin that are also not readily detected by thermodynamic quantities. Metmyoglobin as a model for hemoglobin then suggests conformational rearrangement with little if any abnormality in the enthalpy and entropy changes. However, myoglobin (Fe^{II}) provides quite a different model for hemoglobin. Keyes (10, 11) has worked out a good part of the thermo-dynamic picture for this form of myoglobin, and his results are worth reviewing.

Keyes's initial goal was to explain the nonlinearity of the van't Hoff plots for ligand binding by myoglobin. If real, such curvature indicates a large heat capacity and usually a large change in the folding of the protein during the reaction. If false, such curvature indicates that the system is not adequately controlled to provide reliable data. He found that the curvature was an artifact due to partial contamination of the protein by a very strongly bound small molecule of an as yet unknown species. His thermodynamic changes for oxygen and carbon-monoxide binding to both clean and fully contaminated whale-muscle myoglobin are given in Table 3. In this table are also given Theorell's (12) corresponding results for horse-heart myo-globin. Although Keyes found that the contaminant can be removed without pretreatment by careful gel filtration with a long column, it appears to be relatively easily removed if reducing agents such as ascorbic acid are added before chromatography. Theorell, apparently through an abundance of scientific intuition, treated the protein with dithionite followed by electro-dialysis so that his protein was also contaminant free. The agreement among the thermodynamic parameters is very good, as can be seen in the table. Apparently there is little quantitative difference between O_2 and CO binding to horse-heart myoglobin and to sperm-whale myoglobin.

The entropy changes in Table 3 have been corrected for the entropy of mixing and for the nonideality of the "hydrophobic" interaction of oxygen and carbon monoxide with water. Many ligand-exchange reactions have small entropy changes unless there is a change in charge. There is no change in charge in the reactions of myoglobin with oxygen and carbon monoxide, so that it is highly probable that the unitary entropy changes shown in Table 3 must be attributed to changes in the protein. In contrast to the X-ray differences between acid and alkaline myoglobin, no X-ray differences between deoxy Fe^{II} myoglobin and its oxygenated form have thus far been established (7), but a full-scale X-ray study has not yet been reported. Thus with Fe^{III} myoglobin the X-ray studies demonstrate conformational

TABLE 3

THERMODYNAMIC CHANGES IN LIGAND BINDING TO MYOGLOBIN[a,b]

$X(\text{Alcohol}) + Mb(H_2O) \rightarrow MbX(H_2O)$

Mb preparation	X	$\overline{\Delta F_u^\circ}$ [c] (kcal/mole)	$\overline{\Delta H^\circ}$ (kcal/mole)	$\overline{\Delta S_u^\circ}$ [c] (e.u./mole)	Reference
Horse-heart—clean	O_2	−8.0	−18	−31	Theorell (12)
Horse-heart—clean	CO	−9.6	−22	−41	Theorell (12)
Whale—clean	O_2	−8.0	−18 ± 1	−33 ± 3	Keyes (11)
Whale—clean	CO	−9.8	−21 ± 1	−39 ± 3	Keyes (11)
Whale—fully contaminated	CO	−10.6	−28 ± 1	−58 ± 3	Keyes (11)
Whale—clean	O_2	—	−19 ± 2	—	Keyes et al.[b]

[a] Corrected for hydrophobic interactions of oxygen or carbon monoxide with water; 27°C., pH = 8.0.

[b] Values for the standard enthalpy and entropy of oxygen binding to myoglobin reported by Rossi-Fanelli and Antonini and their co-workers (237) give enthalpy changes about half of those listed in this table. Although there is a number of experimental differences among the methods used by the several groups and no established explanation for the discrepancies, the results obtained with high-precision calorimeter by Keyes and Rajender [last line of table (236)] appear to confirm the van't Hoff results of Theorell and Keyes.

[c] In addition to a correction for the hydrophobic interactions of O_2 and CO with water, the quantities bearing subscript u have been corrected for the ideal entropy of mixing. Even if, as is shown in X-ray structures for Fe^{II} myoglobin, there is no sixth-position ligand before the addition of X, this correction for "cratic" entropy is not significantly different. Of course, the heat of binding would be expected to be somewhat larger for this reason. Without "hydrophobic" or "cratic" corrections the entropy changes are more negative by about 20 e.u. than the tabulated values. The standard state for ligand for this table is unit mole fraction in ethanol.

rearrangements not anticipated in thermodynamic studies, but with Fe^{II} myoglobin the thermodynamic quantities suggest conformational rearrangements that have not yet been detected in X-ray-diffraction studies (cf. Section VI, D) (A).

The corrected entropy of binding carbon monoxide to contaminated myoglobin (Table 3) is still larger in absolute magnitude than that for clean myoglobin, perhaps indicative of the conformational rearrangements associated with linkage of the contaminant site to the CO site. The net interaction free energy at 27° is not large, although the interaction enthalpy and entropy are quite respectable. Thus for the process MbK + MbCO ⇌ MbKCO + Mb, in which K is the contaminant, $\Delta F^\circ = -0.8$ kcal/mole at 27°C with $\Delta H^\circ = -7.0$ kcal/mole and $\Delta S^\circ = -21$ e.u./mole (11). Of more interest is the fact that two or more xenon-binding sites become linked to the CO-binding site when the contaminant is bound (11). A primitive linkage map is shown in Fig. 1. Again the linkage-interaction free energy is not large, since it is limited by the free energy of the linkage between contaminant site and CO site; but there are two or more xenon-binding sites, despite the X-ray studies of Schoenborn and others (*14*), who find only a single xenon atom that is bound immediately behind the heme plane on its proximal side.* Keyes also finds a separate linkage between the CO-binding site and the zinc-ion binding site (*11*). This linkage may not be mediated through the contaminated site and is so shown in Fig. 1. Keyes' quantitative studies of

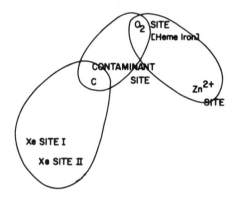

Fig. 1.   Primitive linkage map for a system of binding sites in whale myoglobin [Fe(II) state].   From Keyes (*11*).   See Wyman (*173*) for discussion of linkage maps.   In this map the xenon-binding sites are shown coupled to the oxygen-binding site on heme iron through the contaminant-binding site.   The zinc-ion binding site may or may not be directly coupled to heme iron as shown.   Statistical analysis of the xenon-binding data does not make a significant distinction between two and three binding sites for xenon.

this linkage and of zinc-ion binding indicate that the zinc-ion binding site is probably the same one characterized in X-ray studies of Fe^{III} myoglobin. The latter site lies on the opposite side of the protein from the heme group (*15*).

The linkage system of Fe^{II} myoglobin thus includes the CO site, contaminant site, zinc-ion site, and two or more xenon-binding sites. As such it approaches the complexity of the linkage system of hemoglobin, despite the fact that myoglobin is a single-subunit protein. The contaminant site in myoglobin roughly corresponds to the phosphoacid-binding sites found in

*Schoenborn has recently detected a second xenon-binding site in his X-ray diffraction studies of myoglobin. *Note added in proof.* Wishnia and Pinder (*89*) also find two xenon sites in ferrimyoglobin and ferrihemoglobin. These proteins also bind iodobutane, pentane, neopentane, and butane, and competition experiments show the binding sites for the different molecules to be the same. Detailed studies of pentane binding to myoglobin and hemoglobin showed two sites per heme group and a weak linkage between these sites and the oxygen-binding site in both proteins. Keyes and Lumry also found linkage between cyclopropane- or nitrogen-binding sites and the oxygen-binding site (*90*).

hemoglobin by Chanutin and Curnish (*16*) and by Benesch *et al.* (*167*), though preliminary experiments by Keyes show that phosphoacids effective in hemoglobin are not the contaminants in the myoglobin system. It is quite possible that the linkage system of myoglobin has no relation to physiological activity, but it is important as a model to show that allosteric linkage (*17*) can occur in proteins without multiple subunits. There has yet appeared no reason why multiple subunit proteins are necessary for allosteric linkage, though considerable emphasis has been placed on such a requirement.

The linkage system in myoglobin may be mediated by small conformational rearrangements in the protein and perhaps in nearby water just as in the linkage system in hemoglobin (*18*). Such linkage mechanisms have been called "dynamic rack mechanisms" (*179*), since the close association of heme group with protein conformation introduces stresses that force changes in the geometry of the heme-iron complex to effect rearrangements in the protein conformation and vice versa. The coupling is primarily mechanical— even when it depends on electrostatic interactions among charged groups— and may be influenced by the charge state of ionizable groups near the heme group or quite distant from it. It has been recognized since the work of Pauling, Coryell, and others in the thirties that the electronic properties of heme iron in myoglobin and hemoglobin are unusual as compared with model heme-iron complexes in homogeneous solution (*19*). The small free-energy difference of the order of kT between the high-spin state of heme iron in the Fe^{II} deoxy forms of these proteins and the low-spin state of heme iron in the oxygenated derivatives has particularly impressed most workers in the field. This poised position is the consequence of the rack interaction between the complex ion in the protein, and the conformation of the protein and has long since been explained in a qualitative fashion (*20*). The situation is somewhat more complicated in cytochrome *c*, since the heme-iron complex is not poised at the high-spin—low-spin crossover point. The Fe^{II} form of cytochrome *c* is low-spin, and in contrast to hemoglobin and myoglobin its Fe^{III} form is also low-spin at neutral pH values. Some change in ligand geometry must, of course, accompany the change in oxidation state, but these changes may be even smaller than those in hemoglobin and myoglobin, despite the fact that the oxidation–reduction potential is probably maintained in an unusual state by the rack interaction of the complex ion with the protein in which it is embedded (*21*). There has been reported a number of differences between the Fe^{II} and Fe^{III} forms of cytochrome *c* (*22, 23*), but no unequivocal evidence exists, so far as we know, that drastic changes in conformation occur.

If small changes occur as in hemoglobin and myoglobin, they are not clearly indicated by the thermodynamic changes. By measuring the emf of the cytochrome-*c* half-cell as a function of temperature, Sullivan (*24*) has determined the enthalpy and entropy change for the standard half-cell reaction. With the usual convention that both the half-cell emf and the half-cell enthalpy change for the standard hydrogen electrode are set equal to zero at pH 7.0, the standard half-cell enthalpy $\Delta H°$ and entropy $\Delta S°$ are -18 kcal/mole and -39 e.u./mole respectively if the cell emf is written as $E°_{cell} = E°_{H_2} - E°_{cyt\,c}(pH\ 7)$ and $E°_{cyt\,c}$ at 250°C equals $+0.261$ V. Then

$$\Delta S° = S°_{cyt\,c}Fe^{II} + S°_{H+} - \tfrac{1}{2}S°_{H_2} - S°_{cyt\,c}Fe^{III} \qquad (3)$$

in which $S°_{cyt\,c}Fe^{II}$ is the contribution of the reduced form at standard concentration of 1 M; $S°_{cyt\,c}Fe^{III}$ is that from the oxidized form; $S°_{H+}$ is the contribution due to the protein at 1 M activity [taken as -3.3 e.u. (203)]; and $S°_{H_2}$ is the contribution due to $\tfrac{1}{2}$ mole of H_2 at 1 atm [taken as $+15.6$ e.u. (203)]. The difference in entropy between the oxidized and reduced forms of the protein can be estimated from Eq. (3) using these numbers. It is found to be $+20$ e.u./mole, with the oxidized form having the higher entropy (A). We might have expected the oxidized form to have the lower entropy, since the iron ion gains a single positive charge on oxidation; but there may be counterions bound nearby to influence this effect, so that we cannot draw any conclusions about the significance of the entropy change from these observations.

III. Factors Determining Protein Conformational Stability

In terms of the concepts of interaction and bonding customarily applied to small-molecule systems the list of factors that determine the conformational stability of proteins includes as a minimum the following:

A. Hydrogen Bonding

When proteins are formed by folding a randomly coiled polypeptide, water molecules are displaced from the hydrogen-bonding sites on the backbone and side chains of the protein. The water molecules rejoin bulk water, and the hydrogen-bonding valences of the polypeptide regroup to form intrapeptide hydrogen bonds of one type or another. In this hydrogen-bond exchange process there are changes in enthalpy and entropy. The values of these changes vary from bond to bond, but even the average values are not known accurately. It is probable that on the average the enthalpy change is negative and thus favors the folded state (35). The entropy change cannot now be isolated and thus must be lumped with the other contributions to the configurational entropy of the folded protein and the reference state.

B. Van der Waal's Interactions and "Holes"

At ordinary temperatures hydrogen bonding dominates the geometry of water and the interactions of polar groups of unfolded polypeptides with water. Model studies suggest that the nonpolar side chains of the polypeptide are accommodated in bulk water in large holes formed by collection of small holes from the nearby liquid (59). The effect is quite remarkable. For example, one t-butanol molecule strongly influences the behavior of twenty to fifty water molecules. The nonpolar group interacts through van der Waal's forces with the water surface of its hole. When the polypeptide folds, the nonpolar groups assume new van der Waal's situations in contacts with other nonpolar groups, the sides of α-helical substructures, or the "free" sides of peptide and other groups that form hydrogen bonds. The van der Waal's interactions in the two extreme situations are different, since on the average a given nonpolar group is associated with groups of higher polarizability in the folded protein than when it is accommodated in water, but a sufficiently precise estimate of this difference based on small-molecule studies is probably impossible. If the other contributing factors were clearly of much greater quantitative importance, we should not worry about this inadequacy, but there is no reason to believe that these changes in van der Waal's contributions during folding are smaller than most of the contributions from other factors. In addition, we must realize that the fitting of the polypeptide on folding into a protein will vary from place to place and may be relatively poor in some regions. An interesting example is provided by the region behind the heme plane (proximal side) in myoglobin, which, though apparently well packed as judged by a casual examination of the X-ray structure, is in fact able to undergo rearrangements to accommodate an HgI_3^- ion (25). Estimation of the importance of "free volume" in proteins is rendered difficult by the very strong dependence of van der Waal's forces on interatomic distance. Small separations among oily groups are quite as effective in increasing the van der Waal's potential energy as larger holes.

Holes, in proteins, must in general be located in such a way that their surfaces are predominantly nonpolar. This follows from thermodynamic considerations, since the details of packing by and large are determined by the magnitude of the contribution to the free energy that each detail makes. To form a hole at the expense of one hydrogen bond raises the free energy of the conformation by several kilocalories. In fact, as we shall see, if only two or three hydrogen bonds cannot form in a given folded conformation, that conformation is likely to be unstable relative to the unfolded state. A "hole" of equivalent size made at the expense of van der Waal's interactions, though by no means trivial in the total free-energy balance, is much less unfavorable. One thus expects the mechanical properties of proteins to be dominated by van der Waal's interactions. This is certainly the case in myoglobin, in which nearly 80 percent of the hydrogen bonds formed by the main chain are isolated in α-helical segments. There are not enough data on mechanical properties of proteins to establish the detailed similarity between ordinary roofing pitch and proteins, but the physical strengths, heat capacities, and the temperature dependence of the volume suggest that pitch is mechanically quite similar to proteins. Pitch at ordinary temperatures is not so soft as to be describable as sloppy; and although protein crystals are very soft, the temperature factors in X-ray studies of proteins are not much different from those for any molecular crystal of an organic molecule.

The problem of holes is of considerable importance in protein thermo-dynamics and explains in part why it is virtually impossible to calculate the enthalpies of folded proteins from potential-energy functions. As regards the entropy situation we must expect to find that holes also dominate the entropies of proteins, since the "soft" vibrational modes spread over many atoms make the largest contributions to the configurational entropy of the folded polypeptide. This entropy problem is of course even more difficult than the corresponding enthalpy problem and makes the reliable estimation of protein entropies from the potential-energy approach impossible.

C. Configurational Entropy

We have just discussed the difficulties associated with the estimation of the configurational entropy of the folded states of polypeptides. Even if such estimates were possible, they would be of limited utility in calculating the thermodynamic balance sheet for conformational stability unless the entropy associated with a reference state could also be estimated with equal reliability. Unfortunately, although a true random-coil state of a given polypeptide would be the most convenient reference state, the actual choice is dictated by the intrinsic behavior of the polypeptide when a perturbing variable is changed. It will be possible soon and may even be possible now to use calorimetric measures to provide a random-coil reference state by measuring the transfer of protein to a solvent such as 6.7-M guanidine–water, in which at least some polypeptides approach true random-coil states (26). This procedure introduces considerable complication because the solvent is very different from pure water and may complicate the problem more than simplify it. As we shall see, some proteins under some conditions unfold in simple ways that are convenient for thermodynamic analysis, but they unfold to states that are apparently sloppy nets of partially solvated polypeptide chain, cross-linked erratically in a variety of ways by intrapeptide bonds and hydrophobic group association. The viscosity of solutions of such partially unfolded proteins is often low relative to solutions of random-coil polymers (27, 72). Hollis, McDonald, and Biltonen (28, 29) have shown that the local freedom of (segmental) motion is very high in these "unfolded" sections of polypeptide relative to the motional behavior in folded sections. Their studies and those of Rosenberg et al. (30, 31) and Blears and Danyluk (162) indicate that these nets expand with increasing temperature. At sufficiently high temperature they may actually approximate true random-coil states. This conclusion is important in showing that the so-called "unfolded" macroscopic states we are forced to use as reference states are not much simpler than the folded states. It is also important in showing that the distribution of microscopic (statistical mechanical) states in these "unfolded" macroscopic (thermodynamic) states is very broad. The conclusion is not much help if we seek to calculate the thermodynamic properties of the latter states, but even with a random-coil reference state it is most unlikely that any really reliable estimate of the configurational entropy would be possible. There is, however, little doubt that a major factor working against the formation of a stable folded conformation for a peptide is the loss of the freedom of motion and rearrangement present in the unfolded states. This loss of freedom is measured as the loss of configurational entropy.

D. "Hydrophobic" Bonding

The term "hydrophobic" still needs quotation marks. It is a poor term to use, since it is rather the lipophobic prejudice of water that is responsible for the effect. As discussed above, except at high temperature, water must give up free volume in order to accommodate nonpolar groups. No significant breaking of hydrogen bonds occurs at ordinary temperatures, as a consequence of the relatively large bond strength of these very special secondary bonds. Instead, water gives up free volume only reluctantly, since along with the free volume it also gives up a good share of its own configurational entropy. Whether one looks at this loss in entropy as physically represented by an ordering in the water (clathrate) shell about the nonpolar group, by the "iceberg" of Frank and Evans (32), or by the pictures of Eley (33) and Frank and Evans (32) based on loss of free volume, the thermodynamic consequences are qualitatively the same. Thus, when nonpolar groups cluster together in polypeptide folding, the entropy of the bulk of liquid water increases. Brandts (34, 35) has demonstrated that both the entropy and enthalpy changes resulting from the removal of nonpolar groups from water can be isolated as an individual contributing factor through measurements of changes in the heat capacity of the system. It appears probable that the heat-capacity behavior can be accounted for by two macroscopic states of the subsystem (water + nonpolar group). According to this view, as temperature is increased there is a between-states contribution to the heat capacity due to cooperative transitions from the lower-energy solvation state to the higher-energy state coupled with a slowly decreasing heat capacity in the higher-energy state itself (within-state contribution—see footnote, p. 27). As is always true, the within-states heat capacity of a macroscopic state at any temperature is proportional to the variance of the energy distribution at that temperature over the microscopic states of the particular macroscopic state. We shall shortly show that within present-day experimental errors the heat capacity of this solvation system is independent of temperature (A). It is nevertheless very large, and as a result makes possible a certain factoring of the thermodynamic quantities. There is little question that the "hydrophobic" bonds, which the heat-capacity changes measure, are a major source of folding stability in proteins.

E. Hindered Rotation

It is noteworthy that thus far only one (*43*) peptide bond in the cis configuration has been suggested by X-ray studies of proteins. The peptide bond has a large resonance stabilization that forces it to be planar. The difference in energy between cis and trans configurations is about 1.4 kcal/mole of bonds (*163*). Apparently this cost is usually too high to be accommodated by other factors. On the other hand, the remaining polypeptide bonds as well as most side-chain bonds have lower potential barriers for rotation. In order to effect a tight folding there must be considerable rotation against the potential barriers for hindered rotation. This problem has been overemphasized in relation to the others involved in stability and generally oversimplified. In very simple molecules rotation-barrier heights have been estimated in at least some cases with fair accuracy. The potential functions themselves are usually not well known, and in the case of very low barrier heights, as a consequence of the fact that the Born–Oppenheimer separation does not apply, are not even defined. However, the latter is a trivial matter thermodynamically, since it is the rotations against higher barriers that make important contributions. In complex molecules such as proteins all nearby atoms participate in determining the potential function for hindered rotation. The situation is very complex and must be considered to vary in an important way with the temperature as the protein expands or contracts and local distances change. Nevertheless we should not lose sight of the fact that the potential-energy increase due to the distortions of primary single bonds away from lowest-energy geometry undoubtedly is very important in the free-energy balance and generally must work against stability of the folded polypeptide.

F. Permanent Dipole Interactions

There is a number of groups in proteins with reasonably large dipole moments. The most important of these are the peptide bonds themselves with a dipole stretching from negative carbonyl oxygen to positive imide proton. It is quite probable that the interactions of these large permanent dipoles make important contributions to folding thermodynamics. Flory and co-workers (36, 37) and Brant (38), in particular, have emphasized the possible importance of this type of interaction, but the net magnitude and sign of the effect in amorphous regions of protein are not known and may perhaps average to zero. These effects cannot average to zero in known regular polypeptide structures and must be quite significant for such structures (38).

G. Electrostatic Effects

Estimation of the importance of electrostatic attractions and repulsions among ionized groups on the surfaces of proteins has had a long and troubled history, which we shall not recapitulate (39). Estimates based on pure electrostatic interactions of protein groups actually bearing charges depend in a very sensitive way on the parameters of the model chosen. Experimental studies of salt effects suggest a minor contribution to the thermodynamics of folding—unfavorable at extreme pH values but slightly favorable at the isoionic point. The situation is much more complicated than this, however, because the unfavorable electrostatic interactions at pH values some distance from the isoelectric point strongly encourage the binding of counter-ions to the charged groups. Despite many studies, ion binding is not well understood quantitatively (40), so that the total electrostatic situation is difficult to assess. It seems to us remarkable that X-ray studies of myoglobin crystals precipitated from nearly saturated ammonium sulfate solution show peaks at each surface-charged group of the protein, which undoubtedly indicates a single fixed counter-ion at each such group (41). If this is true in saturated ammonium sulfate solution, how much worse must the situation be in solutions of low ionic strength.

H. Abnormal Acidic and Basic Groups

The pH dependence of protein stability is complicated by the participation of the electrostatic situations of surface groups just discussed and the participation of acidic or basic groups that titrate abnormally and thus are in abnormal environments. However, by holding ionic strength constant, it is often found that a reasonably effective separation is possible. We then detect a pH dependence of the stability of folded and "unfolded" proteins with a relatively weak ionic-strength dependence. One explanation for such behavior is that there are buried charges that are thermodynamically, though not necessarily kinetically, restricted from taking up or giving off protons. However, we can estimate on the basis of low internal dielectric constant that the cost of burying a single charged group deep inside a protein is 20 ± 10

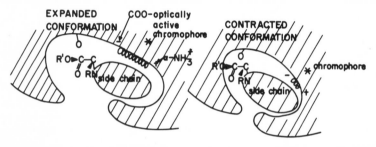

Fig. 2. Buried ion pairs in proteins might function as coulombic springs. With α-chymotrypsin there exist a large optical-rotary-dispersion Cotton effect and a substrate binding process that vary with pH in such a way as to suggest that they are linked to the ion pair formed by the carboxylate group of aspartate 194 and the α-ammonium group of isoleucine 16. This figure provides one possible description of this system. The chromophore is that of the 207-nm Cotton effect (see text).

kcal/mole in free energy. This is rather high and probably can be tolerated only on rare occasions. Instead, the burying of an ion pair formed of carboxylate + ammonium group should be much less expensive. An example of such a grouping has been found by Matthews et al. (42) in tosyl-α-chymotrypsin. It includes the α-ammonium groups of isoleucine 16 and the δ-carboxylate of aspartate 194. It may play a role in the catalytic function of the protein, possibly functioning as an electrostatic spring in the mechanical aspects of chymotryptic catalysis, as shown in Fig. 2.

   Although a simple ion-pair explanation for the pH-dependence of the free energy of folding is possible, the full explanation is probably more complicated and involves many charged groups (Section V) (A). Thus far, however, for the proteins of the chymotrypsinogen A family, the pH dependence of unfolding in the acid region (34, 44–48) plus the low chemical reactivity of three or four lysine side-chain ammonium groups (46) appears to be consistent with the idea that most members of this family have several surface lysine–carboxylate ion pairs. For some of the chymotrypsinogen proteins the contribution of the pH-dependent factors to the thermodynamic balance sheet has already been determined empirically* (see Table 4). The most stable member of this family thus far found is α-chymotrypsin (*47, 48*), for which the maximum instability obtainable by pH variation is estimated to be 14 kcal/mole of protein (cf. Table 4 footnotes). That is, the protein is more stable by 14 kcal of free energy at neutral pH than it is at pH 2. This quantity is not particularly large, but, as we shall see, it is enough to tip the balance between folding and unfolding. As already mentioned, Matthews *et al.* (*42*) have found a buried ion pair in tosyl-α-chymotrypsin, formed between the carboxylate ion of aspartate 194 and the α-ammonium group of isoleucine 16, and since the latter group is produced by hydrolysis of the peptide chain of the zymogen, this ion pair can exist in α-, β-, γ-, π-, and δ-chymotrypsin, but not in chymotrypsinogen. This ion pair is particularly important as the first source of information about the thermodynamic aspects of an ion pair in a protein and because it appears to be closely related to enzymic activity. It is thus useful to try to form some preliminary characterization of the grouping.

Parker (*49*) studied the optical-rotatory-dispersion (ORD) pattern of α-chymotrypsin as a function of pH. Hess and co-workers (*50*) obtained similar results. Biltonen *et al.* (*51*) were subsequently able to show that these variations were due to Cotton effects, one of which is clearly centered near 228 nm and appears to be symmetrical. The larger effect was experimentally difficult to characterize. It is located with a midpoint below 210 nm but may be a complex nonsymmetrical effect due to superposition of individual Cotton effects.† This so-called "207-nm Cotton effect" is larger than that at 228 nm, and its variation with pH appear to be the major source of ORD variation at longer wavelengths found by Parker and Hess and their respective co-workers, although Shiao (*46*) and Kim (*58*) have found the 228-nm effect to vary in a similar way, as have also McConn *et al.* (*201*). The changes in ORD pattern obtained at longer wavelengths, which are very ionic-strength dependent below pH 8 but at higher salt concentrations are simply behaved, show apparent pK_a values of 3 and 9. Parker was led to suggest that the ORD

---

* Brandts (*34, 35*) has estimated the maximum in stability associated with the abnormally titrating groups as about 11 kcal/mole. Shiao's data (given in Table 4) cover a somewhat wider pH range, and he has been led to use a slightly different expression for the extrapolation to neutral pH values than that used by Brandts. Biltonen (*47, 48*) estimates that the net stability from all contributing factors is 7 kcal/mole for α-chymotrypsin at pH 3 and 14 kcal/mole at pH 7, both values corresponding to 27°. In Table 4 the corresponding pH-dependent contribution to the net free energy of unfolding in transition I of dimethionine-sulfoxide chymotrypsinogen is given as 6 kcal/mole. This quantity was measured directly and should be accurate to ±1 kcal/mole.

† Lund and Lumry have recently located the circular-dichroism bands at 202 nm and 230 nm. There is some possibility that they measure changes in character of large amorphous regions of the proteins rather than local cotton effects (*220*) (cf. Section IV, C) (A).

variations are a manifestation of the changes in state of an ion pair with effective carboxylate pK_a of 3 and ammonium-group pK_a of 9 (*52*). Model studies by Rosenberg (*54*) have shown that nearby charges can produce a particularly large enhancing effect on the rotatory strength of amino acid side-chain chromophores, and in view of the large value of the rotatory strength corresponding to the 207-nm Cotton effect, Parker's proposal that the two charged groups of an ion pair are directly responsible for the optical activity in the 207-nm region was reasonable. Oppenheimer and co-workers (*243*) have identified the high-pK_a ammonium group with that of isoleucine 16 and thus, on the basis of the X-ray-diffraction results, with the ion pair of Matthews *et al.* (*42*).

Recent results of Rajender (*55*) suggest that α-chymotrypsin takes up approximately two moles of carbon dioxide at ordinary atmospheric partial pressures of this gas and one additional mole of the gas when the partial pressure is raised to about 1 atm. In view of the work of Roughton (*56*), Chipperfield (*57*), and others, these CO_2 molecules can be presumed to be bound as carbamino compounds of the α-amino groups of α-chymotrypsin. The first two moles of CO_2 bound produce no change in physical properties or catalytic activity of α-chymotrypsin. The third mole bound eliminates the catalytic activity and produces readily detectable changes in ORD, fluorescence yield, and spectrum. These results in turn suggest that destruction of the ion pair by deprotonation of the isoleucine α-ammonium group produces large changes in the organization of the conformation so that a change in 207-nm Cotton effect at basic pH values may be associated with a change in conformation. Thus, the ion pair might influence the chromophore indirectly rather than directly. However, the acid transition $A_b \rightleftarrows A_a$ [specifically $A_b(H_2) + H^+ \rightarrow A_a(H_3)$] (see Fig. 3), which is measured by change in optical rotatory dispersion and is attributed to protonation of the carboxylate groups of the ion pair, can be estimated on the basis of Biltonen's experiments at low ionic strength (*47*) to have a very small entropy change and an enthalpy change of about -4 kcal/mole. These values may be deceptively small if the process is a refolding process, but this is unlikely since it is rapid. Hence, there appears to be no refolding and little conformational alteration in the protonation of the ion pair at acid pH values. Examination of the models of tosyl α-chymotrypsin suggests that no matter where the substrate enters the protein, so long as it is near serine 195, there will be a change in stress on the ion pair, since the carboxylate member of the pair is directly linked to the serine 195 hydroxyl group.

The apparent pK_a values of the two ion-paired groups, according to Parker, are shifted only about 1 pH unit away from their intrinsic pK_a values ($pK_a = 3$ instead of 4 for the carboxylate group). The carboxylate-group behavior, according to Biltonen's estimates, is thermodynamically abnormal, but is not indicated to have associated any large conformational rearrangement. On the other hand, the small upward shift in pK_a observed with the α-ammonium group is misleading insofar as it suggests that only minor changes take place in the protein when this group loses its proton. Since we have quantitative information about no other ion pair in a protein, it is

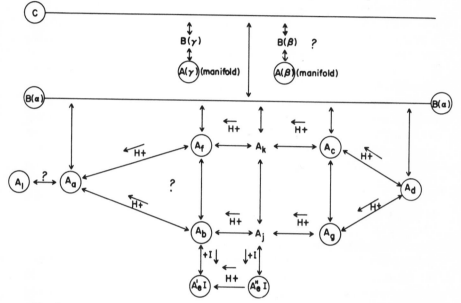

Fig. 3. The states and substates of α-chymotrypsin (bovine). All A substates of α-chymotrypsin are now thought to be "full folded" as judged by the absence of nonpolar groups extruded into water. This conclusion may be in error for some of the A substates in the basic pH region (on the right side of the diagram). All A substates are assumed to have a common B state, B(α). State B is extensively unfolded but retains a large folded core, probably the C chain, which is responsible for the α-helix rotatory-dispersion pattern. State C has not been produced by heating alone. The α-helix ORD pattern is lost in state C. State C is indicated to be a common state of α-chymotrypsin and γ-chymotrypsin and may be common also to β-chymotrypsin as shown in the diagram. All circled states and substates have been detected and partially characterized. A_k and A_j have been detected only through the pH dependence of the enzymic rate parameters and inhibitor binding. This figure is primarily a result of work by Kim (58).

interesting that this ion pair is neither strongly stabilizing nor strongly destabilizing. However, the magnitude of the change in conformation is probably not small. For α-chymotrypsin on the alkaline side the catalytically active species are $A_f(H_2)$, $A_b(H_2)$, $A_j(H)$, and perhaps $A_k(H)$. (See Fig. 3.) The loss of two protons from $A_b(H_2)$ produces the inactive species A_g, and if the loss of activity is due to the disruption of the ion pair, as is quite reasonable on the basis of the findings of Matthews et al. (42) as well as those of

Parker (52), the disruption process $A_b(H) \rightarrow A_g + H^+$ can be considered to consist of two parts: $A_b(H) \rightarrow A_g(H)$ and $A_g(H) \rightarrow A_g + H^+$. Kim ($58$) has estimated the thermodynamic changes for the two-step process as $\Delta F^\circ \approx 13$ kcal/mole, $\Delta H^\circ \approx 28$ kcal/mole, and $\Delta S^\circ \approx 50$ e.u./mole. These total quantities include loss of proton from the isoleucine α-ammonium group, conformational and solvent relaxation, and probably solvation of the aspartate carboxylate as a result of the migration of this group from the inner position out into the solvent. We can estimate the net effect of moving the broken ends of the ion pair, i.e., the ammonium ion and the carboxylate ion, from the solution into the ion-pair position by reprotonating the isoleucine amino group of A_g in water solvent. If its normal pK_a is 8, the thermodynamic changes in reprotonation are about -11 kcal, -10 kcal, and 3 e.u. for the molar free-energy, enthalpy, and entropy changes respectively, so that the formation of the ion pair from the ionized groups extended into solvent $[A_g(H) \rightarrow A_j(H)]$ requires a standard enthalpy change of -18 kcal offset by a standard entropy change of -54 e.u. to produce a net change in standard free energy of about -2 kcal. These numbers are consistent with a pK_a shift from 8 to 9 for the ammonium group and, though rough, are probably of the correct sign and order of magnitude for the formation of the ion pair. They are important numbers, since we could not have anticipated either the sign or magnitude for the enthalpy and entropy changes. Thus there is a large favorable enthalpy change, despite the enthalpy required to desolvate the charged groups. Similarly, although the desolvation process must produce a large entropy increase, the net entropy change is not only negative but quite large in magnitude. However, there are undoubtedly conformation changes as well as solvent and electrostatic changes.

It is generally found that ion pairs formed in polar solvents include especially strong ion–dipole interactions with one or a few solvent molecules. The probability of similar interactions between ion-pairing groups and nearby dipolar groups of the protein and a lack of real knowledge of the solvation state around charged groups extruded into the solvent makes enlightened estimates of ion-pair thermodynamics impossible. Hence, empirical information such as that provided by Kim furnishes the only means we have to assess the importance of the ion pairs in proteins. On the basis of Kim's numbers, ion-pair formation in proteins can be at least slightly stabilizing.

I. Odd Primary Bonds and Electronic Situations

Thus far the only primary bonds that are known to cross-link polypeptides are the disulfide bonds. These bonds are generally supposed to be so well integrated into successful proteins as to introduce no expensive restrictions on folding. They are certainly not essential for folded stability, but, as Kauzmann (59) has pointed out, they must restrict the free motion of the "unfolded" protein and thus the amount of entropy increase on unfolding. This is equivalent to saying that they stabilize the folded form.

Oxazoline and oxazolidine structures have been suggested (60) as possible primary-bond rearrangements in folded proteins, as have the corresponding bonds involving cysteine residues. None of these appears to have been demonstrated to exist in native forms of globular proteins, nor for that matter have any other primary-bond rearrangements or intrapeptide cross links.

Optical-rotatory-dispersion and circular-dichroism studies of the chymotrypsinogen A family of proteins reveal the presence of varying amounts of at least two very abnormal electronic situations in these proteins (2, 51). These have been discussed above as the 228-nm and 207-nm Cotton effects. The sources of these abnormalities have not been established. These sources may be important thermodynamically, but a situation demonstrating abnormal rotatory-dispersion characteristics need not be energetically important.

J. The Interaction of Proteins with Water

It is possible on the basis of available information to form a general impression that water-soluble globular proteins have minimal thermodynamic stability. This will become clear in another section. The impression is quite consistent with the thermodynamic contributions expected from the various contributing factors, with the absence of cis-peptide bond configurations, with the general exposure of charged groups to solvent, and so on, factor by factor. Hence at the present time the hypothesis of marginal conformational stability in globular proteins stands up rather well. It may, however, be quite incorrect for folded polypeptides in the sense that although marginal stability is essential for physiological function, favorable factors for folded stability may under some circumstances heavily overbalance the unstabilizing factors to produce much greater net stability than has thus far been measured. An interesting example of what may appear in some yet-to-be-studied proteins is provided by the core of state B (see Fig. 3) of the chymotrypsinogen family of proteins. This core cannot be unfolded by increasing temperature under any conditions thus far found (47, 51). On the basis of heat-capacity data to be discussed this core can be estimated to include about one-third of the residues. It can be unfolded by high urea or guanidine·HCl concentrations, but this observations gives little idea of the thermodynamic stability. All of the other polypeptide segments of all the chymotrypsinogen family thus far studied, both native and modified types, have low net folded stability even under the most favorable solution and temperature conditions.

Loosely stated, the "marginal-stability" hypothesis requires that whenever structural choices between a more stabilizing situation and a less stabilizing situation exist, the more stabilizing situation will in general be chosen. This simplified form of the hypothesis is insufficient and must be supplemented by the provisos:

(a) The goal of maximum folded stability is achieved only by taking into account the total conformational problem. Bad choices and good choices are thus relative.

(b) Proteins have been elaborated in evolution not necessarily for maximum stability, but rather to carry out some useful function or functions. Thus a bad thermodynamic choice may appear because the resulting situation plays an important role in the function of the protein in which it is found.

The marginal-stability hypothesis requires that all internal hydrogen-bonding valences be reasonably well satisfied. Water molecules will not be found inside proteins unless they make approximately four good hydrogen bonds to neighboring protein polar groups. Internally bonded water is a part of the protein in every sense and should not be grouped with so-called "hydration water."

Among the many quantitative puzzles in protein stability the most interesting and probably the least well understood is the description of the interaction of the protein with water molecules that belong primarily to the bulk solvent. Many descriptions of this interaction have been suggested, varying from simple electrostatic hydration of charged groups at the protein surface to postulations of deep, icelike sheaths of water around the protein. The direct interactions between water molecules and the protein are of course limited to a few water molecules about each charged group of the protein and its associated counter ions, the hydrogen-bonded water held to polar but uncharged groups at the protein surface, and some sort of thin water layer that accommodates itself about the nonpolar groups at the protein surface. However, this cannot be the limit of the influence of the protein on the solvent or the solvent on the protein, since there is a pseudohydrodynamic equilibrium condition due to the cohesive "pressure" of bulk water that must also be satisfied. This cohesive pressure is called the "internal pressure." It provides a major constraint on the conformations available to folded polypeptides. This pressure is in fact responsible for the existence of the protein, since some factors favoring stable folding, especially the hydrophobic bond, do not exist without a liquid water phase. The relative importance of the internal free volume of the protein is derived in large part from this same internal pressure (A). However, no very clear picture of the effects of the internal pressure of the bulk water on the protein has been presented. These effects must depend on the proportion of polar to nonpolar surface of the protein, since this proportion determines the interfacial free energy. However, our own experiments with succinylated chymotrypsinogen and other chymotrypsinogen derivatives in which the net charge is greatly increased (46) do not now appear to indicate that the charge situation is of major importance. These results are puzzling, but may mean that the effective value of the internal pressure depends on the volume of the protein even more than on its surface. We shall have more to say about these very interesting matters at a later point. At present we emphasize that interaction of the folded protein with bulk water is probably of major importance in determining folded stability.

A few summary statements can be made at this point: There is no empirical justification for selecting from among the factors mentioned some subgroup that can be said to make important contributions and a remaining subgroup that can be ignored. There is, in short, no "perfect-gas" model for proteins. Nor for that matter are all the important factors necessarily included in the above list. The conventional approach to thermodynamic stability via approximate potential-energy functions and statistical thermodynamics cannot be expected to have fruitful application to these problems. Even if the potential functions were well known—and no appropriately complex functions of the type required have ever been studied—the complexity of the computing problem is literally of astronomical magnitude. In addition, as Table 4 shows, the net stability in the best folded proteins that have been studied is marginal. In application of *a priori* approaches via potential-energy expressions it is necessary to calculate the absolute enthalpy and the absolute entropy of the macroscopic states and then to compare these states by taking the difference between two very large numbers. The differences are so remarkably small that, quite apart from the problem of accuracy, the precision required in calculation of the absolute entropies and enthalpies would greatly exceed the best anticipated for some years to come. The situation would be perceptibly better if there were "only" the problem of calculating the relative energy of the alternative folded structures, since it might be possible on rare occasions to achieve some success with an energy-minimization procedure, though we find this difficult to believe. However, the data in Table 4 are thermodynamic difference quantities between best-folded proteins and somewhat poorly defined, but nevertheless drastically altered, "unfolded" states. The entropy difference is almost exactly as important as the enthalpy difference in making these comparisons. It is also very important to note that the first thermodynamic state lying higher on the free-energy scale than the native state is *not* a fully folded state, but rather an "unfolded-net" state like state B of chymotrypsin. Comparisons thus must be made among conformational isomers that differ as much in entropy as enthalpy and not among folded conformers of approximately equal entropy.

If we use the solubility of small nonpolar molecules in water to estimate the thermodynamic changes in "hydrophobic" bonding, we would expect to find that "hydrophobic" bonds become more stable at higher temperatures. This is perhaps the result of the more "normal" behavior of water at elevated temperatures (*115*). In any event, at higher temperatures the tendency of oily groups to resist solvation in water becomes an enthalpic rather than an entropic consequence (*34*). If hydrophobic groups of proteins behave similarly, there can be an interval of temperature for some parts of some polypeptides in which the configurational (conformational) entropy increase, which drives the protein from "best folded" state (state A) to state B on increasing temperature, at lower temperature is inadequate to offset the increased stability of the hydrophobic bonds, so that such regions of proteins remain folded. Although at still higher temperatures the entropy must again win out to produce unfolding, small regions of proteins with small configurational entropy gains in unfolding and large proportions of nonpolar residues may show unusual stability against thermal unfolding. This appears to be the case for the folded state-B cores in the chymotrypsinogen family and allows us to speculate that proteins can be constructed that have greater thermodynamic stability against unfolding than those that have been studied.

As we shall see, a distinguishing characteristic of the catalytically active members of the chymotrypsinogen A family is the existence of a number of substates in the general category of "best-folded state"—we call these A-type substates (Fig. 3). The protein readily moves from substate to substate on minor changes in independent variables, and although some of the substates are essentially trivial from the thermodynamic point of view, most can be distinguished from the others by significant enthalpy and entropy differences. The thus-far unique association of catalytic function with substate versatility suggests that globular proteins to be useful functionally must be constructed to have marginal thermodynamic stability. We cannot, of course, prove that this is the case on the basis of the very limited data available, but we suspect that studies of the proteins of thermophilic bacteria and algae will provide a suitable test of this hypothesis. In any event, let us conclude this section by reiterating that if and when thermodynamically abnormal situations such as cis-peptide bonds, buried single charges, and so on are found to occur in proteins, they are probably there because they participate in the specific physiological mechanism of the protein.

IV. Thermodynamic Basis of Protein Folding

A. The "Semiphenomenological" Approach to Protein Thermodynamics

Although in time it may be possible to calculate the thermodynamic quantities for molecular crystals of small organic molecules using *a priori* potential functions, X-ray information on interatomic distances is much less precise for proteins, and there exist very many parameters in even the simplest potential functions that must be specified with high accuracy if similar calculations for proteins are to be more than a formal exercise. Over-optimistic applications of *a priori* approaches have diverted attention from the more realistic and very difficult approaches to theory and experiment that are necessary if quantitative understanding of protein structure and reactions is to be developed. A brief description of the most extensively developed methods for providing quantitative analysis of protein thermodynamics now in use will demonstrate the difficulty of the problem and the severe limitations on quantitative understanding inherent in the limited precision and diversity of experimental information.

B. Two-State and Multistate Transitions of Proteins

For some proteins, at least, the experimental problems associated with obtaining information on thermodynamic stability are considerably simplified by the fact that changes in protein conformation are often so highly cooperative that only two macroscopic states need be distinguished. In unfolding processes these states are the terminal states of most-folding and least-folding. The populations of intermediate species having intermediate degrees of folding are never quantitatively significant and can be ignored. Such processes are called two-state processes and are in fact first-order phase transitions like the melting of ice; but because far fewer atoms are included in the cooperative unit for the protein processes than in the cooperative unit for the melting of normal solids, the transition range of temperature or pH or pressure in which the transition occurs is many orders of magnitude broader for protein two-state transitions than it is for the more familiar two-state melting processes of solids. When a protein transition proceeds through intermediate states that in some range of the independent variable are significantly populated, the problem of relating the variations in thermodynamic quantities measured calorimetrically or by van't Hoff methods to the specific molecular descriptions of the several states becomes very difficult. Such processes are called multistate processes or multistate transitions. These matters have been discussed elsewhere (61), but there are certain possible complications in the analysis of a given protein conformational transition as a two-state or a multistate process that are currently obscuring the two-state concept. These have not been emphasized in the literature, and since they are useful in the present discussion, they should be described here.

Direct calorimetric determinations of heat capacities and enthalpies as a function of temperature, when carried out with adequate precision, are usually the most reliable source of entropy and enthalpy information, no matter what the protein transition. At present, calorimetry is just approaching the requirements in precision and accuracy necessary for studies of protein stability. Enthalpy and heat-capacity changes for a given protein process with equilibrium constant K can also be determined with the van't Hoff equation

$$\ln K(T) = \frac{\Delta H^\circ(T)}{RT} + \text{const.}$$

since the slope of the plot of $\ln K$ versus T^{-1} yields $\Delta H^\circ(T)$ at any T and the variation of ΔH° with T yields ΔC_p° for the process. The ultimate precision of the van't Hoff method for determination of heat-capacity change is obviously much lower than that of the direct determination of heat capacities by calorimetry. In the van't Hoff method only heat-capacity change and not absolute heat capacities associated with the individual thermodynamic state is determinable and then only as the second derivative of the logarithm of the quantity measured, K. In addition, there are possibilities for systematic error in the van't Hoff method that are often very difficult to eliminate. Nevertheless, at the present time the equipment for measuring equilibrium constants is occasionally of such high performance that van't Hoff data rival currently available calorimetric data in quality. The distinction between two-state and multistate processes is easily made from the van't Hoff plot if there are no systematic errors in preparing this plot, but the best test for two-state behavior is to compare the calorimetrically determined heat of reaction with the van't Hoff heat of reaction calculated on a two-state assumption, keeping temperature, pH, and other solution conditions identical. Calorimetry requires somewhat higher concentrations of protein than some van't Hoff measurements, such as spectrophotometric observation, so attention must be given to aggregation and other second- or higher-order protein processes. Thus far, reliable calorimetric versus van't Hoff comparisons of the enthalpy of the first thermally induced cooperative process, transition I between states A and B, have been made only for α-chymotrypsin (46–48, 62) and for ribonuclease A $(63, 64)$.* For α-chymotrypsin transition I was measured directly by Wadsö (62) by adjusting pH from 4 to 2 while the system was held at 40°C. His result, $\Delta H°(40) = 115$ kcal/mole of protein, is to be compared

* Transition I for chymotrypsinogen A has been studied using both solubility (44) and spectral (45) change as observables in the van't Hoff method and in calorimetric studies of high precision (62). Although the tests for two-state behavior that can be applied to van't Hoff data clearly indicate such behavior (61), there is a large discrepancy between the heats of reaction measured by the two methods. The source of this discrepancy has not been established, but in an account of fluorescence studies of the thermal unfolding of this protein, Steiner and Edelhoch (65) show data that suggest that there is a second transition at temperatures below those at which the van't Hoff data reveal a transition. This transition has thus far not been detectable in van't Hoff applications using solubility and indole difference absorption spectra. The situation seems unusual, but it is quite possible that three-state and many-state transitions can give deceptive results in van't Hoff studies if two or more of the states have the same value of the observable at the same set of solvent, temperature, and pressure conditions. Consider the simple three-state process

$$A \rightleftharpoons B \rightleftharpoons C \tag{4}$$

The heat-capacity changes $\Delta \bar{C}^°_{p,A \to B}$ and $\Delta \bar{C}^°_{p,B \to C}$ are assumed to be independent of temperature, and we assume that A and B have the same value of the observable at all experimental temperatures. Then, if the apparent equilibrium constant is $K = [C]/[B] + [A]$,

$$\Delta \bar{H}^°_{\text{van't Hoff}} = \frac{\Delta \bar{H}^°_{A \to B}(T_r) + K \Delta \bar{H}^°_{B \to C}(T_r) + \Delta \bar{C}^°_{p,A \to C}[T - T_r] + K \Delta \bar{C}^°_{p,B \to C}[T - T_r]}{1 + K} \tag{5}$$

$$\Delta \bar{C}^°_{p,\text{van't Hoff}} = \frac{\Delta \bar{C}^°_{p,A \to C} + K \Delta \bar{C}^°_{p,B \to C}}{1 + K} + \frac{\{\Delta \bar{H}^°_{A \to B}(T_r) + \Delta \bar{C}^°_{p,A \to B}[T - T_r]\}^2}{RT^2(2 + K^{-1} + K)} \tag{6}$$

where T_r is the reference temperature. The van't Hoff enthalpy change, Eq. (5), need not demonstrate a maximum within the temperature interval of the transition, and so long as the observables see states A and B as identical, the transition will appear to be two-state. The van't Hoff heat-capacity change, Eq. (6), includes the within-states contribution (first term on the right) and the between-states contribution (second term) and must demonstrate a minimum near the temperature at which $K = 1$. However, this minimum may not be large enough to detect. In the case of van't Hoff study of chymotrypsinogen A neither Brandts (45) nor Shiao (46) has detected a minimum in heat-capacity change within the transition interval. It is possible that there is an alternative explanation of the absence of agreement between calorimetric and van't Hoff data for chymotrypsinogen, but it is nevertheless clear that under some circumstances limited application of two-state tests to van't Hoff data can fail. Of course, there will be no problem in three-state situation if the two transitions are well separated along the independent-variable scale. The two transitions will individually demonstrate two-state behavior. When the transitions overlap, abnormalities will appear as indicated by Eqs. (5) and (6). The latter situation may be appropriate for chymotrypsinogen when studied using spectral changes but not using fluorescence, since the data of Steiner and Edelhoch (65) suggest that there are three states with different fluorescence characteristics. If the transitions between states overlap, no two-state pattern will be detectable. Obviously this matter requires further investigation (A).

with 120 kcal/mole obtained by Biltonen (*47*, *48*) in a spectrophotometric determination of the van't Hoff plot and 117 kcal/mole obtained by Shiao (*46*) independently in the same kind of experiment. The agreement is at the precision level of the several experiments but somewhat better than the absolute accuracy in both kinds of experiments.

The calorimetric enthalpy for transition I of ribonuclease A determined by Danforth *et al.* (*64*) is 86 ± 5 kcal/mole. This can be compared with Brandts and Hunt's (*63*) spectrophotometric van't Hoff value of 88 kcal/mole. These results strongly support a two-state analysis and taken together with the other tests for two-state behavior applied to these same proteins make the two-state interpretation unequivocal for these proteins under the conditions at which transition I was studied. However, although this interpretation is correct for these two proteins, the good experimental agreement may be a fortunate accident. In the case of α-chymotrypsin that this is so is demonstrated by Fig. 3, which includes all the substates of state A that have been detected thus far. All these substates are distinguishable by values of physical or chemical experimental parameters. The enthalpy and entropy changes between those A substates circled on this diagram have been determined on the assumption that these substates are discretely different thermodynamic states. This assumption may not always be correct. For example, the transition $A_b \leftrightarrows A_f$ might involve only these two terminal states, or it might include also a continuous series of intermediate states. In other words this transition itself might be a two-state or a multistate process. However, Kim (*58*) has demonstrated that this particular transition is probably a two-state transition. Be that as it may, it is immediately obvious that the protein can demonstrate strict two-state behavior only if it is initially present as a single pure A-type substate. At pH values where there are at least two A-type substates and when these substates differ significantly in enthalpy, calorimetric and van't Hoff data will correctly demonstrate multistate behavior in transition I even if only one additional substate has been added. The substates in this connection are correctly considered to be separate macroscopic states, and their inclusion as substates of type A in Fig. 2 is an oversimplification, despite the fact that these substates are all much more like each other than like state B. It has not yet been possible to detect substates of state B for this protein or for a number of other members of the chymotrypsinogen A family that have been studied. In fact, the data available now suggest that this partially unfolded state is very similar for all the members of this family (*47*). This is shown by inclusion of a single B state in Fig. 2. Of course, there must be some thermodynamic differences among the B states of chymotrypsinogen, α-chymotrypsin, and δ-chymotrypsin, since these proteins have different numbers of breaks in their peptides. But the differences are small and have not been detectable in state B. They are likely to become so only with precise calorimetric determination of the absolute heat capacities of the different proteins in state B* (A).

Van't Hoff data obtained with α-chymotrypsin at neutral and basic pH values will give different apparent enthalpy changes depending on the parameters studied. In general we cannot expect to find apparent two-state behavior unless the substates available to the protein at a given set of conditions are thermodynamically very similar. For α-chymotrypsin this condition is satisfied by A_a and A_b and approximated by A_g and A_c but not by other pair combination (66). However, once the existence of the type-A substates is known, the analysis of both calorimetric and van't Hoff data can be modified so as to provide enthalpy and entropy information about the transitions between the A substates and between each A substate and state B. The complication is a minor one, so long as there is only a small number of A or B substates and methods are available to measure populations of the substates accurately.

Those derivatives of α-, δ-, and γ-chymotrypsin that are catalytically inactive have been found to be locked into a single A substate in the pH range from about 2.5 to about 9.5 (46, 47, 52, 66). These proteins and chymotrypsinogen A itself (see footnote, p. 27) as well as its derivatives should demonstrate two-state behavior over this range. Shiao (46) has been able to study transition I of dimethioninesulfoxide chymotrypsinogen A up to pH 6.5 and succinylated chymotrypsinogen A from 5.5 to 7. These processes are two-state transitions. On the other hand, transition I of succinylated-α-chymotrypsin appears to show multistate behavior in the pH 6 to 7 region. Confirmation of this finding has not yet been provided.

Less is known about the A-type substates of ribonuclease. Bigelow et al. (67) have interpreted a variety of data for this protein as indicating that there are four denatured species, but it is not clear whether these are thermo-dynamically closely related and thus B-type substates, thermodynamically very different corresponding to several major steps of unfolding, or actually a consequence of transitions among A-type substates not yet detected. Even if the data do correctly apply to differences in the unfolded state, it seems *a priori* quite possible that there are also A-type substates of this protein so that the thermal-transition behavior even in the absence of drastic changes in solvent composition can be complex. Again, however, the van't Hoff data of Brandts and Hunt (63) obtained spectrophotometrically and those of Shiao (46) also obtained spectrophotometrically are in good agreement with the calorimetric enthalpy and heat-capacity changes measured by Danforth, Krakauer, and Sturtevant (64). All of these studies have been in the pH

* Recent studies of the rates of unfolding of the proteins in 8 M urea–water by T. Hopkins reveal significant differences among the B states in this particular behavior (T. Hopkins, unpublished observations from this laboratory).

range from about 1.5 to 3.5. The situation may be complicated when van't Hoff information obtained at higher pH values is compared with low-pH calorimetric data or even with low-pH van't Hoff data. Klee (68) has recently presented studies of the ORD changes and susceptibility to attack by proteases and peptidases as functions of temperature that indicate that a variety of intermediate states may exist in ribonuclease at pH values in the region of 8. His ORD experiments suffer from a lack of precision and his analysis from a lack of knowledge of the temperature dependence of the ORD parameters for the pure states. It does not appear to be generally realized that the maximum precision now obtainable with the best spectrophotometers, spectropolarimeters, etc., is necessary to obtain minimally satisfactory van't Hoff data for protein transitions, and that very small errors in base lines, i.e., temperature dependencies of limit states such as A and B in transition I, make very large errors in van't Hoff parameters.

Klee's (68) studies of enzyme attack on the protein are also less than satisfactory insofar as the data are interpreted as applicable to the free protein. Brandts (69) has pointed out that the catalytic parameters in protein–proteinase reactions have complex pH dependencies further complicated by variations in these parameters with the populations of partially hydrolyzed products. In addition, we note that if measurements of protein transitions are to be meaningful, they must not perturb the system. As was well demonstrated by Linderstrøm-Lang (70), proteolytic enzymes form complexes with their protein substrates in uncommon states of the latter, and in doing so stabilize those states. The enzymic experiments of Klee introduce maximum opportunity for errors from this source. Under the best of circumstances conclusions derived from applications of this method or any method in which the measuring device can perturb the distribution of states of the protein must be considered unreliable. All this, however, does not mean that the thermal unfolding transition of ribonuclease A continues to be of the two-state type as the system is adjusted to neutral and basic pH values. A more reliable test is to be found in the work of Martin (71), who demonstrated that van't Hoff plots obtained at different absorption wavelengths and also by optical rotation were identical. Holcombe and van Holde (72) have also shown that the thermal transition behavior of this protein is the same whether measured by sedimentation or viscosity. These results, obtained in the neutral pH region, strongly suggest that the first reversible thermal transition of ribonuclease is of the two-state type even in this pH range. However, the situation is complicated (61), and we must wait for detailed calorimetric information before forming any conclusions about behavior in the neutral to alkaline pH range.

Although no A-type or B-type substates have been detected for chymotrypsinogen A in the pH range from 1.5 to 10 (see footnote, p. 27), A-type substates of trypsinogen have been reported (73), so that the thermal unfolding processes of this protein should not always demonstrate two-state behavior. However, it is important to note in all these cases that there is as yet no indication of a continuous distribution of states of intermediate folding, each of which in its turn is significantly populated as the temperature is increased. Instead there appears to be a limited set of clearly distinguishable substates. If this is the general situation, it is a fortunate one for the protein physical chemist, since in such cases the effort required to relate thermodynamic changes to molecular details is not overwhelming.

We have emphasized the necessity of precise information about the temperature dependence of the pure A and B states in transition I. Similar precision would be required in studies of this transition produced by pH variation at constant temperature and pressure or pressure variation at constant temperature and pH. This is not a simply attained goal for state B. The degree and type of "unfolding" in this state is very difficult to define except insofar as we can state that it lies intermediate between a tightly folded conformation and a true random coil. Thermal cooperative transitions in our current experience do not produce complete unfolding in the sense that all of the nonpolar groups of the protein are extended into bulk water, nor in the sense that all parts of the polypeptide move with the small local segmental times of unrestricted random-coil polymers. The compromise bonding situation of solvent and protein net is undoubtedly very temperature dependent, a situation already revealed by proton-exchange studies of chymotrypsinogen by Rosenberg and Enberg (31) and by the NMR studies of ribonuclease by Blears and Danyluk (162). This means that the distribution of microscopic states in the macroscopic (thermodynamic) B state is very broad and may be very complex so as to yield unusual heat-capacity behavior for this macroscopic state as well as unusual variations with temperature and pH in the values of the observable used in van't Hoff studies. Thus the extrapolation of the state-B baseline into the temperature range of transition I may be quite difficult to make.

C. Analysis of Thermodynamic Data for Two-State Transitions of Proteins

Let us now proceed with the analysis of thermodynamic data for processes that have been shown to be of the two-state type. The standard free-energy change in such a process, ΔF°, could be factored into what might be called "orthogonal" contributions from the types of interactions that participate in the thermodynamic problem. This factoring is shown in Eq. (7a) in which the variables, ξ, are the members of the orthogonal (independent, non-interacting) set of forces sufficient in number to describe the quantitative conformational situation at some predetermined level of precision. Only in rare instances would these orthogonal forces correspond to the familiar forces by which we describe interactions in systems of small molecules. For example, hydrogen-bond interactions and most types of van der Waal's forces would be mixed and regrouped into several orthogonal forces of quite different and often difficultly describable form. In practice this ideal approach is severely limited by the types of independent variables that can be applied to effect or alter protein transitions. Instead we must use these experimental variables, X_j, themselves as the basis set, regardless of the fact that they are not often likely to be orthogonal so that we must include cross terms in the expression for ΔF°. The formulation for the free energy change then becomes Eq. (7b).

$$\Delta F^\circ = \sum_i \Delta F^\circ(\xi_i) \tag{7a}$$

$$\Delta F^\circ = \sum_j [\Delta F'^\circ(X_j) + \Delta\Delta F^\circ(X_j X_j)] + \sum_j \sum_{k \neq j} \Delta\Delta F^\circ(X_j X_k) \tag{7b}$$

$$\Delta F^\circ = \sum_j \Delta F^\circ(X_j) + \sum_j \sum_{k \neq j} \Delta\Delta F^\circ(X_j X_k) \tag{7c}$$

This equation is essentially a Taylor's series expansion in the variables X_i terminated at the quadratic cross terms, $\Delta\Delta F^\circ(X_j, X_k)$. Even with this procedure the X_j and thus the detail in the description of the transition are limited by the number of independent variables that can be employed in a useful way to influence ΔF°. At present, it is impractical to distinguish independent variables other than temperature, hydrogen-ion activity, nonionic solvent composition, solvent isotope variations, ionic strength, specific ion effects, and pressure. Even in this small but practical list very little is yet known about ionic and pressure effects. For present purposes we can first simplify Eq. (7b) to the form Eq. (7c) and then expand in Eq. (8).

$$\begin{aligned}
\Delta F^\circ = {} & \Delta F^\circ(\text{ion}) + \Delta F^\circ(a_{H+}) + \Delta F^\circ(\text{solvent}) + \Delta F^\circ(T) + \Delta F^\circ(P) \\
& + \Delta\Delta F^\circ(\text{ion}, a_{H+}) + \Delta\Delta F^\circ(\text{ion, solvent}) + \Delta\Delta F^\circ(\text{ion}, T) \\
& + \Delta\Delta F^\circ(\text{ion}, P) + \Delta\Delta F^\circ(a_{H+}, \text{solvent}) + \Delta\Delta F^\circ(a_{H+}, T) \\
& + \Delta\Delta F^\circ(a_{H+}, P) + \Delta\Delta F^\circ(\text{solvent}, T) \\
& + \Delta\Delta F^\circ(\text{solvent}, T) + \Delta\Delta F^\circ(\text{solvent}, P) + \Delta\Delta F^\circ(T, P) \tag{8}
\end{aligned}$$

Further simplification is possible if some of the variables X_j are independent or nearly independent of others or if the process itself can be factored into independent steps. An example of the latter situation is provided by certain aspects of the participation of polar and nonpolar parts of proteins. As has been discussed, the large heat-capacity changes in protein unfolding processes now appear to be due predominantly, if not entirely, to the peculiar inter-action of the nonpolar side chains, extruded into water during unfolding, with bulk water. If this is true, each of the terms in Eq. (8) can be divided into a part associated with this hydrophobic interaction of the oily side chains with water and a part associated with the changes in description of the polar side chains plus the configurational entropy. We shall not rewrite Eq. (8) to take this division into account, since the entire scheme quickly exceeds in detail the current experimentally practical resolution. Instead let us examine the very limited situation that exists today:

Pressure will be considered constant as will be nonionic composition of the solvent. The only small anion is chloride ion in the experiments we will describe. Then we can write the reaction as Eq. (9) and Eq. (7) becomes Eq. (10):

$$A(a_{H+}, a_{Cl-}, T) \rightleftarrows B(a_{H+}, a_{Cl-}, T) \tag{9}$$

$$\begin{aligned}\Delta F^\circ = \ &\Delta F_p^\circ(a_{H+}) + \Delta F_p^\circ(a_{Cl-}) + \Delta F_h^\circ(a_{H+}) + \Delta F_h^\circ(a_{Cl-}) + \Delta F_p^\circ(T) \\ &+ \Delta F_h^\circ(T) + \Delta\Delta F_p^\circ(a_{H+}, T) + \Delta\Delta F_h^\circ(a_{H+}, T) \\ &+ \Delta\Delta F_p^\circ(a_{Cl-}, T) + \Delta\Delta F_h^\circ(a_{Cl-}, T)\end{aligned} \tag{10}$$

The subscript p refers to all details of the unfolding process except the hydrophobic interaction between bulk water and the oily groups that become immersed in water as a consequence of the unfolding process of Eq. (9). The subscript h refers to the latter "hydrophobic" interactions only. We may think of the total process, Eq. (9), as divided into the unfolding itself, which takes place without rearrangement of liquid water around the newly exposed oily groups, and a second step of water rearrangement as the solvent adjusts to the situation of lowest free energy about these oily groups. The first part of the process is labeled p and the second h, since the latter is assumed to be uniquely associated with hydation of hydrophobic groups. This division is, of course, arbitrary.

The approximation that assigns the heat-capacity changes in process Eq. (9) to changes in hydrophobe–water situations as first postulated by Brandts (*34*) is supported by a considerable body of data obtained from small-molecule systems. Nevertheless, this division must still be considered to be no more than a working hypothesis until the important aspects of water–protein unfolding processes become considerably better understood than they are at present.

Although the studies of Brandts and Brandts and Hunt (*63*) show that $\Delta\Delta H_p^\circ(a_{H+}, T)$ and $\Delta\Delta H_h^\circ(a_{H+}, T)$ in the cross terms $\Delta\Delta F_p^\circ(a_{H+}, T)$ and $\Delta\Delta F_h^\circ(a_{H+}, T)$ probably cannot be ignored for transition I of ribonuclease in the acid region, Brandts (*34, 45*), Shiao (*46*), and Biltonen (*47*) have found them to be negligible for a variety of chymotrypsinogen A proteins including α-chymotrypsin. This situation was first pointed out for chymotrypsinogen A by Brandts, who also first emphasized the complexity of ionic dependencies of transition I. However, for the chymotrypsinogen proteins $\Delta\Delta H_p^\circ(a_{Cl^-}, T)$ and $\Delta\Delta H_h^\circ(a_{Cl^-}, T)$ in the cross terms $\Delta\Delta F_p^\circ(a_{Cl^-}, T)$ and $\Delta\Delta F_h^\circ(a_{Cl^-}, T)$ have also been shown by the same authors to be small in the acid pH region. As a consequence $\Delta F_p^\circ(a_{H^+})$ and $\Delta F_h^\circ(a_{H^+})$ and the other remaining terms in Eq. (10) that are not explicitly dependent on T are adequately approximated as pure entropy changes and may now be written as such: $\Delta F_p^\circ(a_{H^+}) = -T\Delta S_p^\circ(a_{H^+})$ and so forth. To simplify the discussion we will refer all experiments to a constant (reference) chloride-ion concentration (0.01 N) so that Eq. (10) becomes Eq. (10a):

$$\Delta F^\circ = T\,\Delta S_p^\circ([Cl^-]_{ref}) - T\,\Delta S_p^\circ(a_{H^+}) - T\,\Delta\Delta S_p^\circ(a_{H^+}, [Cl^-]_{ref})$$
$$+ \Delta F_p^\circ(T) + \Delta F_h^\circ(T) \tag{10a}$$

Using Brandts' (*34*) explicit sets of hydrogen-ion equilibria for molecules in state A with acid dissociation constants $K_{A,i}$ and in state B with $K_{B,i}$ we write Eq. (11), which is suitable for the chymotrypsinogen family of proteins but not for ribonuclease:

$$\Delta S_p^\circ(a_{H^+}) + \Delta\Delta S_p^\circ(a_{H^+}, [Cl^-]_{ref}) = R\ln\prod_i\left(\frac{1 + a_{H^+}/K_{B,i}}{1 + a_{H^+}/K_{A,i}}\right) \tag{11}$$

The choice of standard state for hydrogen-ion activity is arbitrary, but it would be convenient to choose some pH near neutrality, since there is little or no pH dependence of the process in the pH region from about 5 to 8, and this is the pH range of maximum stability. At lower pH values the influence of pH on the process is to decrease stability with increasing proton uptake. Nevertheless, it is often experimentally difficult to make direct measurements at pH values in the 5-to-8 region so that choice of reference state for pH in the acid or alkaline region where transition-I experiments can be made is necessary. This is the case for many members of the chymotrypsinogen A family of proteins.

In a complete study of pH effects the acid titration behavior of the protein in state B and state A must be determined to validate the form of Eq. (11). For the chymotrypsinogen proteins it has not yet been possible to establish that the form of Eq. (11) is correct, and even if it is correct, we can anticipate that it may prove difficult to distinguish between a small set of acid groups strongly perturbed to abnormal pK_a values and any larger set experiencing smaller perturbations (*47*).

Since simple measurements of the process of Eq. (9) as a function of pH at constant anion composition yield the correction factors that allow the data to be adjusted to any arbitrarily chosen reference state for hydrogen-ion activity within the experimental range, it now remains to determine the form of the $\Delta F_\mathrm{p}^\circ(T)$ and $\Delta F_\mathrm{p}^\circ(T)$ in Eq. (10).

The development to this point has been strictly phenomenological, since the form of the pH dependence given in Eq. (11) should be considered provisional and the assumption of separability of the free energy terms into polar terms and hydrophobic terms has not yet been fully established. Although Brandts's choice of the form $\Delta F_h^\circ(T) + \Delta F_p^\circ(T)$ shown in Eq. (12) was guided by expressions for the free-energy of interaction of small nonpolar groups with water, Eq. (12) nevertheless may be considered to be chosen for its ability to fit the experimental transition-I data from several proteins. In this sense it is also phenomenological.

$$\Delta F_\mathrm{p}^\circ(T) + \Delta F_\mathrm{h}^\circ(T) = \alpha + \lambda T + BT^2 + CT^3 \tag{12}$$

From Eq. (12):

$$\Delta H_\mathrm{p}^\circ(T) + \Delta H_\mathrm{h}^\circ(T) = \alpha - BT^2 - 2CT^3 \tag{13}$$

$$\Delta S_\mathrm{p}^\circ(T) + \Delta S_\mathrm{h}^\circ(T) = -\lambda - 2BT - 3CT^2 \tag{14}$$

$$\Delta C_\mathrm{p}^\circ(T) + \Delta C_\mathrm{h}^\circ(T) = -2BT - 6CT^2 \tag{15}$$

The coefficients α, λ, B, and C are constants for a given process; B and C are related only by the hydrophobic solvation effect. According to Eq. (15) the heat-capacity change (at constant pressure) is temperature dependent. However, Shiao (46), using a technique suggested by Biltonen (84) to minimize systematic errors in van't Hoff data associated with extrapolations of state-A and state-B temperature dependencies (base lines) into the transition range, finds a negligible temperature dependence for the heat-capacity change for a variety of the chymotrypsinogen proteins in transition I over a total experimental range of temperature from 10 to 50°. Thus within current experimental error the heat-capacity change can be considered constant. Indeed, Shiao finds that this choice gives a better statistical fit to his enthalpy and free-energy data than does the choice of Eq. (15) (A). The assumption of constant heat-capacity change leads to a simpler equation than Eq. (12). Shiao's equation is Eq. (16), from which the enthalpy, entropy, and heat-capacity changes can be written. (In the following, $\Delta C_\mathrm{p}^\circ \equiv \Delta C_{P,\mathrm{p}}^\circ$ and $\Delta C_\mathrm{h}^\circ \equiv \Delta C_{P,\mathrm{h}}^\circ$).

$$\Delta F_\mathrm{p}^\circ(T) + \Delta F_\mathrm{h}^\circ(T) = \Delta H_\mathrm{p}^\circ(T_\mathrm{r}) + (\Delta C_\mathrm{p}^\circ + \Delta C_\mathrm{h}^\circ)(T - T_\mathrm{r})$$
$$- T\,\Delta S_\mathrm{p}^\circ(T_\mathrm{r}) - T\,\Delta S_\mathrm{h}^\circ(T_\mathrm{r}) - (\Delta C_\mathrm{p}^\circ + \Delta C_\mathrm{h}^\circ)T\ln\frac{T}{T_\mathrm{r}} \tag{16}$$

$$\Delta H_\mathrm{p}^\circ + \Delta H_\mathrm{h}^\circ = \Delta H_\mathrm{p}^\circ(T_\mathrm{r}) + \Delta H_\mathrm{h}^\circ(T_\mathrm{r}) + (\Delta C_\mathrm{p}^\circ + \Delta C_\mathrm{h}^\circ)(T - T_\mathrm{r}) \tag{17}$$

$$\Delta S_\mathrm{p}^\circ + \Delta S_\mathrm{h}^\circ = \Delta S_\mathrm{p}^\circ(T_\mathrm{r}) + \Delta S_\mathrm{h}^\circ(T_\mathrm{r}) + (\Delta C_\mathrm{p}^\circ + \Delta C_\mathrm{h}^\circ)\ln\frac{T}{T_\mathrm{r}} \tag{18}$$

Equation (16) can be fitted to transition-I data by a suitably weighted least-squares procedure to provide values of

$$[\Delta H_p^\circ(T_r) + \Delta H_h^\circ(T_r)], \; [\Delta S_p^\circ(T_r) + \Delta S_h^\circ(T_r)], \; \text{ and } \; [\Delta C_p^\circ + \Delta C_h^\circ]$$

for any choice of reference temperature T_r. A representative sample of the thermodynamic quantities is given in Table 4. The temperature at which the protein molecules are equally divided between state A and state B at the particular choice of reference pH and reference chloride-ion concentration is T_0. The temperature at which the maximum standard free-energy difference between states A and B occurs, T_m, is also given, together with the maximum free-energy difference, $\Delta F_p^\circ(T_m) + \Delta F_h^\circ(T_m)$.

Thus far no assumptions about the separation of polar and water–hydrophobe interactions have been made, so that the data in Table 4 are still essentially phenomenological. The most important conclusions to be drawn for this table are the following:

(1) The maximum stability, which is measured by $\Delta F_p^\circ(T_m) + \Delta F_h^\circ(T_m)$, is small at pH 2. The most stable protein of the chymotrypsinogen group thus far found is α-chymotrypsin at neutral pH. Even at pH 6, where the unstabilizing effect of pH is completely absent, the maximum stability is only about 14 kcal/mole for this protein. The instability due to pH change from 6 to 2 is about 12 kcal/mole. At pH 2 chymotrypsinogen is the most stable.

(2) The standard heat-capacity changes vary over a wide range.

(3) The variations in the enthalpy change at T_0 among the proteins are not given in the tables, since they are deceptively large; the T_0 values vary, and the heat-capacity changes are large. Consequently for comparisons among the proteins the standard enthalpy and entropy changes are given in Table 4 at a common temperature value, which, for convenience, has been chosen as 314.5°K, the T_0 value for chymotrypsinogen. However, even with this choice, large variations in enthalpy and entropy change are found. Since the enthalpy precision is not poorer than 2 kcal/mole, the quantitative thermodynamic differences among the proteins are easily detected.

The thermodynamic quantities refer to one mole of cooperative units and in these cases to one mole of protein. The variations in heat-capacity change, enthalpy change at 314.5°K, and entropy change at 314.5°K show that the size of the region of the protein undergoing cooperative unfolding, the cooperative unit, varies from protein to protein over a considerable range of values.

(4) The parameter most useful in comparing the characteristics of cooperative units among proteins is the ratio of standard enthalpy change to standard entropy change for a cooperative unfolding process, but even this "thermodynamic phase relation" (2) is not of much use as a measure of the

TABLE 4

PHENOMENOLOGICAL INFORMATION FROM TRANSITION I OF SOME MEMBERS OF THE CHYMOTRYPSINOGEN A FAMILY[a,b]

Protein	$(\Delta C_p^\circ + \Delta C_h^\circ)^e$ (kcal/mole-deg)	$(\Delta H_p^\circ + \Delta H_h^\circ)$ (kcal/mole) at 314.5°K	$(\Delta S_p^\circ + \Delta S_h^\circ)$ (e.u./mole) at 314.5°K	T_0 (°K)	T_m (°K)	$(\Delta F_p^\circ + \Delta F_h^\circ)(T_m)$ (kcal/mole)
α-chymotrypsin	4.0	118	390	303.5	285.6	2.2[g]
γ-chymotrypsin	4.1	106	350	305.1	288.7	1.8
δ-chymotrypsin	3.0	90	290	306.3	284.7	2.4
Diphenyl carbamyl-α-chymotrypsin[c]	4.2	123	403	304.1	285.8	2.3
Chymotrypsinogen A[d]	3.2[f]	93	295	314.5	286.7	4.0
Dimethionine-sulfoxide-chymotrypsinogen	2.3	73	246	297.7	283.3	0.8

[a] From Shiao (46).

[b] pH_ref = 2.0; (Cl⁻)_ref = 0.01 N.

[c] For dimethionine-sulfoxide-chymotrypsinogen at pH 6, $(\Delta F_p^\circ + \Delta F_h^\circ)(T_m) \approx$ 6.0 kcal/mole.

[d] For chymotrypsinogen A at pH 6, $(\Delta F_p^\circ + \Delta F_h^\circ) \approx$ 11.5 kcal/mole.

[e] Heat-capacity changes at constant pressure.

[f] From Shiao (46); Brandts' value (45) is somewhat lower if calculated at lower temperatures;   at higher temperatures his values are the same.

[g] According to Biltonen (47, 48), for α-chymotrypsin at pH 6, $(\Delta F_p^\circ + \Delta F_h^\circ) \approx$ 14 kcal/mole.

quality of folding in a cooperative unit at arbitrary pH, temperature, and salt concentration, since it is dependent on these three variables as well as the amino-acid composition of the protein. In comparing proteins with different amino-acid compositions this measure should be applied using enthalpy changes and entropy changes that have been experimentally corrected to remove contributions from ionic-strength effects and abnormally titrating groups. Comparisons can then be made with or without the water–hydrophobe contributions and at a constant temperature or at the respective T_0 values for the proteins, since each comparison gives additional information about the character of the cooperative unit. In making comparisons among proteins of a single family, i.e., having the same or nearly the same amino-acid composition, useful information can be obtained at any common reference state of pH and solvent salt composition, although we may be forced to choices that are not the most informative. For comparisons of the phase relation variations among the chymotrypsinogen proteins we are at present limited to such conditions. Specifically, we choose $T = T_0$, since for this choice the phase relation is simply T_0 so that the comparison is among T_0 values.

Major differences in T_0 are found in Table 4 only with chymotrypsinogen and dimethioninesulfoxide-chymotrypsinogen. The "thermodynamic character" of the cooperative process as judged using this parameter is quite different in these two proteins, and in turn these processes are quite different from those of the other proteins listed. Alternatively, the sizes of the cooperative unit of δ-chymotrypsin and chymotrypsinogen estimated from the heat-capacity change are very nearly the same. This would have been expected on the basis of chemical information, since δ-chymotrypsin is formed from chymotrypsinogen by removal of the dipeptide at positions 14 and 15 of the latter. The fact that under the conditions of Table 4 chymotrypsinogen is the most stable protein of this group may be attributable to its intact peptide chain, which, as judged by the value of the phase relation, requires a larger enthalpy input per unit entropy gain than the other proteins. On the other hand, dimethioninesulfoxide-chymotrypsinogen requires a considerably smaller enthalpy input per unit entropy gain than the other proteins, despite the fact that its peptide chain is intact. This situation requires that the cooperative units in these two proteins be not identical.

(5) It is thought that the diphenylcarbamyl group of diphenylcarbamyl chloride acylates the hydroxyl group of serine 195 (74), which is also the site of acylation during normal chymotryptic catalysis. According to the data in Table 4 this acylation process improves the "folding" of α-chymotrypsin by about 5 kcal/mole in enthalpy and 13 e.u./mole in entropy. It would be interesting to know if this enthalpic "improvement" is due to the hydrophobic interaction of the diphenyl carbamyl group with water in state B.

(6) Although it is not shown in Table 4, in which only the high-temperature value of T_0 is given, the heat-capacity change is such that as the temperature is lowered, both enthalpy and entropy changes pass through zero, change sign, and then increase in magnitude with the new sign. As a result, there is a lower temperature, not always available, at which $\Delta F^\circ(a_{H+,\text{ref}}, [Cl^-])$ equals zero. The folded state, state A, becomes unstable relative to state B at temperatures *lower* than the lower-temperature T_0. The lower-temperature T_0 values at different pH and ionic-strength conditions all lie below 0°C for chymotrypsinogen and ribonuclease, but the other proteins in the tables, by suitable adjustment of a_{H+} and $[Cl^-]$, can be made to undergo transition I by cooling as well as by heating. An unfolding transition produced by cooling is called an "inverted transition" (75). The existence of inverted-transition behavior for transition I of chymotrypsinogen A was proposed by Brandts (34).* It was demonstrated by Biltonen (47) in α-chymotrypsin and its dimethioninesulfoxide derivative. Some "cold-precipitating" proteins may manifest this behavior, since inverted transitions are probably quite common.

The existence of two values of T_0 explains the occurrence of a condition of maximum stability already discussed and illustrated by data in Table 4. The reason for the constancy of the temperature T_m for the proteins in Table 4 is too complicated to be described here, but is largely a consequence of their common amino acid composition and the similarity of the amino acid composition of the cooperative units of these proteins. The value $T_m = 285°K$ has no general significance; T_m can be calculated to be lower than 273°K for ribonuclease A in water.

Before moving on to the next step of this thermodynamic analysis we note first that Rosenberg and Enberg (31) have shown with chymotrypsinogen A that exchange of some protein protons with those in solution as measured using tritiated proteins and hydrogen oxide solvent can occur through a transition-I mechanism, which can be given in simplified form in Eq. (19):

$$A_{\text{tri}} \rightleftarrows B_{\text{tri}} \xrightarrow{\ H_2O\ } B_{\text{hyd}} \rightleftarrows A_{\text{hyd}} \tag{19}$$

This mechanism of exchange applies to a single family of protons that exchange as a group, but this type of cooperative exchange occurs only when solution conditions and temperatures are such that there exists some, as yet not established, minimum proportion of the protein molecules in state B. At lower temperatures or in pH regions of greater thermodynamic stability of state A a quite different mechanism assumes major importance (30, 31).

* Brandts demonstrated this "inverted-transition" behavior and found the lower-temperature T_0 for chymotrypsinogen by using urea–water rather than water solutions (45). The first clear example of an inverted transition was produced by Schellman, using ribonuclease (75).

The quantitative agreement between proton-exchange data and transition-I thermodynamics for chymotrypsinogen A under some sets of conditions provides a means whereby the thermodynamic changes in transition-I can be determined without producing any appreciable concentration of the state-B species. Most proteins aggregate in what amounts to an irreversible process from the point of view of transition-I experiments when in state B, if the temperature is high or the concentration in state B is high. As a result of this rapid aggregation it is not possible in proteins such as carbonic anhydrase to study reversible unfolding reactions by ordinary methods that require directly measurable concentrations of unaggregated state-B species. Unfolding thermodynamic behavior in such cases can be studied through convenient and highly precise proton-exchange methods at least under some sets of circumstances. Proton-exchange data can provide an estimate of the sizes of cooperative units as well as unique information about conformation change.

A number of the chymotrypsinogen proteins for which transition-I data have been obtained have also been studied by Hollis, McDonald, and Biltonen (*28, 29*) using bandwidths of the proton NMR bands for non-exchangeable protons. A fortunate pattern of bands corresponding to different nonexchangeable protons of assignable chemical shifts provides "NMR fingerprints" that are characteristic of different sections of the peptide chain of the chymotrypsinogen proteins. When a given section of polypeptide is moving with the slow tumbling motion of the protein or a slow fluctuation in conformation, the correlation time is long and the NMR bands are broad. When this segment is released from tight confinement by the protein, local segmental motion, which is at least three orders of magnitude faster, decreases the correlation time and as a result increases the lifetime and decreases the energy uncertainty, i.e., narrows the bands. Thus fast-moving protons tend to reduce bandwidths so that analysis of the shape and area of the proton bands can yield the relative populations of fast-moving and slow-moving protons of the several "chemical shift" classes. Using this method these workers have been able to relate the different cooperative units suggested by the data in Table 4 to the different segments of the polypeptide, and to estimate with certain assumptions the amount of the polypeptide unfolded in state A as well as the probable segments of the polypeptide in the fast-moving category (see Fig. 4). Estimates of the sizes of the cooperation units are based on the assumption that α-chymotrypsin is fully folded, an assumption suggested by the calorimetric heat-capacity data for α-chymotrypsin now available (*76*), and the absence, real or apparent, of narrow-band contributions in the NMR spectrum for this protein. The quantitative agreement between heat-capacity data for transition I and the estimates of cooperative-unit size by NMR is good. The heat-capacity results are an independent source of information about the folding of proteins but do not in themselves lead to any specific molecular detail. We shall examine the heat-capacity analysis presently. The analysis of NMR data leads to

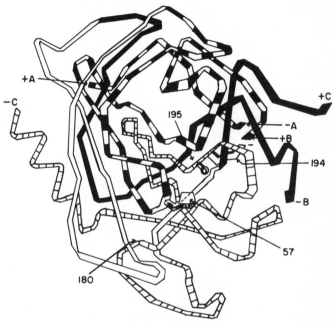

Fig. 4. A provisional dissection of the X-ray model of α-chymotrypsin in terms of partial unfolding in *A* and *B* states. The figure is closely based on the structure for tosyl-α-chymotrypsin reported by Matthews *et al.* (*42*) with interpolation of residue side chains for serine 195 and aspartate 196 (*4*). The assignments of partial folding are due to Hollis, McDonald, and Biltonen (*28, 29*). The cross-lined polypeptide that runs from 168 to 245, nearly all of the *C* chain, appears to be the stable core of state *B*. The solid black section that runs from 109 to 151 is "unfolded" in chymotrypsinogen and in δ-chymotrypsin. The peptide section marked by alternate filled and unfilled sections is "unfolded" in dimethionine-sulfoxide-α-chymotrypsin but not in chymotrypsinogen. It runs from 1 to 78. The point of acylation of α-chymotrypsin by substrate acid-groups is the hydroxyl of serine 195. The base catalyst is the imidazole group of histidine 57 and the ion pair directly related to catalysis and optical rotatory dispersion may be that formed between the sidechain carboxylate group of aspartate 194 and the α-ammonium group of leucine 16, which is marked +B. The oxidation of methionine 180 is thought to be directly responsible for the unfolding of the section from 1 to 78 (see text).

suggestions not only about the size of the cooperative unit in transition I but also about the specific sections of the peptide chain that are folded or unfolded. This technique promises to become a very powerful source of information about folding and dynamic characteristics of specific regions of proteins, but it should be noted that the work of Hollis, McDonald, and

Biltonen is the first of its kind, and their analyses, which we quote in the following paragraphs, have not been fully substantiated. Lack of agreement between NMR and X-ray diffractions results may be due to error in analysis of data. Alternatively, since the NMR method looks at the protein in solution, lack of agreement may mean that the protein in solution is not conformationally identical with that in the crystal.

TABLE 5

EFFECT OF DIPHENYL CARBAMYL CHLORIDE (DPC)
REACTION ON SOME SPECIES OF CHYMOTRYPSIN[a,b]

	$\overline{\Delta C_p^\circ}$ (kcal/mole-deg)	$\overline{\Delta H^\circ}$ (kcal/mole)	$\overline{\Delta H^\circ}/\overline{\Delta C_p^\circ}$	Ref.
α-CT	4.0	100	25	(46, 47)
DPC-α-CT	4.2	104	25	(46)
δ-CT	3.0	78	26	(46)
DPC-δ-CT	3.2	80.5	25	(46)
DMS-α-CT	1.8	52	29	(47)
DPC-DMS-α-CT	3.4	82	24	(46)
CGN	3.2	79	25	(45, 46)

[a] From Shiao (46).

[b] $\overline{\Delta C_p^\circ}$ and $\overline{\Delta H^\circ}$ are standard heat-capacity and entropy changes in transition I at 310°K. DMS = dimethionine sulfoxide.

Among the chemically modified proteins and unmodified proteins of this family, the descriptions of the cooperative units are quite different. We cannot explore all the results of this study, but in Table 5 are listed the transition-I parameters for α-chymotrypsin, dimethionine sulfoxide-α-chymotrypsin, diphenylcarbamyl-α-chymotrypsin, diphenylcarbamyl-δ-chymotrypsin and diphenylcarbamyl-dimethioninesulfoxide-α-chymotrypsin. According to the NMR data dimethioninesulfoxide-α-CT is extensively unfolded in its "best folded state," state A. Furthermore, according to the NMR analysis, the unfolding, which is a consequence of oxidation of methionine 180, disrupts the peptide region around histidine 57, the imidazole of which plays a crucial role as a base catalyst (77) in the on- and off-acylation steps of chymotryptic catalysis. When this protein is treated with diphenyl-carbamylchloride to produce DPC-DMS-α-CT, a large fraction of the unfolded region refolds. This pattern of events is also reflected in the transition-I data shown in Table 5. In addition, however, a comparison of the thermodynamic data for DPC-α-CT with those for DPC-δ-CT strongly suggests that the DPC reagent, in producing DPC-DMS-α-CT, has forced refolding into a conformer that resembles δ-CT much more than it does α-CT or γ-CT. The most interesting implication of these findings is that some ordinary substrates must act like the DPC reagent to produce refolding before primary bond rearrangement steps of catalysis become possible. This implication is an immediate consequence of the fact that DMS-α-CT is nearly as efficient a catalyst for small ester substrates as α-CT itself (78).

D. The Approach with Models

Despite the promising nature of transition-I studies with the variables that have been employed, much greater power is necessary if a reasonably complete description of protein stability is to be developed. Addition of variables such as pressure, solvent composition, and chemical modification provide additional power and so provide increasingly detailed analysis of the problem. However, it is also desirable to make use of data from small-molecule systems whenever possible, bearing in mind that this is a treacherous undertaking, the results of which must be viewed with considerable caution. When such data are used, the analyses of protein-transition data are properly termed semiphenomenological to indicate that the protein itself is no longer the only model employed (2, 35). The first major step toward a semi-phenomenological description of the conformational thermodynamics of proteins has been taken by Brandts (34, 35). His method is based on an observation first made by Edsall (79) that the nonpolar groups of the type found in proteins have a remarkably large heat-capacity change associated with their solubility in water. Brandts, following a computational procedure used by Tanford (80), estimated the heat-capacity, enthalpy, and unitary free-energy changes that might be expected if a nonpolar sidechain buried in the folded protein is exposed to water on unfolding. A polar organic solvent is assumed to simulate the solvation of the nonpolar side chain inside the protein. Polar organic solvents resemble at least roughly the inside of the protein, but close simulation proves to be unnecessary, since the large heat-capacity effects are due primarily to the water–hydrophobe interaction so that the unitary thermodynamic changes in the transfer of the side chain from polar organic solvent to water are relatively insensitive to the polar organic solvent chosen. A typical computational procedure for estimating these unitary free-energies of transfer is as follows:

leucine (polar organic solvent) → leucine (water); $\Delta f_l(T)$
glycine (water) → glycine (polar organic solvent); $\Delta f_g(T)$

---

leucine side chain (polar organic solvent) → leucine side chain (water)
$\Delta f_{leucine} = \Delta f_l(T) + \Delta f_g(T)$

Brandts used expressions of the form $\Delta f_i^\circ(T) = a_i T + b_i T^2 + c_i T^3$ for the various subgroups of nonpolar amino-acid side chains and then, by summing these expressions over the hydrophobic groups, i, of a protein of known amino acid composition he obtained the coefficients for the average nonpolar residue:

$$\frac{1}{N_h}\sum_i a_i, \qquad \frac{1}{N_h}\sum_i b_i, \qquad \frac{1}{N_h}\sum_i c_i$$

in which N_h is the total number of hydrophobic side chains. The details of this procedure have been presented elsewhere (35), but it is important to note that the validity of this procedure rests on the assumption that nearly all of the heat-capacity change in a protein unfolding reaction is due to an increased number of nonpolar groups exposed to water (A).

We have followed Brandts in our previous separation of $\Delta F^\circ(T)$ into $\Delta F_p^\circ(T)$ and $\Delta F_h^\circ(T)$, but for additional progress we now assume that $(\Delta C_p^\circ)_p$ is zero within the errors of the method, as is suggested by the side-chain transfer measurements of polar amino acids. However, since we cannot yet in any case describe in adequate detail the amino-acid composition of any cooperative unit, this assumption is still insufficient and we must also adopt Brandts's concept of an average nonpolar residue. This means that we assume that the cooperative unit of the transition has the same average nonpolar residue as the total protein, regardless of the size of this unit. This last assumption is obviously a poorer and poorer approximation as the cooperative unit becomes a smaller and smaller part of the total molecule.*

There are 245, 243, and 241 amino-acid residues in chymotrypsinogen A, δ-chymotrypsin, and α-chymotrypsin, respectively, of which approximately fifty percent are classified as hydrophobic by Brandts. His classification is an empirical one based on the enthalpy and entropy change when amino-acid side chains are transferred from a polar organic solvent to water (cf. Section III, D). At body temperatures and below, a hydrophobic side chain on immersion in water produces a small reduction in enthalpy but a relatively large reduction in entropy. This behavior is manifested by tyrosine, norleucine, isoleucine, proline, phenylalanine, tryptophan, valine, methionine, lysine (through its methylene chain), alanine, and arginine (nonpolar part), but not by glycine, aspartic acid, glutamic acid, asparagine, glutamine, histidine, cystine, serine, or threonine. Ribonuclease A has 46 hydrophobic residues out of a total of 124 according to this classification. In Table 6 Shiao's data for several proteins have been analyzed using Brandts's method but not his temperature polynomial, Eq. (12). Instead, the heat-capacity changes on

---

* A detailed consideration of the significance of the empirical quantities obtained by Brandts' analysis in terms of polar and nonpolar residues has been given elsewhere (2).

TABLE 6

SEMIPHENOMENOLOGICAL PARAMETERS FOR TRANSITION I OF SOME REPRESENTATIVE PROTEINS[a,b]

Protein	$\dfrac{\Delta C_h^\circ}{\gamma}$	ΔH_p° (kcal/mole)	$\Delta H_h^\circ(T_r)$ (kcal/mole)	ΔS_p° (e.u./mole)	$\Delta S_h^\circ(T_r)$ (e.u./mole)	f_R[e]	Δh_p° (kcal/mole)	Δs_p° (e.u./mole)
Diphenyl carbamyl-α-chymotrypsin	80[c]	166	−43	990	−590	0.67	1.01	6.0
Chymotrypsinogen A	60[c]	126	−33	730	−440	0.51	1.01	5.9
Dimethionine sulfoxide chymotrypsinogen	44[c]	96	−24	540	−320	0.37	1.05	6.0
Ribonuclease A	50[d]	118	−22	590	−280	0.71	1.34	6.7

[a] From Shiao (46).

[b] Estimates based on zero temperature dependence of ΔC_h°; $\overline{\Delta C_h^\circ} = \overline{(\Delta C_p^\circ)_h}$. Mean hydrophobic heat-capacity change is γ. $pH_{ref} = 2.0$; $(Cl^-)_{ref} = 0.01$; $T_r = 314.5°K$.

[c] Based on average nonpolar residue approximation with standard transfer heat capacity per mole of side chains $\gamma = 53$ cal/deg-mole.

[d] Based on average nonpolar residue approximation with standard transfer heat capacity per mole of side chains = 50 cal/deg-mole; ΔC_h° for transition I of ribonuclease is 2.0 kcal/deg-mole using Shiao's (46) data and is independent of temperature within error. At pH 2.8, $T_0 = 314.5°K$ and $\Delta H^\circ = 83$ kcal/mole from Shiao (46).

[e] Fraction of total residue experiencing change in transition I.

transfer of side chains are considered to be independent of temperature. Then the heat capacity of transfer for the average hydrophobic residue of any protein is estimated by the appropriate average of the transfer heat capacities for the hydrophobic residues of the protein using heat capacities of transfer obtained from amino-acid transfer data. In the case of the chymotrypsinogen proteins this average is 53 cal/mole-deg. The experimental heat-capacity change for transition I divided by this number provides an estimate of the number of average hydrophobic side chains exposed to water, and since we have assumed that the cooperative unit of transition I has the same average composition as the total protein, this estimate is also an estimate of the number of hydrophobic residues that become exposed to water as a result of the unfolding in transition I. These estimates are given for four proteins of the chymotrypsinogen family and ribonuclease A in Table 6. Also given is the fraction of the total residues affected by the transition I as estimated on the assumption that the proportion of nonhydrophobic ("polar") residues involved in the transition is the same as the proportion of these residues in the total protein. This is not an unreasonable assumption insofar as nearly all of the charged ends of the lysine, aspartic acid, glutamic acid, and arginine are already exposed to water in state A and may not greatly change their solvation in state B. Except for chymotrypsinogen, these fractions, denoted as f_R in Table 6, are considerably lower than the fractions reported by Brandts and Hunt (63) and by Biltonen and Lumry (48). The differences are due to the method of analysis, since the experimental data obtained by Shiao are in excellent agreement with those obtained by the other workers.

By definition Brandts's empirical classification of side chains includes in the nonhydrophobic or polar group all those whose amino-acid models have a negligible heat capacity of transfer. This group has already been listed. According to Brandts's interpretation the enthalpy, entropy, and free-energy changes in transition I for the hydrophobic interactions produced by unfolding can be estimated from these quantities calculated for the average hydrophobic residue. The enthalpy and entropy changes in transition I from this source, ΔH_h° and ΔS_h°, are strongly dependent on temperature. The remaining contributions from hydrogen-bond exchange with water, configurational entropy, and so on for all residues affected by transition I are essentially temperature independent. These quantities are labeled ΔH_p° and ΔS_p° in Table 6, and the contributions per residue to these quantities are designated Δh_p° and Δs_p° in this table. A few observations based on this analysis are now in order.

(1) It is clear that α-chymotrypsin as represented by its slightly "better folded" derivative diphenylcarbamyl-α-chymotrypsin ($\Delta C_h^\circ = 4.2$ kcal/mole-deg against 4.0 kcal/mole-deg for α-CT itself) by no means fully unfolds in

transition I. This is consistent with the observations of Biltonen *et al.* (*51*) that that part of the optical-rotatory-dispersion pattern characteristic of the α-helix remains unchanged in the transition from state A to state B and disappears only in state C. Ribonuclease is also incompletely unfolded in transition I. This appears to be consistent with the analysis of ribonuclease unfolding as a stepwise process by Bigelow and co-workers (*67*).

(2) We have previously remarked on the unusual stability of the folded core of state B. If we accept state A as the reference state for heat-capacity analysis—and it is consistent with the limited absolute heat-capacity information on state A of α-chymotrypsin—then according to data in Table 6 the cooperative unit for transition I of this protein contains 67 percent of the protein, and the state B core is 33 percent or 80 residues. The NMR data of Biltonen *et al.* (*29*) for a wide variety of members of the chymotrypsinogen family suggest, though they do not prove, that the residues from 152 to 245, the bulk of the C chain, are never unfolded in any of these proteins in state A. This section of 93 residues is thus a candidate for the state-B core, especially since it also contains the only α-helix segment thus far detected in this protein by X-ray-diffraction work (see Fig. 4). It will be interesting to see whether or not isolated C chains of α-chymotrypsin are able to preserve a folded conformation in the absence of the A and B chains.

(3) Biltonen (*47*) and Biltonen and Lumry (*48*) have previously argued that the B states of chymotrypsinogen, α-chymotrypsin, and dimethioninesulfoxide-α-chymotrypsin are identical by several criteria. If so, the smaller numbers of residues in the transition-I cooperative units for chymotrypsinogen and DMS-α-chymotrypsin must be due to incomplete folding of these proteins in state A. This conclusion receives support from the work of Hollis *et al.* (*28*), which suggests that there is rapid motion in states A in the peptide chains of the methyl and methylene protons from residue 109 to 151 in chymotrypsinogen and from 1 to 78 in DMS-α-chymotrypsin (*84*) (see Fig. 4). The latter protein, DMS-α-CT, is stable at pH 2 only in state B, regardless of temperature and salt composition, and thus serves as a very useful state-B reference for ORD studies and other comparisons of state B and state A. It folds to state A at pH 3 with a T_0 value of 310°K compared with 314.5°K for DMS-chymotrypsinogen at this pH. The temperature-independent heat-capacity change can be determined using Biltonen's data for the pH-3 condition as 1.8 kcal/mole-deg. Using these data we find that DMS-α-chymotrypsin has about 83 residues "unfolded" in state A as compared with the 78 residues in section 1–78, indicated to be fast moving by the NMR method. A similar comparison for chymotrypsinogen finds 40 residues unfolded in state A by the heat-capacity method and 42 residues in the section from 109 to 151, found fast moving by Biltonen and co-workers (*29*) (A).

The agreement between the two methods is remarkably good—too good in fact—and thus probably is due in part to a cancellation of errors in the heat-capacity method. This probability follows from the use of the side chains of amino acids as models for estimating the heat-capacity change when an average hydrophobic side chain of the protein is removed from the polar-organic environment within the protein and extruded into bulk water. Such evidence as is now available from studies of Rosenberg and Enberg (31), of Tanford and co-workers (26), and from viscosity studies (27) on proteins in unfolded conformation in pure water solvent suggests that the protein is still a net with much random and intermittent cross linking within the poly-peptide. There is undoubtedly incomplete solvation in the sense that the water molecules grouped about hydrophobic side chains only partially resemble the solvation of hydrophobic side chains completely surrounded by bulk water. We expect to find that the heat-capacity analysis built on amino-acid-transfer heat capacities underestimates the number of residues that become exposed to solvent during transition I. Wetlaufer et al. (81) and Krescheck and Benjamin (82) have criticized the application of the amino-acid-transfer free energies to fully solvated hydrophobic side chains of true random-coil polypeptides. These criticisms are reasonable, and it seems possible that for quite accidental reasons the transfer free-energies are a better model for the partially exposed side chains in the net conformations of state B than they would be for the water–hydrophobe interactions in true random-coil polypeptides (A).

The reliability of the two methods of analysis remains to be firmly established. These initial applications are of considerable interest, but may be in error. For the present, we can consider the use of side-chain-transfer heat capacities, enthalpies, and free energies obtained from amino-acid-transfer reactions as a potentially sound pragmatic tool in thermodynamic analysis. In other words, regardless of the considerations that led to their use, the amino-acid-transfer quantities may prove to be a reliable means for estimating the number of residues participating in unfolding transitions from thermodynamic data.

(4) As is well known (83), δ-chymotrypsin, which is formed from chymo-trypsinogen A by removal of a dipeptide from positions 14 and 15, is as active in chymotryptic catalysis of small ester substrates as α-chymorypsin. The conversion of δ-chymotrypsin to α-chymotrypsin occurs following hydrolytic removal of the dipeptide at positions 146 and 147. According to the assignment by Hollis et al. (28) the conformational changes that mark the conversion of chymotrypsinogen to α-chymotrypsin should not occur with removal of residues 14 and 15, but rather with removal of 146 and 148, which are within the peptide section from 109 to 151. This conclusion receives strong support from Shiao's thermodynamic studies, as can be seen in Table 5. Within the experimental errors the heat-capacity, enthalpy, and entropy

changes in transition I for chymotrypsinogen and δ-chymotrypsin are identical. Hence, the thermodynamic data suggest that the conformation of δ-chymotrypsin is very different from that of α-chymotrypsin and probably very similar to that of chymotrypsinogen.

(5) It should be noted that although the unfolded parts of chymotrypsinogen and DMS-α-chymotrypsin now appear to be parts of the cooperative unit for transition I of α-chymotrypsin, if the interpretation of Hollis, McDonald, and Biltonen is correct, the "unfolded" residues in the two proteins share no common members.

Just what actually occurs when a protein unfolds remains to be established. Despite the fact that a relatively minor displacement of a section of the polypeptide from its neighboring sections in the compact conformation is all that is required to convert slow-moving protons to fast-moving protons, the resulting freedom of local motion, if maintained in the crystal, might blur detail in the X-ray pictures. It remains to be seen whether or not this blurring actually is observable. If it is, additional testing of the assignments of unfolding sections should be possible in this way. If it is not, crystallization may have restored the folding. In addition, we note a similar complication due to the existence of numerous A substates of the active chymotrypsin species. Each of these substates may have its own crystal structure and thus confuse the comparisons of a given active form of chymotrypsin in solution and crystal phases.

(6) There is considerable difference between the Δh_p° values for the chymotrypsinogen proteins and ribonuclease A, as well as between the Δs_p° values for these species (Table 6). It is possible that the analysis of Shiao, which we have used, exaggerates these differences, since Brandts and Hunt (63) compute considerably smaller differences. Nevertheless, we might expect significant differences. Not only the sizes but the relative compositions of the ribonuclease and chymotrypsinogen proteins are quite different. Both proteins have the same relative serine and threonine content, 20 percent, but ribonuclease has a 19-percent content of charged groups compared with 12 percent for chymotrypsinogen and a much lower content of hydrophobic residues: 35 percent versus 43 percent for chymotrypsin (85).

V. Entropy-Enthalpy Compensation and Its Relation to the Participation of Water in Physiological Processes

Nonpolar or weakly polar solvents accommodate solute molecules in holes formed by the rupture of the weak intermolecular van der Waal's interactions between solvent molecules. Except for the entropy of mixing, the thermodynamic consequences of such solvation are usually slight. As is well known, water as a solvent is quite different. It already has a large amount of unfilled space, which can be made available to solutes by rearranging hydrogen bonds and by breaking hydrogen bonds (*32, 33*). Polar solutes pay for their space at least in part by the bonds they make to water molecules. The low solubility of hydrocarbon molecules and honorary hydrocarbons such as oxygen, nitrogen, and the rare gases is always due to the strength of the water–water hydrogen bond, but its thermodynamic manifestation at lower temperatures is an unfavorable unitary entropy change and at higher temperatures an unfavorable enthalpy change. Many descriptions of the molecular rearrangements in these processes have been given, but none has been established and no one is clearly more correct than any of the others. The clathrate or "iceberg" description of Frank and Evans (*32*) is the most generally accepted description, but is too simple to fit the data and may be incorrect in its major details (*211*). At the present time a phenomenological approach is necessary to avoid deception, as Lumry and Rajender have discussed in another place (*205*).

If the solute molecule contains strongly polar groups as well as hydrocarbon parts, the solubility process depends on the ratio of the polar to nonpolar "surfaces" of the solute molecule, its shape, and volume. If the number of "polar valences" on the solute surface are such as to duplicate the thermodynamic situation that existed at the water surface of the cavity in the pure solvent, the unitary free-energy change in solvation should contain a very small enthalpy gain due to pressure–volume work and a small entropy loss due to configurational restrictions on water near the interface. However, protein molecules appear to have a less polar surface than is required to duplicate the thermodynamic interaction of water with water at the surface of the water cavity formed to contain the protein. The solvation situation that results from this inadequacy of matching of protein surface to water cavity surface is a major area of uncertainty (cf. Section III.J) in protein chemistry (*205*). The adhesive force between water and protein is certainly less and probably much less than the cohesive force within water, so that the protein is exposed to a large "internal pressure" tending to compress it so as to reduce the cavity surface particularly in ways that will increase the relative hydrophilicity of this surface. Using the Gibbs–Duhem equation we find the thermodynamics statement of the surface–volume relationship to be $-A\,d\bar{\sigma} = V_{w}\,d\overline{W}_{w} + V_{p}\,d\overline{W}_{p}$ in which A, V_{w}, and V_{p} are the total protein area, free volume in water, and protein volume, respectively (*205*). The interfacial free energy per unit area between water and protein is $\bar{\sigma}$. $\overline{W}_{w}$ and $\overline{W}_{p}$ are the charges in free energy per unit change in the water free volume and protein volume respectively. We see that any changes in the character of the interface, the internal structure of the protein as it affects the resistance of the protein to compression, and the ease with which water accommodates the change in protein volume will generally change the area and volume of the protein. A protein with a highly hydrophobic surface can be expected to have a higher density than a protein with a relatively more hydrophilic surface, though conventional partial-molal-volume experiments cannot be used to measure the difference. The contribution to the free energy from the compression process may be considerable, but at present it is difficult to make an educated guess as to what the quantitative consequences of this interaction between protein and bulk and "hydration" water are. The matter is of great current concern in protein study, but as shown in particular by Lauffer and his co-workers (86), it is very complicated. In addition, even greater complications in the water–protein relationship are suggested by the "kinks" (slope discontinuities) of Drost–Hansen (87), which are frequently found when physical and solvation properties of water are plotted as a function of temperature. Any reader who has developed a complacency toward water should see Ref. (88) by Kortum and Steiner in which the calorimetric heats of dilution of ethanol–water solutions by water are shown as displaying 27 discontinuous changes in slope when plotted against the mole fraction of ethanol. This is perhaps the world's champion collection of kinks in water behavior and, if experimentally correct, far exceeds our current ability to explain. However, the appearance of slope discontinuities when properties of water or water solutions are measured as a function of temperature must be accompanied by heat-capacity spikes, and these have not yet been observed.

We shall avoid any entanglement with the remarkable kinks. However, there have been some recent developments in the general area of water behavior that provide a clear-cut experimental approach to the problem of "protein hydration," although they do not give any resolution of the current confusion. In this section we will consider these matters, which can be closely grouped around the term "enthalpy–entropy compensation," that is, around phenomena that can be characterized by a compensation of enthalpy change by entropy change. In particular we shall discuss a specific process of this type that we suspect must be a feature common to many and possibly to most reactions in liquid water. Let us begin by implicating this special compensation process in the reactions of proteins.

The facts though quite remarkable, are easily presented. In Fig. 5 are plotted experimental enthalpy and entropy changes for several reactions of α-chymotrypsin with small molecules. In order to understand such "compensation" plots the reader should first examine Fig. 6, which contains a statement of the way in which compensation between ΔH_i° and ΔS_i° for the part process i in a total process shows up in the total enthalpy and entropy change. The compensation plot is $\Delta H_{\text{tot}}^\circ$ versus $\Delta S_{\text{tot}}^\circ$ as shown in this figure. When this plot contains a straight-line segment, there is entropy–enthalpy compensation in one or more of the part processes of the total process. The slope of the straight-line segment is called the "isoequilibrium" or "isokinetic" temperature, β, in the discussions of Leffler and Grunwald (91). We prefer the term "compensation" temperature, T_c. At $T = T_c$ the enthalpy change in the part process, ΔH_i°, is exactly balanced by the $T_c \Delta S_i^\circ$ term, so that there is no change in free energy due to this part process.

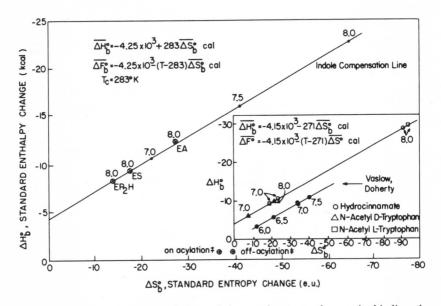

Fig. 5. Compensation plots of the enthalpy- and entropy-changes in binding the "side-chain" type of inhibitor to α-chymotrypsin. The solid points in the inset are for the binding of N-acetyl-L-dibromotyrosine studied by Doherty and Vaslow (93), using equilibrium dialysis. The open points in the inset figure were obtained by Yapel (94), using a temperature-jump method (95) (see text). The small solid points in the larger figure were obtained by Yapel using indole. The points marked ES, EA, and EP$_2$H were obtained by Rajender, Han, and Lumry (96) from steady-state-kinetics studies of chymotryptic hydrolysis of N-acetyl-L-tryptophanethylester at pH 8. These are defined in Eq. (20). The cross-in-circle points at the bottom of the larger figure are for the "on-acylation" and "off-acylation" activated complexes of this substrate. The numbers beside the points are the pH values of the experiments.

It is important to emphasize that enthalpy–entropy compensation is a thermodynamic characteristic of reacting systems in water or in any other solvent only in rare circumstances [cf. Ref. (205)]. Many examples of compensation have been reported; these matters as well as the nature of compensation phenomena have been discussed by Leffler (113), Leffler and Grunwald, and by Ritchie and Sager (234). It is often possible to attribute compensation to solvation effects; but no single explanation for the existence of compensation has been found, and such examples as are at all understood suggest that there are many different explanations. However, in the present discussion we are proposing that many if not all examples of compensation in processes occurring in water solution have a common origin in the properties of bulk water as a solvent.

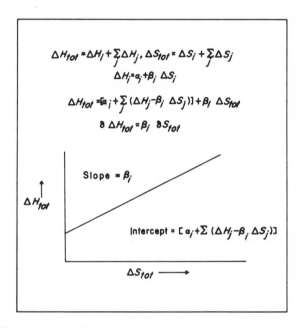

Fig. 6. The simple thermodynamic description of entropy–enthalpy compensation. An overall reaction is considered to be divided into partial processes j. Of these partial processes process i manifests compensation as defined by the equation in the second line of the figure. β_i is the compensation temperature (also called T_c). δ is an operator defined as "the variation in enthalpy (entropy) due to change in some independent variable" (91).

Enthalpy–entropy compensation can be detected if the independent variables of the system can be varied so as to produce different pairs of $\Delta H : \Delta S$ values for a given reaction. This variation is most commonly produced by small chemical variations in reactants to produce a homologous series of similar processes. Alternatively it may be possible to change the pH, salt composition, or concentration of co-solvents such as alcohols to produce variations in the $\Delta H : \Delta S$ pairs. As shown in Fig. 6, if a plot of ΔH_{tot} versus ΔS_{tot} yields a straight line, compensation is present in at least one part-process—such as the specific solvation of hydrogen ion and acid anion in the total dissociation process of carboxylic acids (92). Complex compensation patterns may be difficult to analyze if more than one part-process has its own compensation mechanism. Compensation may be somewhat more widespread than is now thought, but has escaped detection because of the quantitative confusion produced in this way.

Figure 5 is a consequence of work carried out by Doherty and Vaslow in 1952 (*93*). In the small box of this figure the solid circles represent data for the binding of *N*-acetyl-*L*-dibromotyrosine to α-chymotrypsin obtained by them using the dialysis-equilibrium method. The data are fit by a straight line with a compensation temperature of $271°K$. The open circles, squares, and triangles in the same box are data obtained by Yapel (*94*), using a temperature-jump method with which he measures directly the accessibility of the imidazole groups of this protein to a pH-indicator molecule freely dissolved in the solution (*95*). The indicator does not bind to the protein, but the binding of those competitive inhibitors that resemble normal substrates in having a bulky uncharged aromatic or aliphatic side chain prevents the proton transfer between this imidazole group and the indicator. Yapel was able to obtain standard enthalpy and standard entropy changes in binding by this method with a precision of one or two percent. As can be seen in this figure, although he did not study binding of the same competitive inhibitor studied by Vaslow and Doherty, he found with very similar inhibitors a very similar compensation plot in the sense that the compensation temperature is the same. We suggest that the best way to compare compensation processes is in terms of the compensation temperature. If so, Yapel finds the same process in his work as that found by Vaslow and Doherty. The lack of congruency of the lines obtained by these workers may be a result of the use of an impure protein preparation by the early workers, or it may be due to difference in the inhibitors. It has only been in the last few years that the contamination problem in α-chymotrypsin has been given adequate attention (*95*). Impurity errors would give an apparent low enthalpy of binding and would shift the compensation curve to the right, since the entropy axis shows negative values in this figure.

The affinity of the protein for these inhibitors changes little with pH. However, the enthalpy and entropy changes become progressively more negative with increasing pH. If, as we shall suggest, the Vaslow–Doherty compensation process is a common part-process in many protein reactions, the near constancy of the free energy of binding despite the very large changes in enthalpy and entropy provides an object lesson for the methodologies used in studies of protein mechanisms, since it is apparently necessary to recognize that conformational changes can produce large enthalpy and entropy changes that compensate each other to produce a minor change in free energy. We can expect to find that for all types of reversible conformation processes produced by changes in temperature, the enthalpy and entropy changes will have the same sign and thus tend to compensate. This expectation, suggested some years ago (20, 93), is in fact the major reason why the Vaslow–Doherty compensation process has been a matter of unusual interest.

In the main diagram of Fig. 5 are given data obtained by Yapel for the binding of indole to α-chymotrypsin. The straight line is drawn through his very precisely measured points. The value of T_c in this plot is 283°K rather than 271°K, so that we may conclude that the compensation phenomenon is quantitatively different with different inhibitors, but is nevertheless qualitatively the same in the two cases. This may mean that in inhibitor binding there is at least one other part-process in which weak compensation occurs to explain the variation in T_c. In any event, the qualitative identity of the binding of the two classes of inhibitors requires that we broaden the range of T_c values we will use to characterize the Vaslow–Doherty compensation phenomenon to the range from 270 to about 290°K, though in the absence of better understanding the range is quite arbitrary.

Along the indole compensation line in Fig. 5 are plotted points labeled ES, EA, and EP_2H. These points were determined by Rajender, Han, and Lumry (96) for the metastable intermediates in the chymotryptic catalysis of N-acetyl-L-tryptophanethylester analyzed according to the mechanism of Hartley and Kilby (97), Eq. (20):

$$E + S + H_2O \rightleftarrows ES + H_2O \rightleftarrows EA + H_2O + P_1 \rightleftarrows EP_2H + P_1 \rightleftarrows E + P_1 + P_2H$$

$$(20)$$

where   S = N-acetyl-L-tryptophanethylester;   ES = enzyme-substrate Michaelis–Menten complex; P_1 = ethanol;   EA = N-acetyl-L-tryptophanyl-α-chymotrypsin; EP_2H = Michaelis–Menten complex for acid product and enzyme; P_2H = N-acetyl-L-tryptophan in free acid form. EP_2H, ES, and EA binding thermodynamics were obtained by steady-state studies of the catalytic reaction. As has been discussed elsewhere (2), the thermodynamic changes in forming the activated complexes for "on-acylation" and "off-acylation" bear no similarity to corresponding quantities estimated for acidic or basic catalysis of ester substrates in homogeneous solution. There is a clear pattern of protein involvement in the process, since the enthalpy and entropy values for formation of the metastable intermediates and the activated complexes from separated enzyme and substrate are always negative. Indeed, relative to this reference state the activation energies for both on-acylation and off-acylation are zero or slightly negative.

In summary, the binding of certain types of inhibitors to α-chymotrypsin is marked by a compensation behavior that also characterizes the metastable intermediates ES, EA, and EP_2H. The inclusion of EA in this list is particularly interesting, since in this metastable intermediate there is a primary bond between the acid part of the substrate and the hydroxyl group of serine

195 of chymotrypsin (*77, 98*). Probably as a result of the restraint provided by this bond the magnitude of change in the compensation process is less than that obtained with inhibitors at higher pH values (see Fig. 5). These matters are discussed in detail elsewhere (*2*), where it is suggested that the compensation process is an anisotropic shrinking of the protein and expansion of water that mechanically forces ES, EA, and EP_2H into their activated complexes for bond rearrangement.

Whatever the specific compensation process involved in the binding and catalytic reactions of chymotrypsin, compensation is not unique to chymotryptic catalysis. Likhtenshtein (*99*) has collected entropy-change and enthalpy-change data for the rate parameters of a large number of enzymic reactions and finds a common pattern of compensation among them. The extreme variety of his examples is disconcerting, since we do not expect the limiting velocity at saturating substrate concentrations or the Michaelis–Menten constants to measure similar rate phenomena in the different examples. We suspect that many of the examples he quotes are artifacts and thus not members of the Vaslow–Doherty family of compensation processes. There are several experimental pitfalls in determining whether or not a process contains a compensation subprocess and even greater possibilities for error in determining the value of the compensation temperature [Cf. Ref. (*205*)]. For example, if the experimental method for measuring the equilibrium position or the rate of a type of process is limited, only those reactions with equilibrium or rate values in the experimentally available range can be studied. Then the free energies measured are limited to a narrow range, so that the enthalpy and entropy changes must approximately compensate. This type of error is more likely to appear in studies of homologous series of chemically different reactants than in the variety of experiment in which pH, solvent additive, and similar alterations are made to vary ΔH and ΔS. However, even in the latter type of experiment, errors in the determination of enthalpy change—which may be quite common in proteins because of unusual heat-capacity effects—will generally lead to compensatory errors in computing the entropy from the free-energy change and the enthalpy change. The first type of error is a natural consequence of poor experimental design and technique. The latter error is somewhat more subtle and might be expected to be common in protein unfolding processes when there are very large enthalpy changes. In any event, Likhtenshtein's presentation of a compensation pattern common to a variety of enzymes, though less than convincing, does suggest that entropy–enthalpy compensation may be common in enzymic processes.

In Figs. 7 and 8 are given additional examples of compensation processes for proteins. Figure 7 is a compensation plot for the activation enthalpy versus activation entropy in the "irreversible" thermal denaturation of

hemoglobin carried out with varying amounts of ethanol added to predominantly aqueous solutions of this protein (*100, 101*). The compensation
temperature is 290°K in this figure, and the range of activation enthalpy and
entropy change is very large—so large in fact as to make it quite improbable
that a compensation process of this magnitude with unchanged compensation temperature can be ascribed entirely to the protein. The only material
of uniform properties present in sufficient amount to support this large degree
of compensation change is obviously water. A similar conclusion can be

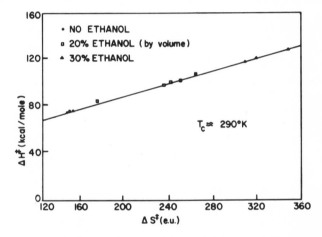

Fig. 7. Compensation plot of the activation enthalpy and activation entropy for thermal
denaturation of hemoglobin in water and ethanol–water solutions. From Refs. (*100*)
and (*101*).

drawn from Fig. 8, which is a compensation plot of changes in the standard
entropy and standard enthalpy for the first thermally produced unfolding
transition, transition I, of ribonuclease A when various amounts of ethanol
are present. The data are taken from Brandts and Hunt (*63*). The
compensation temperature is 285°K.

There is a number of additional examples of protein denaturation
reactions that manifest compensation part-processes with T_c value in the
Vaslow–Doherty range. Some and perhaps many of these are untrustworthy
for the experimental reasons described above, and a complete reexamination
of each of these is necessary before any unequivocal assignment of compensation is made. Nevertheless, taken in aggregate these examples are impressive.

Thus far we have presented evidence suggesting that the Vaslow–Doherty
compensation phenomenon is widespread among the reactions of proteins.
Now we should like to provide evidence that suggests that it is widespread
among reactions of other types in water. We shall not be able to describe

the examples in detail [the reader is referred to Refs. (2) and (205)] but can list a few of the more reliable ones (A).

One of the most important of the small-molecule examples is the quenching of the fluorescence of indole in water-alcohol systems. This reaction, studied by Walker, Bednar, and Lumry (102), is unusual for fluorescence quenching

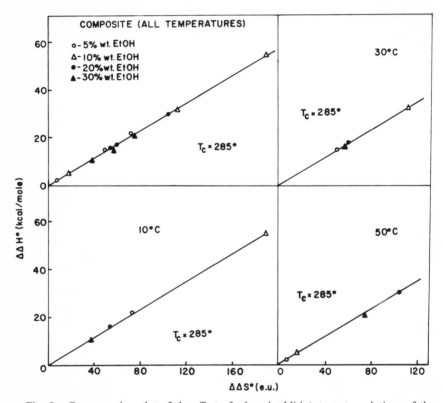

Fig. 8. Compensation plot of the effect of ethanol addition to water solutions of the ribonuclease *A* on the standard enthalpy and entropy changes in the reversible unfolding (transition I). Computation (by D. S. Shiao and G. Pool—unpublished) based on data from Brandts and Hunt (63). Note that as the alcohol concentration is increased the thermodynamic differences first increase along the compensation line to a maximum and then reverse direction and move back down the line. Although the dependence on alcohol concentration varies with temperature, the compensation temperature is always found to be 285°K.

in that in pure water the activation energy is very high, nearly 9 kcal/mole. In addition, when various proportions of methanol or *t*-butanol are present, the pairs of activation-enthalpy and the activation-entropy values form points on a compensation plot with a T_c value of 285°K. Thus we are again led to believe that the Vaslow–Doherty compensation phenomenon is being measured.

The large temperature dependence of the quenching reaction, though not unique, is very unusual, particularly if quenching proceeds in the usual way via intersystem crossing to the triplet state and from thence to the ground state. However, Grossweiner and Joscheck (*103*), using flash photolysis, have found that the excitation of indole in water and alcohols produces hydrated electrons.* This is particularly interesting, since electron ejection into the solvent requires the provision of large holes for the electron. The de Broglie wavelength of the electron is large. Approximately 60 ml per mole of free volume must be provided by the water to accommodate the electron (*104*), although the net amount will be less than this as a result of electro-striction in the solvent (A).

A second and more extensive category of small-molecule reactions that appears to manifest the Vaslow–Doherty compensation phenomenon are the solvation rearrangements that occur when simple carboxylic acids of all common types dissociate in water into proton plus acid anion. Divisions of the total ionization process into a part-process sensitive to variations in the internal electronic characteristics of the acids and a part-process that is the solvation of the proton and acid anion once formed in the first part-process have been carried out by a number of authors. However, the most extensive work of this type has been that of Hepler and his co-workers (*92*). These workers find a compensation temperature in the 280–290°K region, as do Ives and Marsden (*105*), whose discussion of the problem of compensation in these reactions is very illuminating.

The aggregate of these ion-solvation processes suggested to be of the Vaslow–Doherty type by the similarity of their compensation temperatures is large. However, there is a still larger and a much more varied group of processes that may also fall into the family of Vaslow–Doherty compensation processes. We have selected our examples for discussion of compensation in this chapter to emphasize the special behavior of alcohol–water solvents. In indole quenching and the reversible unfolding of ribonuclease a consistent pattern of alcohol effects can be detected. Starting from zero alcohol the effect of alcohol rises with its mole fraction to produce some maximum extent of enthalpy and entropy change. Then the degree of advancement of the compensation process falls with a further increase in alcohol concentration and becomes less than the degree of advancement in alcohol-free solutions. Although the data are insufficient to make complete quantitative comparisons, there is a clear indication that the larger the nonpolar part of the mono-hydroxy alcohols, the smaller the mole fraction of alcohol at which the

---

* Hopkins and Lumry have confirmed this observation using a scavenging method (*208*).

maximum in advancement of the compensation phenomenon occurs. The influence of the alcohol is considerable. For example, since the maximum in enthalpy and entropy of activation for the quenching of indole fluorescence occurs at about 0.025 mole fraction in t-butanol–water solutions, it is clear that a single molecule of this alcohol influences in a very important way the behavior of about 40 water molecules. At 10°C the maximally effective mole fraction for ethanol in the ribonuclease studies is about 0.04.

This pattern of compensation behavior of alcohol–water solutions is already well known as a unifying feature of a large number of processes that have been studied in water–alcohol solutions. Some of these processes are the rate of solvolysis of t-butylchloride (106, 107), the solubility of strongly nonpolar ions such as sodium tetraphenylborate (108), and the displacement of the maximum in the charge-transfer absorption band of iodide ion (109). The reader is referred to the excellent reviews by Franks and Ives (110) and by Arnett (108). Many workers have made contributions to the list, but the essential unity of these processes as regards their common sensitivity to alcohol has been most emphasized by Franks and Arnett. We shall label these processes the "alcohol-perturbed" group for future discussions. Our examples of Vaslow–Doherty behavior in alcohol–water solutions with small mole fractions of the alcohol fall into the alcohol-perturbed category as well as the Vaslow–Doherty group and thus suggest the possibility that both kinds of process have a common unifying feature and are to be grouped together. This does not mean that the Vaslow–Doherty process is the most important subprocess in the alcohol-perturbed group. This cannot be the case, since the latter group generally does not demonstrate either a simple compensation behavior or the same apparent compensation temperatures found in the Vaslow–Doherty group. Nevertheless, compensation of enthalpy change by entropy change is the important manifestation of alcohol dependence, and we suspect that the processes in the group are simply more complicated because they contain other subprocesses in which compensation also occurs, but with their own compensation temperatures different from the 270–290° range. The analysis of the rates of solvolysis of t-butanol by Arnett and co-workers (111) have shown that the overall processes can be broken down into part-processes in somewhat the same way that Hepler and co-workers (92) have dissected the ionization processes for carboxylic acids. Thus, we suggest that there is a common dependence on the properties of predominantly water solvents that unites all the reactions discussed in this section.

In order not to overemphasize the role of alcohol or perhaps the better to emphasize the fundamental importance of liquid water it should be noted that the transfers of small organic molecules such as ethane from pure water to 8-M urea–water or 6.7-M guanidine hydrochloride–water solutions as determined by Wetlaufer and co-workers (*81*) are characterized by compensation with compensation temperatures near 270°. Thus it would appear that the alcohols are not alone in their effect on water properties. In this example it is important to note that we do not suggest that urea or guanidine hydrochloride is equivalent in its effect to the alcohols; it is well known that there are very significant differences. We do suggest however, that one of the effects of both kinds of solvent additive is to control the degree of advancement of the Vaslow–Doherty compensation process. The quantitative manifestation of this control is complicated by the participation of the other part-processes, particularly in many of the alcohol-perturbed examples.

It remains to be seen whether or not the rather sweeping generalizations suggested above are indeed true or, if true, provide remarkable information about the behavior of bulk water on the one hand and proteins on the other. In any event, the occurrence of compensation apparently attributable to water in protein reactions requires that a more searching analysis of the interaction of water with proteins be carried out. These observations suggest that the point of view most effectively championed by Klotz (*112*) that there is organized water about globular proteins and perhaps about all polymers having a large proportion of hydrophobic groups deserves more serious consideration. We believe that the accommodation by water of large protein molecules with generally less polar surfaces than the surface of the water cavity in which they lie may induce drastic reorganization in nearby bulk water out to rather large distances from the protein surface [see also Likhtenshtein (*99*)]. The collected observations presented in this section may ultimately prove to have been misleading. Nevertheless, at present they suggest that possibility of an entirely new and unexplored aspect of protein chemistry and they provide small-molecule processes that may be used as guides and controls in exploring this new aspect.

It is probable that the Vaslow–Doherty phenomenon and probably the alcohol-perturbed phenomena are manifestations of some sort of free-volume bookkeeping required by the hydrogen bonds of water. At 273°K there exists an exact compensation process in water–ice mixtures such that volume can be made available to a system of solutes by melting of the low-density form of water, ice. It is possible that this melting process occurs at temperatures significantly above the true melting point as a continuation of normal melting but in a much less cooperative way than melting at 273°K because the remaining ice crystallites are much smaller than the domains of cooperation at 273°K. If normal icelike microcrystallites or similar low-density forms of water are the so-called "low-temperature water" as is suggested by Jhon *et al.* (*114*) and by Horne and co-workers (*115*), then their melting to provide space for the increasing volume of a reacting system may be the basis for the Vaslow–Doherty phenomenon. The application of this idea to an enzyme system is shown in Fig. 9. As shown in Fig. 9 we may hypothesize that an enzymic protein can borrow free energy from nearby bulk water by increasing or decreasing its volume, depending on whether we believe that the predominant "structured" species is less dense or more dense than the liquid species. More specifically, we might suggest that an enzymic protein is a machine programmed in evolution to couple the substrate mechanically to bulk water so as to provide a suitably directed force for chemical change in the substrate system; e.g., so as to distort a substrate into

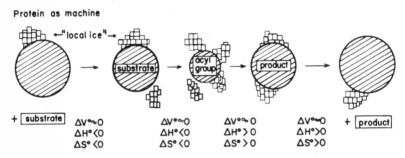

Fig. 9. A "sample hypothesis" of an expansion–contraction relationship of protein and small ice domains, which might be the Vaslow–Doherty compensation process. Protein serves as a machine to "apply" internal pressure of liquid water to the substrate–product system in chymotryptic catalysis in such a way as to facilitate chemical reaction. Ice domains of 20–40 H_2O molecules act as enthalpy reservoirs in this process. (See text.)

an important activated complex. Specific mechanisms based on anisotropic swelling and shrinking of the protein can be proposed for chymotrypsin and are consistent with some of the characteristics of chymotryptic catalysis. However, it is important to point out that if the Vaslow–Doherty phenomenon is the manifestation of this process, the protein cannot borrow sufficient positive free energy from water in this way to produce the total observed catalytic efficiency. The evidence that a compensation process located primarily in water does accompany the reactions of chymotrypsin is strong, but it is quite possible that we have misplaced the emphasis. Thus the Vaslow–Doherty phenomenon may be a universal concomitant of changes in the volume of a reacting solute system in water but not a process that nature has been able to exploit effectively* (A). If so, the role of the compensation

---

* In chymotryptic catalysis with small ester substrates the effect of compensation is to replace large activation enthalpies for chemical steps that are rate limiting by large negative entropies of activation. This type of substitution would appear to have importance for cold-blooded organisms, since it can reduce the sensitivity of the organisms to variations in ambient temperature. For further discussion of this possibility see Ref. (2).

process may be minor or negligible and thus in a sense a red herring. Perhaps it does properly indicate that changes in protein volume take place during specific physiological reactions of proteins and that it is to these volume changes that we should direct our attention. In any event, the compensation phenomenon or phenomena in at least some protein systems must be taken into consideration in any quantitative study of these proteins, whether it be concerned with unfolding kinetics, unfolding thermodynamics, acid and base ionization, protein association reactions, solubility, ion binding, or physiological functions.*

VI. The Lilliput Principle

A. Rack Mechanisms

The diversity of the specific physiological reactions of globular proteins tends to emphasize the differences among proteins rather than the similarities. In fact, the essential similarity is much more important than the differences. This similarity is the ability to provide an extraordinary delicacy of balance in the electronic situation at one or more local sites where the specific chemical reactions take place. The catalytic efficiency of enzymes, like that of all catalysts, is determined by the ease with which electron density is transferred between the specific catalytic groups of the protein or protein-bound coenzyme and the atoms of the substrate undergoing reaction. Thus the problem faced by nature in developing catalytic proteins has been to establish electronic states with energy, polarizability, electronegativity, etc. most appropriate to the specificity and efficiency of a desirable substrate transformation. This is essentially the same problem that has been solved to establish the oxygen affinity of hemoglobin and myoglobin in a given organism and to establish the oxidation–reduction potentials of the cytochromes. All these proteins share the common feature of precise electronic balance. Furthermore, it is to be noted that there exist one or more mechanisms common to proteins that make species adjustment and natural selection of electronic properties possible. For example, the various mammalian hemoglobins support the same elementary reaction, which is the binding of oxygen to iron; yet the binding constants vary widely among the species of

---

* Since this chapter was written, a number of articles describing enthalpy–entropy compensation in protein systems has been found. In all cases the T_c values are near 285°K. The most extensively studied cases are the binding of sixth-position ligands by methemoglobin, studied by Anusiem, Beetlestone, and Irvine (209), and the binding of inhibitors to acetylcholinesterase, studied by Belleau and co-workers (210). These are discussed at length in Ref. (205).

hemoglobins, and each hemoglobin appears to have been evolved to have the most functionally suitable oxygen affinity for the organism in which it is found. Similar situations exist for enzymes. In some instances the quantitative catalytic parameters are not much different when enzymes with the same catalytic function, but obtained from two different organisms, are compared. In other cases, however, the kinetic parameters are different, and the catalytic mechanisms can be quite different. Thus, in contrast to the behavior of reactions of simple catalysts in inorganic and organic chemistry, proteins possess the property of chemical mutability. In this section we shall discuss the possible mechanisms by which electronic balance and chemical mutability may be achieved in proteins.

We must begin by acknowledging the fact that there is a number of ways in which detailed chemical balance might be achieved, and undoubtedly a number of these are put to use in one protein or another. A particularly interesting example is the finding by Chanutin and Curnish (16) and by Benesch et al. (167) that the affinity of hemoglobin can be greatly altered by phosphoacids, particularly 2,3-diphosphoglyceric acid. In hemoglobin, oxygen affinity is in part controlled by this regulatory substance, even though only one diphosphoglycerate ion is bound to a four-subunit hemoglobin molecule (118). As has been mentioned already, Keyes (10, 11) finds a contaminant with similar regulatory ability in myoglobin preparations. In addition, a number of other examples of so-called "allosteric regulation" by small molecules have been reported (119–121, 174–176). It has been the contention of the author and his co-workers for some years that this regulation is usually a consequence of adjustments in protein fabric that, through mechanical forces, cause alterations in the electronic behavior of the protein at its immediate functional sites. The mechanical process by which electronic properties of functional groups of the protein and the chemically reacting atoms or bonds of substrates are supposed to be altered is generally called a "rack mechanism" (20, 133) to emphasize the importance of strain and distortion in the mechanism. In another place (2) [see also Refs. (122, 116, 179)] we have discussed at length the manner in which rack distortions associated with conformational readjustments can produce increased reactivity in bound substrates and thus a lowering of the free energy of activation for the chemical steps of rearrangement through distortion of substrate and protein. In the present discussion we will concentrate on a different group of proteins and a different aspect of the mechanical–chemical interaction of protein and substrate. Specifically we shall discuss how the functional groups of proteins as well as coenzymes and metal ions bound to proteins can be caused to take on functionally useful, but often very unusual, electronic characteristics. Most of the discussion will use one or another of the hemeproteins as models. For a more thorough discussion of these and other cases of rack mechanisms the reader should consult the original references in which are described the application of rack mechanisms to hemoglobin (*18, 20, 116*), myoglobin (*18*), cytochrome *c* (*21*), chymotrypsin (*2, 116, 124*), oxidative phosphorylation (*124*), and other physiological systems.

B. Definition of the Reaction Part

It is convenient to think of any protein that supports a specific physiological reaction as being divided into two parts. The first of these is the small part containing the atoms directly involved in the electronic rearrangements of the reactions. The second part is the remainder of the protein. The first part we will call *the reaction part* and second *the protein remainder.* Since the evidence firmly precludes the participation in protein ground states of low-lying, thermally excited states linking electronically any large region of the protein, this division is always possible. For heme proteins the division is easily made, at least in hemoglobin and in the simple cytochromes. The reaction part is the complex iron ion consisting of the iron ion; the prophyrin ring, whose nitrogen atoms form the base plane of the roughly octahedral complex; and the ligands at the fifth and sixth coordination positions. In such cases the complex ion can often be constructed free of protein, and many relevant small iron ions have been studied. In fact, further progress toward an understanding of the mechanism of the heme proteins requires that appropriate small models be well understood, both theoretically and experimentally, in the absence of complications due to the remainder of the protein.

The properties of the reaction part can be considered to be developed in three stages. The zero-order properties are those revealed in studies of the isolated reaction part, the complex iron ion. On incorporation into the protein remainder these properties, specifically the electronic properties, are altered by the major influences of the protein to a set of first-order properties. Final adjustment is achieved under the aggregate effect of the weaker influences of the protein. This final set of properties may be simply called the second-order set.

C. Influence of the Protein Remainder

(1) The major factor determining the range of obtainable first- and second-order properties is the zero-order set. The first-order set arises as the result of a major perturbation of the zero-order state, and by definition the differences between first-order and second-order properties are due to small

perturbations of the first-order state, as occur from species to species. We note already the possibility that the reaction part is not thermodynamically stable when removed from the protein. A major "first-order" role of the protein in such cases is to stabilize the reaction part. More specifically, we note that coordination number, ligand selection, and ligand configuration about the metal ion are under the control of natural selection through the construction of the polypeptide chain. Simple complex ions that are unstable and thus can be neither synthesized nor easily studied in the laboratory can be and probably are common as reaction parts of proteins. This ability to construct and stabilize complex ions not found in the laboratory is undoubtedly the major first-order role of the protein conformation in proteins such as hemoglobin and carboxypeptidase A, in which a metal ion participates in function as a center for electronic rearrangements.

The list of additional ways by which the protein remainder might influence the electronic properties of the reaction part in such metal proteins is as follows:

(2) A change in Born-charging free energy relative to water solution for a reaction part bearing one or more charges. This change is inevitable for charged reaction parts and is due to the fact that the protein provides an environment of different and probably very complex effective dielectric constant.

(3) Specific positioning of charged and dipolar groups about the reaction part. If the reaction part is charged, this factor cannot be dissociated from factor (2).

(4) Restriction on changes in ligand geometry. For example, oxygen does not easily oxidize hemoglobin iron, although it is generally supposed that the zero-order complexes with iron in the II state will be oxidized by oxygen very readily. The protein apparently holds the proximal imidazole group and the porphyrin plane so that the distortions necessary to achieve the activated complex configuration for electron transfer require a large activation energy (*125*). It has been suggested that the distal imidazole aids in this protective mechanism by preventing the oxygen molecule from making an end-on attachment to the iron ion (*126*). The situation is interesting for several reasons, not the least of which is the probability that when oxygen does oxidize iron in hemoglobin and myoglobin, it is not the bound oxygen molecule that is the specific oxidant. Electron transfer then takes place via a less direct pathway, and deoxyhemoglobin is much easier to oxidize than oxyhemoglobin—so also myoglobin. On the other hand, the complex-ion configuration of cytochrome *c* appears to be adjusted by evolution so that the protein holds the complex in a geometry corresponding closely to the geometry of the activated complex for electron transfer. In this way the

oxidation and reduction of heme iron in cytochrome c can take place much more rapidly, though it is not clear that this behavior has been the major result of the evolution of this protein.

(5) Provision of supplementary oxidation–reduction groups on the protein or coenzyme to facilitate multiple-electron oxidation–reduction processes. As is well known, most organic oxidation–reduction processes require a two-electron change and when forced to go through two one-electron steps usually become sluggish because of the high free energy of the one-electron intermediates. In some of the flavin enzymes, in addition to the flavin redox group the protein contains a suitably positioned and remarkably reactive disulfide group, so that when oxidation of amino acids occurs, for example, there is simultaneous electron transfer to the disulfide group and the flavin (*127*). The one-electron intermediate stage of the amino acid is thus effectively bypassed.

This is but one example of chemical cooperation in protein systems. In such cases it is perhaps most useful to group together the two chemically participating systems, i.e., the flavin and the reactive disulfide group, as the reaction part.

We should include under this heading the possible provision of displaceable ligands, by which we mean that one and possibly two protein-held ligands may be displaced by other protein-held groups when the oxidation state of the metal changes or the nonprotein ligand changes. Thus, there is a possibility that the sixth-position ligand of Fe(III) cytochrome, possibly a methionine side chain (*128*), is displaced by another protein-held group, perhaps a phenol, as a step in the total reduction process.

(6) Pi bonding to protein groups and in general inductive and resonance interaction between the reaction part and protein or coenzyme groups (see Fig. 10).

(7) Provision of what might be called charge-adjustment sites. For example, the distal imidazole of myoglobin has been shown to serve as a local binding site for negative ions whenever myoglobin is in the Fe(III) state and does not carry a negatively charged sixth-position ligand (*235*). As was apparent in discussing the effective dielectric constant inside proteins, it must usually be true that the production of a charge at an inside point of a protein is very expensive in free energy.

(8) Distortions of the reaction part away from the geometry of its lowest energy. Almost an infinite variety of geometries can be produced in this way to provide very delicate adjustment of the electronic state or states of the reaction part. This adjustment may be primarily a static matter, as is the case in the adjustment of the heme complex to establish the evolutionarily useful oxygen affinity of a given species of myoglobin, or it may be dynamic in the sense that coordinated fluctuations in protein geometry produce a cyclic

pattern of oxidation–reduction potential values in a cytochrome, for example.

The dynamic cooperation between reaction part and protein remainder is probably very important. Keyes (*11*) has provided some evidence by comparison of the entropy of sixth-ligand binding to myoglobin with the nephelauxetic parameters (*202*) for the ligands that there is a continuous

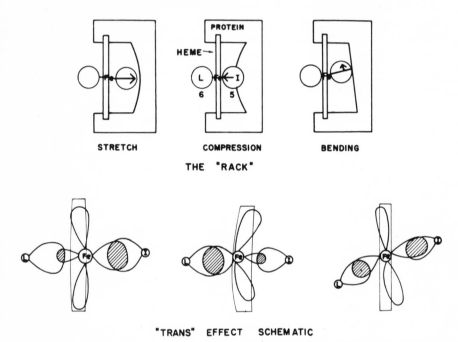

Fig. 10. An early pictorialization of the rack mechanism in heme proteins showing some of the ways the electronic properties of iron could be altered by positioning of the proximal imidazole group under partial control of protein conformation. The lower series emphasizes in an approximate way the "trans" effect linking electronically the ligands at the fifth and sixth positions of heme iron and the porphyrin. From Refs. (*18*, *116*).

adjustment of the protein remainder to the increasing net binding energy of ligand and iron ion if, as is generally supposed, this interaction is adequately estimated by the nephelauxetic parameter. If this suggestion is correct (see also Table III) the ligand-binding process is governed by a net free energy of binding containing important contributions from the protein remainder as well as from the reaction part. A situation of this type is responsible for the close balance between orbital-energy separation and electron-pairing energy in ferric myoglobin and ferric hemoglobin and in ferrous myoglobin and ferrous hemoglobin. The easy conversion of high-spin to low-spin state in these proteins is not a zero-order property of the reaction parts, but rather the consequence of delicate free-energy balance effected by balancing the free-energy changes associated with conformational displacements in the protein remainder against the energy and entropy changes in the reaction part. According to Keyes' suggestion, this is a continuously variable process, the extent of which is determined primarily by the interaction between the sixth-position ligand and the heme group. The mutual distortions of reaction part and protein remainder that accompany ligand binding may well produce changes in symmetry leading to changes in geometry of the porphyrin not related to Jahn–Teller effects or enhanced atom rearrangements leading to lower symmetry through the participation of the Jahn–Teller effect. Clearly the element of adjustability lies in the protein remainder once the reaction part has been chosen.

(9) An equally important, but more complex case of item (8) is the provision of coupling or, as it is more commonly called, "linkage" between reaction parts. In the best known example the heme groups of hemoglobin are connected to coordinate oxygen binding at these four sites. In addition, heme iron ions of hemoglobin are coupled to a "regulatory" controlling site where a controlling chemical species is bound (16, 167).

Although one or more examples of most of these first- or second-order roles of the protein conformation can be provided, it is not possible to discuss all the roles in detail here. It is obvious that some of these must be common in proteins with metal-containing reaction parts. Thus, the selection of ligands for a metal ion, the alteration of the dielectric situation at charged reaction parts, and the restrictions on major changes in geometry while the reaction part is fixed in the protein are always present and always important. In addition, the protective feature provided by burying the reaction part in the protein may be important, not only to prevent destruction of the reaction part, but also to limit the variety of nondestructive reactions. Pauling and Lein (129) have shown that the protein structure about the sixth-position face of the heme group in hemoglobin and myoglobin can greatly restrict the binding of large ligands such as tertiary butylisocyanide at the sixth-position. George and co-workers have emphasized this feature of the sixth-position site of the heme group (130). It is possible that the absence of any sixth-position ligand in deoxy-Fe(II)-myoglobin, as is suggested by X-ray studies (13), is also a consequence of the conformation about this face of the heme. However, it is equally possible that the affinity for water is so low at the sixth-position site as a result of the geometry enforced on the porphyrin and the proximal imidazole group that water occupancy is greatly reduced.

The latter explanation is another example of the application of item (8) from the list, and it is this item and item (9) that we should like to emphasize, particularly in connection with the known peculiarities of hemoglobin and myoglobin. For myoglobin these include delicate adjustment of affinity for oxygen and prevention of oxidation of iron by bound oxygen; for hemoglobin, linkage among the heme groups, between heme groups and proton sites, and between the heme groups and the control site. In addition, there are many nonphysiological properties of these proteins produced as the result of laboratory modifications, which fall under the heading of linkage, but which show a variety of specific linkage patterns (*131*).

Despite the general occurrence of a distal imidazole group (α^{58}, β^{63}, or His 65 in myoglobin) in normal hemoglobins and myoglobins, the oxygen-binding reactions in the Fe(II) states appear to be insensitive to its presence. It is possible that this group plays a steric role in oxygen binding, but even this is not established. In addition, since neither oxygen nor Fe(II) heme is charged, the dielectric effect of confinement to the interior of the protein and the use of charged groups to influence the electronic properties of heme iron are second-order effects, unless there is large delocalization of iron electrons on the oxygen molecule to set up a large electric dipole. The existence of a large $Fe-O_2$ dipole is not consistent with the independence of the oxygen-binding process of myoglobin on pH, that is, on the charge state of the distal imidazole, nor with the magnetic state of oxymyoglobin and oxyhemoglobin. Aside from steric restrictions on the sixth-position side, the only important influences on the heme group in oxygen binding are those mediated through protein groups in direct contact with the heme group. Sensitivity of oxygen binding in hemoglobin or myoglobin to structural details and changes in structural details at distant parts of the protein would seem to require the "passage of information through the protein fabric" and the contact groups to the heme group. See Note added in proof, p. 106.

Of the contact groups the more important would appear to be those that are ligands of iron. There are possibilities for charge-transfer "binding" between aromatic groups in contact with the porphyrin, but we have too little information about these interactions to discuss them in a useful way here. On the other hand, it is quite clear that the protein can and does exert considerable control over the electronic state of heme iron as it is manifested in the oxygen-binding reaction or in the iron oxidation–reduction process, and does so, at least in part, by adjusting the geometry of the liganding groups. Such adjustments must, in almost all cases, involve a distortion of the zero-order complex ion—that is, the reaction part—away from its state of lowest free energy. Conversely, the directed-valence forces of the iron ion tending to adjust the ligand geometry to achieve the state of lowest free energy for the zero-order complex ion must produce distortion of the protein fabric away from its conformational state of greatest stability. There is thus established through mutual adjustment of these two opposing forces a compromise geometry that has associated with it the physiologically desirable electronic properties of heme iron. Even though studies of model Fe(II) complexes are not yet adequate to provide a quantitative description of the results of such ligand distortion, it is obvious on the basis of the information obtained from other small complex ions that a wide variety of quantitative changes in the electronic state of iron is possible. The specific adjustments, all generally describable in the conventional language of complex ions, should perhaps be listed: distortion of the porphyrin plane—now shown most thoroughly by Fleischer and co-workers (132) to be both common and versatile; distortion of the substituents at the periphery of the porphyrin; compression or stretching of the bond between the proximal imidazole nitrogen atom and the iron ion; bending of the latter bond; rotation of the plane of the proximal imidazole. Jahn–Teller distortion, control of orbital symmetry, magnetic susceptibility, and in fact all aspects of the electronic state of the heme group can be adjusted by these means, though only within certain quantitative limits set by the zero-order choice of fifth- and sixth-position ligands. In the first section of this chapter we estimated that the contribution of the protein conformation to physiological processes is probably not greater than about 15 kcal per mole of reaction part. This may not be a bad guess for the upper limit of the adjustability of the heme-protein system; but the sixth-position of heme iron in deoxymyoglobin (13) is unfilled, and if it is also unfilled in deoxyhemoglobin, and in some states of partial oxygenation of hemoglobin, a much larger unfavorable free-energy change may take place in the protein to keep the ligand affinity low.

D. Rack Mechanism in Heme Proteins

The general picture of distortion and strain established between heme iron and protein conformation for the purpose of chemical adjustment is a special example of a mechanism called "the rack" by Eyring, Lumry, and Spikes (133). In fact, it was the first example of a rack mechanism in which a specific molecular mechanism was proposed (20). In this original proposal the oxygenation behavior was attributed to unusual electronic characteristics of iron due in turn to distortions of the porphyrin and the interactions between proximal imidazole group and the heme group produced by protein folding about the heme group. There was both a static and a dynamic aspect to this proposal. (See Fig. 10.)

The static rack picture is perhaps best described in terms of the behavior of $IrX(Co)(P(C_6H_5)_3)_2$, which in benzene solution binds oxygen if X is Cl^- or I^-. As shown by Vaska (134), oxygen binding is readily reversible when $X = Cl^-$, but recently McGinnety, Doedens, and Ibers (135) have found oxygen binding to be irreversible when $X = I^-$. The iodide compound is shown in Fig. 11, taken from the work of the latter authors. The O—O bond is 1.30 Å in the chloride compound and 1.47 Å in the iodide compound. As suggested by McGinnety *et al.* (*135*), the enhanced inductive displacement of electrons from the iodide ion to the metal and thus into the metal–oxygen bonds is the probable explanation for the relative increase in oxygen affinity. Many years ago Williams (*136*) pointed out that in heme proteins different fifth-position ligands from the protein should behave in just this way, since each would have different electronegativity and must thus alter the interactions of iron and oxygen. The point of view taken in the rack proposal is qualitatively different. In the heme proteins for which structural information exists the fifth-position ligand is always an imidazole group of the protein.

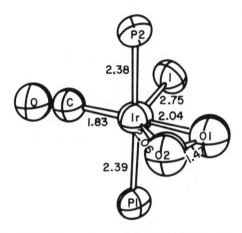

Fig. 11. The inner coordination sphere of IrIO₂(CO)(PPh₃)₂ . The errors in the distance are 0.02 Å or less. (This figure was taken from *Science*, Vol. 155, Feb., 1967, p. 709 in an article by J. A. McGinnety, Robert J. Doedens, and James A. Ibers. By permission of authors. Copyright 1967 by the American Association for the Advancement of Science.)

Small variations in the geometry of this group relative to the heme plane as a consequence of even minor variations in amino-acid composition must alter not only π-bonding but also σ-bonding and must thus change the affinity of iron for the sixth ligand, though not usually in a very large way. Different geometric adjustments must produce quantitative differences in the properties of heme iron, despite the fact that the protein-controlled ligand is always an imidazole group. The most obvious mechanical connection between heme group and protein conformation is the bond to the imidazole nitrogen formed at the fifth position of heme iron. Hoard and co-workers (*137, 138*) have presented impressive arguments to the effect that the iron ion, which is found about 0.4 Å out of the porphyrin plane toward the fifth-position ligand in model high-spin heme complexes (*137*) and in the high-spin forms of myoglobin (*13*), should move toward the porphyrin plane as the iron ion goes into the low-spin state. This motion might, for example, draw the imidazole group toward the porphyrin plane with a large concomitant distortion of the protein conformation. Unfortunately no relevant X-ray studies of low-spin heme complexes have been reported to test the proposal of Hoard *et al.*, and the X-ray studies of oxymoglobin, which is a low-spin complex, necessary to determine whether or not the iron ion lies closer to the porphyrin plane than it does in the deoxymyoglobin form have not been experimentally possible (A). High-spin to low-spin changes occur in Fe^{III} myoglobin when cyanide ion or hydroxyl ion replaces water in the sixth ligand position. In these cases the evidence thus far reported does not show the drastic movement of the iron ion anticipated by Hoard and co-workers, but distortions of the porphyrin occur, apparently with little change in the position of the imidazole nitrogen with respect to the lattice coordinates (7, 8). It is thus not at all clear whether or not a change in protein conformation is produced via the iron-to-nitrogen bond when deoxy-myoglobin binds oxygen. It is quite possible that a major role of the protein is to prevent the iron from moving closer to the porphyrin plane so as to reduce the rate of oxidation of iron as well as reducing the affinity for oxygen to a useful value.

  In the sense that the imidazole group is rigidly fixed in position relative to the heme group so as to establish the desired oxygen affinity, we speak of a static rack mechanism. Small geometric changes in the heme complex that must occur during oxygen binding as well as those resulting from replacement of one kind of sixth-position ligand by another need not spread into the protein fabric. However, it appears probable that those geometric effects do in fact spread from the ligands of iron into the protein. For example, "trans-effects" between fifth- and sixth-position ligands may force a repositioning of the proximal imidazole group so as to produce geometric changes. If so, in hemoglobin, these could become amplified to produce a cooperative set of atom displacements extending not only to the limits of a single subunit but also on into the other subunits of the molecule through the coupling at the contact regions of these subunits. Alternatively, there is evidence (*131, 139*) that suggests that the immediate mechanical stimulus for these rearrangements is not the imidazole repositioning, but rather changes in porphyrin geometry, perhaps with some interesting electronic consequences, since some β substituents of porphyrins are known to be far from inert in their effects on the metal ion in metal–porphyrin complexes. The oxidation–reduction potentials of such complexes are a good example of this effect (*140*).

  It remains to be established just how the geometric rearrangements in hemoglobin and perhaps in myoglobin are triggered, but some dynamic rack mechanism appears to be required to explain hemoglobin behavior and other examples of allosteric mechanisms. Wyman and Allen (*141*) in 1951 specifically suggested that heme–heme interaction is associated with conformational changes in hemoglobin, and, shortly thereafter a dynamic rack mechanism was proposed (20) as the vehicle for connecting the conformation of the protein to the electronic properties of the heme group. There is still uncertainty about this connecting mechanism, but there is no longer much uncertainty about the conformational basis of heme–heme interaction, although in adequate descriptions of this mechanism we shall have to consider the relationship of the conformation changes to changes in protein volume and "hydration water." This undertaking requires precise thermodynamic information not yet available. For the present, we take the position that the weight of evidence in favor of a conformational basis for the interactions among the heme groups in ligand binding is sufficiently great that we may accept it as a working hypothesis with a high probability of being correct. The reorientation of the subunits of hemoglobin during oxygenation must either rest on conformational changes within the subunits—probably coupled to changes in nearby solvent and solvent in the pronounced cavities between the subunits—or on some as yet undetected cooperative interaction among charged groups (A).

The EPR signals of McConnell and co-workers (4) and the immunological data obtained by Reichlin et al. (5) come very close to proof for conformational changes within the hemoglobin subunits, although they can still be explained by changes in solvation or perhaps changes in electrostatic situations at the surfaces of the subunits. In addition, there is of course a very wide variety of less easily interpretable and thus less convincing evidence for atom rearrangements within the single subunits [partially reviewed in Ref. (131)]. If the X-ray precision can be significantly improved, we anticipate that some intrasubunit atom displacements will be detectable by this method, though it is doubtful that a reliable description of these rearrangements can be provided by this method, since the smaller atom motions are unlikely to be detectable. We thus face a decision that is also becoming apparent in studies of several other proteins. Specifically, what methods and what criteria can be used to describe and confirm the existence of conformational rearrangements in protein function? It is true that we do not now have a rigorous proof that such intrasubunit rearrangements occur during oxygenation of hemoglobin, but it seems certain that even if X-ray studies can detect some of these rearrangements, they can neither fully describe nor fully prove a conformational mechanism. This point of view may prove to be overpessimistic in some cases—hopefully in that of hemoglobin—and in any event, the X-ray method is by far the most powerful source of information about protein reactions where it is applicable. In general, however, X-ray information must be strongly buttressed by quantitative information from other types of investigation. The importance of this matter is particularly great for hemoglobin, since among proteins for which significant structural information is now available, only hemoglobin is a suitable model for much more complicated physiological systems such as the mitochondrion.

The phenomenological behavior of hemoglobin requires a rather complicated conformational explanation. For example, to a good first approximation the Bohr effect and heme–heme interaction are found to be independent of each other. Similarly, the binding of one 2,3-diphosphoglycerate ion according to Benesch and Benesch (*118, 167*) decreases the affinity for oxygen, but does not significantly change the Bohr effect or the heme–heme interaction pattern. Wyman (*117*) has shown that in principle, under certain conditions, a single linkage system can produce independent behavior among linked pairs in a larger linkage system. It is probable that the explanation of the independence required to explain the observations just mentioned is to be found in a special case of Wyman's theoretical analysis and depends on the details of conformational behavior. Three such details are seen to be required, and to illustrate what is meant by the phrase "details of conformational behavior," we shall now develop a sample hypothesis.

Winterhalter (*142*) and Beychok (*143*) and their respective co-workers have shown that a four-subunit molecule having optical-rotation and hydrodynamic properties like normal hemoglobin can be produced by adding two heme groups to two dimeric globin molecules. Thus, two hemes return the very different conformation of globin to something very similar to that of normal hemoglobin. One heme group per dimer controls the first step of conformational change. The oxygenation behavior is not quantitatively restored in this way, but Parkhurst and Gibson (*139*) have shown that the addition of two protophorphyrin molecules free of iron to fill the remaining two heme-binding sites on these half-filled molecules restores ligand binding kinetics for the two heme groups to their quantitative behavior in normal hemoglobin. There has been much work in this direction by the Rome group (*144*), all of which shows that the β substituents of the prophyrins are important in oxygenation.

These studies appear to show that heme–heme interaction is predominantly controlled by the way in which the protein folds about the porphyrin plane and not by the iron-to-nitrogen bond at the fifth position. The first two heme groups produce refolding and association of subunits into tetramers. The second two heme groups or porphyrins are tightly bound, but in the binding process the favorable free energy of association of the subunits is decreased, as is required by the heme–heme interaction in the tetramer. Thus, distortions of the protein fabric that link the heme groups and decrease the affinity for oxygen require a positive free-energy change, which is derived from the large negative free energy of heme binding. This is characteristic of rack mechanisms found in enzymic catalysis where substrates are bound in ways that produce the distortions in protein and substrate necessary for catalysis rather than the most favorable binding of substrate to protein (*2, 179*).

Heme–heme interaction is manifested in the electronic characteristics of the heme group (A). For example, Mizukami and Lumry (*147*) find that the dissociation of horse hemoglobin from dimers to single subunits is accompanied by an unusually large spectral change in the 265–270-nm region. The effect is much too large to be attributed to the chromophores of the protein remainder and is due to the heme group.* Urry (*145*) and Beychok *et al.* (*146*) find strong circular-dichroism bands in this region in cytochrome *c*, myoglobin, and hemoglobin and attribute them to the heme group. Antonini *et al.* (*148, 149*) find a distinct spectral change in the Soret region in the dissociation of human hemoglobin into single subunits, and since they also have shown that heme–heme interaction is lost in this process (*148*),there is clear evidence that heme–heme interaction produces changes in the electronic states of the heme groups. There is also a considerable bulk of older literature that relates changes in visible spectra in a quantitative way to oxygen affinity [see Ref. (*150*) for an interesting example, Barcroft's "span"]. More specific electronic information is provided by the work of Hayashi *et al.* (*204*), who have demonstrated using EPR that the electron-spin state on the two ferric-iron atoms of hemoglobin $M_{\text{Hyde Park}}$ is altered when oxygen becomes bound at the remaining two ferrous heme groups. It will be recalled that this hemoglobin M has the distal imidazole of the α chains, 92, replaced by a tyrosine, which stabilizes the ferric form of the β heme groups. Oxygenation of the α quarters that remain in the ferrous state produces a large change in the EPR spectrum of the ferric spin system (A).

A similar support for a mechanism of heme–heme interaction depending at least in part on changes in the electronic states of the iron ions has been provided by Shigu, Hwang, and Tyuma (*151*), who report that EPR signals from the unpaired electron of nitric oxide bound as a ligand at the four heme

---

* In a recent article Perutz *et al.* (*212*) report the contacts between α_1 and β_1 and α_1 and β_2 subunits of hemoglobin. There is a single tryptophan residue, C3β, located in the $\alpha_1\beta_2$ interface, and since the indole group of this residue must undergo a considerable change in its environment if dissociation of the subunits occurs at this interface, a changing indole spectrum can be expected in such a dissociation. Mizukami and Lumry (*147*) find such a spectral change in the first dissociation step of oxy-horse hemoglobin in 1.0 *M* NaCl at pH 7.0 and 25°C. Their analysis indicates that the change occurs as a result of the dissociation of the four-subunit protein into two two-subunit products. This finding suggests that the process is $\alpha_1\alpha_2\beta_1\beta_2 \rightleftarrows \alpha_1\beta_2 + \alpha_2\beta_1$, in agreement with the proposal of Perutz *et al.* As pointed out by these authors, this mode of dissociation does not appear to be consistent with the idea that the major intersubunit cooperation responsible for heme–heme interaction takes place between the subunits of the two-subunit products. Confirmation of this proposal will require studies of the dissociation equilibria of oxy- and deoxyhemoglobin and a comparison of the thermodynamic changes in heme–heme interaction obtained in this way with those obtained from oxygen-binding isotherms.

groups are distinctly sensitive to the aggregation state of the subunits, the specific types of subunits studied, and chemical modifications of the subunits. The EPR signal would be expected to be a good indicator of the electronic state of the Fe^{III} ions, or at least of changes in this state. However, it should be mentioned that there is at least an apparent disagreement between their results and those of Alben and Caughey (152), who find variations in the CO stretching frequency of carbonyl hemoglobins only with those M-type hemoglobins that have substitutions for the distal imidazole. In light of our attempts to associate electronic changes at heme iron with conformational events, this is a disturbing finding. However, it is quite possible that Caughey and co-workers would have seen a difference at intermediate states of CO binding, and perhaps the results they obtained with the M hemoglobins are equivalent to what would be observed with half-saturated normal hemoglobins.

To return to our description of a conformational mechanism for hemoglobin, we note that despite any important role of the porphyrin in heme–heme interaction, the imidazole nitrogen-to-heme-iron bond is of major importance in establishing the necessary quantitative electronic characteristics at heme iron. Is it possible that the electronic state of heme iron can be modified by these two pathways and yet give phenomenological independence of the two in experimental studies of ligand binding? The answer is probably yes. Consider, for example, a protein in which the mechanical coupling among the heme groups passes through the porphyrins and has only two states of iron adjustment, determined by the terminal states of a cooperative conformational process in the protein. We label these states + and −. Within certain limits cooperation is maintained regardless of the quantitative adjustments made in these terminal states. Suppose, then, that the Bohr effect is associated with a continuous set of adjustments of the imidazole–iron geometry that alters the electronic behavior of iron and thus its oxygen affinity in the + state. Similarly, suppose that the effect of 2,3-diphosphoglycerate ion is on the − state. Heme–heme interaction will then occur in substantially the same way regardless of the adjustments made in the + and − states, but the affinity for oxygen is the net result of the participation of all three factors. To a first approximation all three factors will be phenomenologically independent. This does not mean, as we will discuss, that the specific "allosteric model" of Monod, Wyman, and Changeux (17) is necessary, since only two subunits need be cooperatively linked to produce the all-or-none (two-state) process. It is clear that our illustrative model requires such a two-state process and is thus consistent with recent analyses of experimental results. The mechanism is based only on distortion and strain and thus on the mechanical aspects of linkage, since each factor operates by this mechanism in our model. These matters will be discussed in the following section.

E. Generalization of the Lilliput Principle

The central feature of importance in rack mechanisms is the regulation of electronic properties in primary bonds of substrate, coenzyme, and protein, which is effected mechanically by strains set up in a structure held together only by secondary bonds. The protein conformation is sufficiently strong to support these strains only because the relatively weak secondary interactions are so numerous. The title of this section is given as the "Lilliput principle" by no means facetiously, but rather to emphasize a somewhat more sophisticated and more fundamental aspect of rack mechanisms than is immediately obvious in enzymic processes, although the principle is the same. The term "Lilliput principle" emphasizes the ability of many weak bonds to control one or more primary bonds, and of course the analogy is to Gulliver and the Lilliputians. The original choice of the term "rack" emphasizes stretching, although it is probable that compression and twisting of primary bonds is more common and more easily effected by protein conformations. The new term conjures up better molecular descriptions of the conformational basis for physiological function of proteins and would have been a better original choice of name. Alternatively, the rack mechanism could have been called the "muscle mechanism," since the interconversion of chemical and mechanical free energy as occurs in muscle function is still another way of describing the fundamental feature of the rack. In fact, with the emergence of proton pumping through contractile membranes as the possible direct mechanism of photophosphorylation (*153*), a finding that may be duplicated in oxidative phosphorylation, it begins to appear possible that most physiological processes of protein systems can be grouped under the single heading of contractility, or more specifically, chemical–mechanical free-energy interchange (*122*). Insofar as the function of single proteins and small systems of proteins are concerned, it appears possible to explain under such a heading the following important characteristics of physiological reaction:

(1) The production of unusual electronic properties in small groups of atoms, as is necessary to establish unusual reactivity. In this group falls the delicately adjusted system of hemoglobin, complete with pH control and metabolic control mediated through 2,3-diphosphoglycerate binding.

In the cytochromes, although major changes in oxidation–reduction potentials from cytochrome to cytochrome are probably due to the selection of fifth- and sixth-position ligands in the zero-order complex, fine adjustment, perhaps within about 100 mV and apparently delicate enough to make 1-mV discriminations, may be a manifestation of rack control. Indeed, in cytochrome *c*, as in hemoglobin, the constancy of redox potential or delicately adjusted oxygen isotherm from species to species, despite wide variations in amino-acid composition, may rank among the most remarkable consequences of rack control.

Additional examples of unusual reactivity of functional groups are easy to find. The easily reduced and oxidized protein disulfide groups of some flavoproteins may derive their special characteristics from distortion. This is an old idea and still an attractive one. The zinc atom in carboxypeptidase A appears to be the focus of the bond-rearrangement steps in enzymic hydrolysis (154), just as does the remarkably effective hydroxyl group of serine 195 in chymotrypsin (155) and the functional sulfhydryl of papain (156). In the case of the zinc ion the picture seems particularly easy to understand, since the characteristics of the complex zinc ion in the reaction part of this protein are probably established in the same manner as are the properties of heme iron in the heme proteins; that is, nature, through evolution, not only selects a useful set of ligands but also determines their positions around the zinc and the characteristics of the bonds they make to zinc, as is required for efficient catalysis. This proposal for metal ions in proteins was first made by Smith (157), who emphasized the unusual geometries and consequent reactivities possible under the control of protein conformation.

The special reactivity of the serine hydroxyl group in chymotrypsin is considerably more difficult to explain. Current explanations favor co-operation of this group with one or two imidazole groups, appropriately positioned for concerted action (158, 159). This suggestion is attractive, and certainly the required geometry would seem to be attainable in polypeptide evolution. There are some mysterious Cotton effects in the ORD spectrum of this protein (51), at least one of which appears to be related to both catalytic ability and the isoleucine–aspartate ion pair (cf. Section III, H) and may indicate an acid–base chemistry more complicated than those to which we are accustomed (2). There is evidence that the substrate is forced into a distorted activated complex by the protein, and it may prove that this is the only rack mechanism involved in chymotryptic catalysis (2). The serine hydroxyl reactivity would then be due to group positioning, the local electrostatic field, and perhaps some of the other useful variants of protein control listed at the beginning of this section.

(2) Specificity in enzyme catalysis and other chemical processes of proteins is obviously a major accomplishment of protein evolution. Not only substrates but chemical transformations are selected with delicate precision. According to Bernhard (160) there is a factor of 10^{10} in the speed of chymotryptic hydrolysis of some L substrates relative to the speed of hydrolysis of their D enantiomorphs. We believe this achievement to be the result of a complex pattern of dynamic changes that relate speed and specificity, and have recently discussed the matter in another place (2). It is not possible to go into detail here, but Parker (52) noted that each substrate side chain

important as a focus of substrate selection by chymotrypsin produces a different protein-spectral-difference pattern in the "aromatic" region near 290 nm. This observation suggests a wide range of adjustability in the protein in a mechanism that may be identical with the early and very important mechanism of "configurational (conformational) adaptability" proposed by Karush (*161*). It further suggests an infolding of substrate, as was first proposed by Doherty and Vaslow (*93*), such that the total process of substrate binding and bond rearrangement forms a unit operation coordinated by the mutual conformational adjustments of protein and substrate.

(3) Strain and distortion mechanisms for the specific physiological function of proteins obviously manifest the necessary characteristic of chemical mutability through genetic alteration of amino-acid sequence, and they do so to a remarkable degree. In the heme proteins the number of degrees of freedom for such adjustments by changes in amino-acid composition at points distant from heme iron as well as close by is remarkably large. We have mentioned the adjustments in geometry of the porphyrin nucleus and its β substituents as well as the bending, stretching, and rotational degrees of freedom of the proximal imidazole group. Indeed, there exists very nearly a continuum of electronic properties for heme iron available in this way to nature through direct mutation. In addition, of course, there are residue changes that alter solubility or aggregation state in such a way as to alter physiological function at the heme group indirectly via these solubility or aggregation changes. Mechanisms of this sort are already well known from studies of abnormal hemoglobins (*131*). Then there is also a large class of mutations that might be labeled "contact mutations," since they alter the local environment around substrates. The positioning of functional groups such as imidazole, phenol, or hydroxyl groups of side chains to fix in a very delicate fashion the orientation of these groups to substrates and thus to establish chemical behavior with equivalent delicacy is certainly a major mechanism in the mutation of chemical behavior in proteins. This type of genetic control is particularly obvious for enzymic proteins in which a large substrate is enfolded by the protein, and one might be tempted on the basis of such examples to assume that this type of positioning of local functional groups, perhaps passive positioning along the lines of Koshland's "induced-fit" proposals (*164*), is the only important method of genetic control, at least as applied to direct changes of the functional properties of the catalytic site.

It is certainly to be expected that the exhaustive searching process of evolution has found and developed a number of unusual electronic mechanisms in the chemistry of protein functional groups. These will surface and become understood one by one as protein chemistry develops. But such mechanisms are specialized to a variety of chemical tasks and cannot represent the most fundamental level of protein chemistry, or so we believe. Instead, we seek a fundamental and general property of functionally active proteins that makes such mechanisms possible, and hemoglobin plays a unique role in this search, since of the well-known proteins no other is at the same time so complex as to manifest the most sophisticated aspects of protein chemistry and yet so simple as to be within range of our understanding at the present time.

Heme–heme interaction and the control over oxygen affinity in hemoglobin effected by a single diphosphoglyceric acid molecule cannot be explained by passive positioning of functional groups, nor in general can they be altered with sufficient delicacy and variety by changes of contact groups alone. All or very large parts of a chemically functional protein operate as a single integrated net of interactions, producing by cooperation a whole considerably larger than the parts. In principle, any modification by amino-acid substitution, no matter where it lies in this functional unit, must cause modifications in properties of the unit, though of course in some instances of nonlethal mutation the changes are undoubtedly too small to be detected or they manifest their effect indirectly and are detectable only under special circumstances. We believe that the heme proteins provide some outstanding examples of the delicacy of adjustment possible. Bypassing hemoglobin for a change, let us examine cytochrome c, which is especially interesting because the mammalian forms of this protein, insofar as they have been studied, all have standard half-cell potentials of about -256 to -260 mV at 25°C (140, 165), despite the fact that there are often large variations in aminoacid composition (166). The zero-order properties of the reaction part, the heme-complex ion, are determined by choice of fifth and sixth ligands, and the first-order properties are determined in some degree by the local electric field, most important when heme iron bears an uncompensated charge. The second-order properties, best exemplified by the high precision of adjustment of the standard half-cell potential, are the place to look for subtle control, and this control appears to be so delicate that 1-mV differences in half-cell potential can be selected by mutation. However, it must be shown that the half-cell potential is indeed a second-order rather than a zero-order property. That this is the case is strongly suggested by studies of Sullivan (24) and Yue (168), who were able to produce changes in the half-cell potential of 100 mV or more by additions of small amounts of ethanol to water solutions of horse-heart cytochrome c. This sensitivity to solvent composition appears to reveal a strong interaction between the conformation of the protein and heme iron, but there are serious hazards in drawing conclusions from limited experiments of this type. Recent work by Benesch and Benesch (118) indicates that the intrinsic oxygen affinity of the subunits of hemoglobin in the absence of "regulatory substances" such as diphosphoglyceric acid is nearly identical to that of myoglobin. Perhaps after all, oxygen affinity is not so sensitive to amino-acid changes as we have imagined. A systemmatic study of "clean" myoglobins and hemoglobins of different species is very much needed to answer fundamental questions of this type.

In connection with the problem of cleanliness we will end this section with an important practical remark on protein contamination. In proteins cleanliness may not be next to godliness, but it is certainly next to wisdom. The very important finding of Chanutin (16) and Benesch (167) and their respective co-workers of a "contaminant" site controlling oxygen affinity in hemoglobin is paralleled by Keyes's finding (10) of a contaminant-dependent linkage system in myoglobin. One wonders how many additional dependencies on contaminants, specific-ion effects, and so on will be revealed before hemoglobin becomes a reliable material for serious investigation. Myoglobin and hemoglobin are typical in respect to these complications and not rare examples. Despite many years of study of chymotrypsin, it was found only a few years ago that the best preparations of this material were often seriously contaminated (95). The internal consistency now observed among quantitative experiments on this protein suggest that it is now worthy of serious study, but there may yet be some rude surprises even in this case. Since protein chemistry is beginning to advance into its early middle age of quantitative study, the problems of contaminant control and accidental chemical modification have become the most important among the experimental problems.

VII. Vignettes

A. Allohysteria

Although the term "allosteric interaction" signifies nothing more exotic than the heme–heme interaction phenomenon of hemoglobin, well known before the invention of this term, the term itself has acted as a remarkable catalyst for growth of interest in linkage systems in individual proteins and protein complexes. The sudden growth has produced erratic emphasis and quite unnecessary rigidity in the use of the term. Among the more extreme examples of this rigidity has been the restriction to limiting cases of linkage and the intrusion of symmetry into considerations of protein complexes and multisubunit proteins such as hemoglobin (17, 171). The original specific application of linkage concepts to protein systems is due perhaps more to Wyman than to any other author, but Wyman's formulation (170) was general and did not require, for example, that all the cooperative processes linking all the sites take place in the binding of one ligand (the generalized "allosteric model"). In fact, the general equation for linkage among identical sites is simply the famous equation of Adair [see Ref. (*172*), for example] recently put in much more sophisticated garb by Wyman using the "binding potential" (*117, 173*) (grand partition function). The so-called "allosteric" model and the "Koshland" model for hemoglobin (*171*) are simply special cases of the general linkage mechanism, and so far as we know, no one has come forth with unequivocal evidence that the one or the other or any intermediate linkage situation is preferred in nature. It is now thought that the "allosteric" model applies to *D*-glyceraldehyde-3-phosphate dehydrogenase (*175*) and aspartate transcarbamylase (*11*), but an intermediate model applies to hemoglobin according to latest evidence (*117, 176*).

There does not yet appear to be any reason to suppose that multiple subunits are necessary for linkage, and as has been discussed, Keyes (*10, 11*) has shown that a complex linkage system can exist in a single molecule of myoglobin. Similar considerations apply to the requirements of symmetry, at least in allosteric systems thus far examined in detail. Such symmetry as does exist in multiple-subunit systems is very likely an evolutionary accident. There is no apparent symmetry in function in hemoglobin that is a consequence of symmetry in structure. Indeed, the intermediate states of oxygenation have lost any functional symmetry existing in the fully deoxygenated or fully oxygenated protein.

Allosteric interaction in relatively simple systems can be described with the usual highly simplified electronic potential-energy surfaces of reaction kinetics. Figure 12 is such an example, and it provides a useful basis for a qualitative consideration. In Fig. 12 the conformational changes are indicated on the ordinate and the progress along the chemical coordinate is indicated by motion along the abscissa. The single figure does not adequately provide for changes in cratic (concentration) free energy associated with different partial pressures of oxygen. However, we may think of the potential-energy wells as retaining the same depth, which is associated only with the enthalpy, but increasing in total size (volume in phase space) as the cratic entropy of oxygen in solution increases. This follows from the fact that this size is a measure of the entropy. As oxygen concentration is raised, no expansion of the hemoglobin well occurs; the HbO_2 well increases with the logarithm of the first power of the oxygen concentration, the HbO_4 well expands with the logarithm of the second power, and the system moves progressively from Hb through HbO_2 and so on, finally coming to rest with HbO_8 having such a small free energy that all the systems find themselves in this well.

We see no reason why any aspect of this diagram should be required to be fixed by nature. The size of the wells (area in the figure), which is roughly proportional to the entropies of the several states, certainly need not show monotonic decrease, nor need the depth of the wells, which measures the

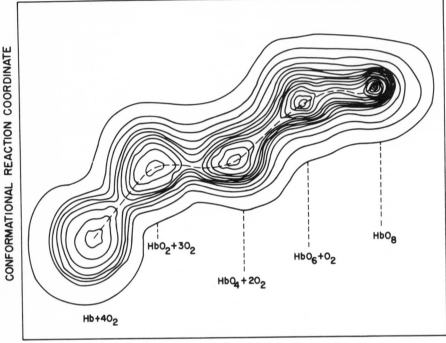

CHEMICAL REACTION COORDINATE

Fig. 12. A sample potential-energy diagram for the oxygenation of hemoglobin. The conformational rearrangements responsible for linkage among the heme groups are very roughly described in terms of a single conformational reaction coordinate, which is the ordinate of this figure. The chemistry of oxygen binding progresses along the abscissa. The contour lines are lines of constant electronic potential energy with decreasing potential energy as the lines move in closer to the potential "wells." The stable or meta-stable wells are described on the figure as the states of Hb, HBO_2, HbO_4, HbO_6, and HbO_8. The dashed path is the course of the oxygenation reaction. Since it is desirable to consider the variation of oxygen concentration rather than the fixed standard state of oxygen required for a single diagram of this type, the size of the well (area on the plane of the figure) can be considered to increase with the logarithm of the oxygen concentration with the following rates: no concentration dependence for Hb well; increase proportional to $\log [O_2]$ for HbO_2; increase proportional to $\log [O_2]^2$ for HbO_4; increase proportional to $\log [O_2]^3$ for Hb_6; and increase proportional to $\log [O_2]$ for HbO_8. Thus as oxygen concentration increases the most common species at equilibrium moves from the Hb well toward HbO_2 and so on until at sufficiently high pressure the most common species is HbO_8. No attempt has been made to represent the data in a quantitative fashion.

enthalpy of the states, be fixed absolutely or relatively. Vertical and horizontal positions are similarly adjustable, or so it now appears. In the Monod "allosteric" model the HbO_2 well would lie along the abscissa approximately where it now lies but raised on the ordinate to the relative positions shown for the HbO_6 and HbO_8 states and the HbO_4 well would have this same ordinate value. This is obviously only one of an infinite number of quantitative adjustments possible by changes in amino-acid composition, solvent variables, or regulatory substances. Cooperative behavior is a matter of the relative depth and size of the different wells. If those for HbO_2, HbO_4, and HbO_6 were all small and shallow, only the states Hb and HbO_8 would ever be present in significant concentration. Thus the degree of cooperation is relative and can take any one of a number of patterns dependent on the relative quantitative characteristics of the wells. According to some current analyses of human-hemoglobin behavior (*117*) the HbO_4 and HbO_8 wells are larger or deeper or both than the HbO_2 and HbO_6 wells, since the latter two states have higher free energy, regardless of the oxygen pressure (*170*). This pattern is probably not invarient for hemoglobin, since horse hemoglobin appears to conform to a quite different potential-energy diagram (*177, 178*).

Diagrams of the type of Fig. 12 also make it easy to see how the Bohr effect, for example, can be independent of heme–heme interaction, though of course they do not give the molecular detail that is responsible for the independent behavior. So long as the free energies of the states bear the same semiquantitative relationship to each other, the cooperative pattern of behavior that appears as the all-or-none process of heme–heme interaction in hemoglobin remains unchanged. However, variations in pH, diphosphoglycerate concentration and many other changes in hemoglobin solutions or in hemoglobin itself will shift the positions of wells in an almost random manner on the plane of the figure, keeping only the required ordering along the chemical coordinate. HbO_2 may lie near the top of the figure but HbO_4 near the bottom and so on, with no apparent restriction placed on these constructions by nature. Well depth can also change without change in heme–heme interaction at the first-order level of precision. So long as pH and regulatory substances do not interact strongly with each other, their influence on behavior will also be independent. As we have already mentioned, the best way to avoid such interaction and certainly the only generally adequate way is to have pH control one well and the Chanutin–Benesch substances control another, though only within the limits required to maintain the heme–heme interaction pattern within its own required limits. For hemoglobin we see that the total number of phenomena that can be independent to a good first-order approximation is equal to twice the number of two-state (cooperative, all-or-none) processes plus one. If hemoglobin has two two-state processes, the maximum number of independent processes having an influence on oxygen binding is five. If there is only one two-state process, the number is three. If the separation into two two-state processes is only approximate, there can be three independent processes and one that is only weakly coupled to heme–heme interaction, though it will be independent of the remaining two processes.

Figure 12 is very useful even with little molecular information and becomes progressively more useful as molecular information appears. The particular version of this figure chosen for illustration has obviously not been carefully constructed and should not be considered to have any quantitative relevance.

B. Energy Transfer in Proteins

It should be realized that the migration of vibrational energy from normal vibrational mode to normal vibrational mode is so rapid in small molecules at room temperature as to make the concept of normal modes quite inappropriate for purposes of describing unimolecular reactions. Furthermore, in a molecule with such a large and weakly bonded structure as a protein there would seem to be no way to insulate one part from another. Thus, we expect that every small domain of a protein including only a few atoms is in separate thermal equilibrium with its neighboring domains and consequently with the solvent. Fluctuations in local temperature will occur and of course are required at points where chemical reaction takes place to provide the thermal energy for activation. On the basis of a detailed examination of some of the more precise Arrhenius plots for enzymic processes, we have found (179) no indication that special attention need be given to the participation of vibrational modes of atoms near such chemical reaction centers or elsewhere as contributing sources of thermal energy. Of course the energy does migrate to the points where chemistry occurs via these modes, and the modes of the local domain do contribute thermal energy for activation. The point to be stressed, however, is that enzymes appear to function primarily as devices for reducing free energies of activation for chemical processes relative to their behavior in homogeneous solution, and they do so by increasing the number of elementary steps as a necessary means for reducing the effective activation free energies for the difficult steps. There seems to be no evidence that they act as thermal reservoirs to any extent more important than, say, the vibrational modes of ethane in a liquid solvent. It is not improbable that there are aperiodic conformational rearrangements having both the amplitude and the energy to produce desirable distortions of substrate or protein reacting subsystems, but this type of behavior is a typical example of the dynamic rack mechanism already discussed at length in this chapter. It seems quite possible that such aperiodic fluctuations in conformation could come into periodic resonance with a suitable oscillatory perturbation and may be detected in this way in time. Ultrasound is perhaps the most promising perturbation to use in this search, but if the resonant conformation frequency is less than about 10^7 Hz, as one might guess at this point in history (though Eigen and co-workers (176) suggest a time of about 10^{-7} sec for the conformational rearrangements of hemoglobin), the experimental problem will be severe.

In any event, returning to the thermal energy in vibrational degrees of freedom, it is very unlikely that sufficient vibrational energy frequently accumulates in a single domain to equal the energy of the electronic quantum necessary for local electronic excitation. If this did occur, although the energy would ultimately have to be returned to the thermal reservoir, electronic excited states might become populated for a brief time at the expense of the vibrational energy and then could provide semiconducting pathways or a particularly effective local catalyst state for electron rearrangements. Although there has been considerable study of the semiconduction properties of dry and wet proteins and nucleic acids (180) and it is quite possible that hydrogen-bonded regular arrays of polypeptide or nucleic acid have semiconduction states, the existence of such states in globular proteins as aspects of normal mechanism has not yet been detected and certainly can in no way be said to be established. Their detection should not be difficult. Semiconduction via an n-type mechanism may occur over some limited domain when proteins are treated with sufficiently strong reducing substances, and there is some evidence that this does take place on attack by hydrogen atoms (181). The high energy required to excite a protein into a conduction level is not the only reason that semiconduction is in general improbable. Of equivalent importance is the need for long pathways of orbitals having the correct symmetry. This is certainly impossible in the amorphous regions of proteins. Electrical conductivity has been observed in wet protein samples at a threshold potential of 46 kcal and higher (180), but it is more probable that the water provides the conduction pathway than the protein.

No evidence suggesting the participation of electronic excited states in normal enzymic function existed in 1961 when we last reviewed this matter (18), and to the best of our knowledge no evidence requiring the slightest commitment to this point of view applied to proteins has appeared since then. Even for the process of vision light excitation seems to be used to produce chemical changes in prosthetic groups controlling enzyme function rather than to provide semiconducting states for energy migration, though this is a complicated matter still subject to considerable argument. In photosynthesis, despite a large number of attempts covering a period of many years, the evidence is strongly unfavorable to the participation of crystal-like semiconductor processes in normal photosynthesis (*182*). Instead, electronic energy moves as electronic quanta via the resonance-transfer process (sensitized fluorescence), the theoretical and experimental basis of which owes much to Förster (*183*), though its original development goes back to the Perrins (*184*), and Kallman and London (*185*), or through electron transfers once the electronic quanta have been trapped at the sites where they are converted to the electronic potential energy of ground-state molecules, probably of the free-radical type. We suspect that there may also be important pathways for transfer of energy in photosynthesis as mechanical free energy and will discuss this matter shortly.

In less complicated and less exotic protein systems, although the conditions for resonance transfer of electronic quanta produced by photon absorption via phenol or indole chromophores should be excellent in the absence of special quenching situations, the implications of such transfer for enzyme chemistry are trivial, just as they are trivial in oxidative phosphorylation, in photosynthesis, and in vision. When there are coenzymes, prosthetic groups, or nearby chromophoric molecules bound or unbound and having the necessary overlap between their absorption spectrum and the emission band of one of the protein chromophores, particularly indole which is itself a natural collector of migrating quanta, then quanta absorbed by the protein chromophores can be transferred to the special molecule to effect fluorescence from that molecule, triplet states, or chemistry. We need cite only one example to establish the accuracy of these statements. It is well known that CO can be dissociated from carbonyl-myoglobin by light absorbed into the two indole groups of this protein (*186*). The light absorbed into both indole and heme chromophoric systems is used with almost equal efficiency, despite the fact that neither indole group is in contact with the heme group. Energy migrates by resonance transfer from indole to the metal porphyrin, and there, probably at a charge-transfer band in the far red, excitation to a state in which the CO–Fe interaction is repulsive occurs and the CO molecule is driven off.

Other resonance-transfer processes to foreign molecules have been studied [some examples are given in Refs. (*187*) and (*188*)], and it is quite clear that indole is a good sensitizer, given the appropriate overlap of emission and absorption bands. The efficiency of phenol in tyrosine residues of proteins as a sensitizer, i.e., transmitter, in resonance transfer is still not well established (*195*). Recent studies by Longworth (*189*) show that the efficiency can be considerably higher than previously indicated (*195*). There is still much to be learned about the possible paths of energy migration and degradation from photoexcited phenol groups, and despite what appears to be a somewhat simpler behavior, the alternative paths of migration and loss of electronic excitation by indole groups are by no means fully understood (cf. Section VII, D).

There is no doubt that resonance transfer is intrinsically interesting and very useful both at the present state of our knowledge and potentially as a source of information about proteins. Fluorescence has been much exploited as a means for following chemical reactions and as a device for studying changes in protein conformations or the binding of quenching molecules able to accept quanta from protein chromophores and then to fluoresce. The potential power of the fluorescence and phosphorescence methods, even when restricted to protein chromophores, is very considerable. Nevertheless, from the point of view of energy transfer in normal physiological processes where no optical photons are available, resonance transfer is quite without importance.

It is commonly believed that the major processes for energy transfer in particulate systems such as the mitochondrion or the thylacoid unit of photosynthesis are by electron transfer. This is undoubtedly true at least in the sense that electron transfer is the chemical indicator of energy transfer. However, there is more to the story than this, as is obvious from the difficulties that have been encountered whenever a direct connection for free-energy transfer between the electron transfer acts at the cytochromes, for example, and the phosphorylation processes has been sought. We will not review the alternatives that have been suggested, but simply state that the most promising possibility seems to us to be the rack mechanism whereby free energy is stored and transferred as the "mechanical" free energy of macromolecular distortions and conformational rearrangements (124). This is why hemoglobin stands as a special protein in the history of physiological mechanisms, since hemoglobin is the prototype for musclelike interconversion of chemical and mechanical free energy in small proteins systems that have no obvious mechanical function. Hemoglobin thus establishes the principle that this type of interconversion is possible and, in its own well-known characteristics of behavior, establishes the existence of mechanisms that now appear to be sufficient to explain even the most complex aspects of the behavior of the mitochondrion and similar chemical factories. Thus it was necessary to realize that free energy is transferred from one heme group to another by conformational rearrangements in order to attract attention to the possibilities for mechanical free-energy exploitation in both simple enzymic function and in macromolecular complexes that support synthesis. The free-energy changes associated with conformational rearrangements, whether due to outright strain or to changes from one "unstrained" structure to another of different free energy, are what has been meant throughout this chapter by the term "mechanical free-energy changes," since in both cases the free-energy state varies with geometric parameters and there is a potential for doing mechanical work, even if the processes are restricted by high free energies of activation for refolding, etc.

As an illustration of an extreme case in which mechanical forces would be the major means for the transfer of free energy, consider the case of four cytochromelike proteins encased together in a rigid matrix as shown in Fig. 13. These molecules have different shapes in oxidized and reduced forms. If molecule I is reduced, part of the chemical free energy of reduction is converted to mechanical free energy, as indicated by the "stress" arrows. Electron transfer from II to III in this figure does not eliminate the stress, but the

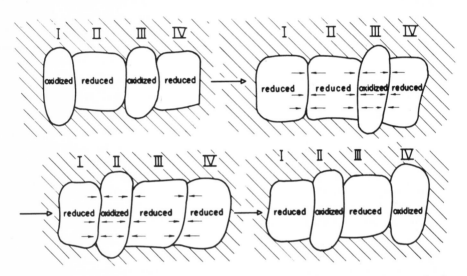

Fig. 13. A model rack mechanism for a system of four different cytochromes confined in a rigid structural matrix. Changes in shape on change of reduction state produce internal mechanical stresses that force cooperation among the proteins. This picture shows the basic picture of stress generation and change when an electron is introduced into I from the left and finally removed from IV at the right. Various alternative uses for such a mechanically coupled system in phosphorylation and in other physiologically useful processes are discussed in the text. No attempt has been made to discribe a known electron-transport process.

subsequent loss of an electron from IV to the acceptor on the right (not shown) does reduce the stress. This process could appear in photosynthesis, where the electron entering from the left is the high-free-energy product of a previous photochemical step. In any event the free energy migrates through this system of four proteins primarily as mechanical free energy, and the mechanically coupled protein system provides the following possibilities:

(1) Electrons need not move from one cytochrome to the next. In order to reduce the stress after the first step, an electron would enter II from some donor in the matrix, and another electron would leave III, going into the

matrix. Thus the chemical balance would be maintained, but direct electron-transfer from cytochrome to cytochrome would not be necessary. The stress situation temporarily raises the electron affinity of the oxidized cytochrome and reduces that of the reduced cytochrome. This is just what happens in the oxygen-binding process of hemoglobin, though it is oxygen affinity rather than electron affinity that fluctuates.

(2) The oxidation–reduction potentials of all four cytochromes could be linked together and would then vary with the state of reduction. Thus it would appear to be possible to produce by mechanical means a wide variety of electronic properties starting with a single small set of molecules. Since these altered electronic properties would be reflected in spectral changes, the spectral variations often interpreted as being due to different cytochromes could instead be due to different electronic states of a small set of cyto-chromes. A similar possibility exists for the many spectrally different chlorophyll "species" in green plants.

(3) If the system also contains sites for phosphorylation and the phos-phorylation process, no matter where it is located on the proteins, allows conformational rearrangements that reduce stress, then some of the stored mechanical free energy would be reconverted to the chemical free energy of ATP.

(4) By suitable construction of the protein molecules, chemical free energy associated with oxidation–reduction processes could be pooled and re-distributed so as to provide the specific amounts required for photophos-phorylation steps without loss. Chemical processes requiring large amounts of free energy could be driven from a mechanical supply accumulated in several electron-transfer steps. Problems of electron balance in which a pathway of one-electron processes, such as is found in the simpler cytochromes, meets a more conventional two-electron oxidation or reduction step, could be solved either by abstraction of the extra electron from a nearby donor or by the delivery of electrons two at a time through the mechanical coordination of at least two cytochromes.

(5) Chemical changes along pathways of proteins linked as a single mechanically coupled unit could be coordinated for purposes of control, efficient use of chemical free energy, and so on. If long-range correlation of chemical processes is necessary for function in physiological processes, mechanical coupling would appear to provide an efficient and highly adjust-able vehicle for this kind of information transfer in particulate systems.

(6) The use of mechanical pathways for free-energy transfer would tend to reduce requirements on the translocation of small-molecule participants. In the extreme case discussed under (1) above, there would be no direct transfer of electrons or small electron carriers from one cytochrome to another.

Although we may have been overingenious in describing what might be achieved by rack mechanisms in the multistep processes of particulate structures, it does seem unlikely that nature has ignored all these desirable characteristics; it is encouraging that such mechanisms are steadily becoming more popular postulations in both oxidative phosphorylation (*190*) and photophosphorylation (*191*).

C. Electron-Transfer Processes

So long as it remains unnecessary to consider long-range migration of electrons or positive holes via semiconduction mechanisms in protein systems, the mechanisms of electron transfer in biology take on no unusual properties relative to current information based on small-molecule oxidation–reduction reactions. The mechanism of these processes is discussed in some detail elsewhere (*192*), so that we need review only a few of the details that may be particularly important in connection with oxidation–reduction processes involving protein groups. We note that electrons can migrate from reductant to oxidant via "jumps" which are quantum-mechanical tunneling processes, or as parts of nuclear fragments which move from one reactant to another. Obvious examples of the latter are hydrogen atoms or hydride ions. Since electron-transfer processes of the latter type are no different in principle from any other chemical process requiring strong interaction between reactants in the activated complex, that is, any other process involving the breaking and making of chemical bonds, we shall concentrate on the "electron-jump" processes. [The topics and references in this section are to be found in Ref. (*192*).]

(1) To ensure a high probability of electron migration during the lifetime of the activated complex (about 10^{-13} sec) the oxidant and reductant must interact through overlap of their electronic orbitals. However, the minimum magnitude of the interaction necessary is about 100 cal/mole and thus very small indeed. For this reason casual encounters between reactants usually provide sufficient interaction energy for efficient electron transfer, provided there are no selection rules to reduce this probability. Activated complexes for electron-transfer reactions then require little interaction energy and relatively slight distortions of the nuclear framework as compared with the extensive reorganization necessary in activated complexes for primary bond rearrangement. Hence the potential-energy problems associated with the application of quantum-mechanical theories to complicated activated complexes of unknown geometry are bypassed to a considerable degree. Marcus and Hush as well as Levich and his comrades have already developed

theories of electron-transfer processes that give good agreement with experiment in simple cases. These simple cases are usually those in which the activated complex is the result of casual contact of reactants without preceding chemical rearrangements. Such activated complexes are called "outer sphere" activated complexes. Electron transfer between ferrocyanide and ferricyanide ions in water is an example, though even this apparently simple reaction proved to have somewhat unusual features. However, if this reaction required the removal of one cyanide ion so that a bridge consisting of a single cyanide ion could be formed between the iron ions, the activated complex would be of the "inner-sphere" type and the reaction could be more complicated. For example, the rate-limiting step might be that of bridge formation or bridge destruction, rather than the elementary step in which electron transfer takes place.

(2) Electron migration down conjugated pathways can be very rapid, carotene or porphyrin molecules being as good conductors as metals. Even in saturated systems, migration along a few bond lengths can occur with adequate probability, and transfer processes down a peptide chain are efficient over at least two residues and perhaps three. On the other hand, the degree of overlap of the $p-\pi$ orbitals of nitrogen and oxygen across interpeptide hydrogen bonds, though certainly adequate for electron migration if only one such hydrogen bond appears in an otherwise efficient pathway, is not great enough to support electron migration over any long pathway of hydrogen bonds. The probability of transfer falls off rapidly with increasing number of hydrogen bonds. Hence electron migration through helical or pleated-sheet structures of hydrogen-bonded polypeptide is very unlikely to provide a useful path of significant length for electron transfer between oxidant and reductant bound to a protein. Although oxidant and reductant groups may be spaced a short distance apart on or in a protein with the protein providing a short link for transfer (probably no more than three residues as mentioned above), it is probable that most electron-transfer actions in the biology of electronically unexcited molecules involve direct contact between electron-donor and -acceptor groups.

(3) Electron-transfer processes are much like photoexcitation processes in the sense that their special characteristics are due to Franck–Condon restrictions. More specifically, since the "electron jump," if and when it occurs, does so in a typical valence-electron time of motion, about 10^{-15} sec, nuclei heavier than hydrogen cannot move during the jump. Even hydrogen nuclei lag well behind the electron. In its fastest "chemical" process (which is the movement from one potential well of a hydrogen bond to the other, as occurs in proton migration through water and ice) the proton requires ten times the period of the electron jump. It is thus obvious that to a firm first approximation the activated complex for electron transfer must be doubly degenerate

in a very specific way, such that the two quantum states of equal energy differ from each other only insofar as the migrating electron is on the reductant reactant in the "reactants" activated complex and on the oxidant reactant, now become reduced product, in the "products" activated complex. A consideration of the statistical mechanics of an ensemble of such degenerate activated complexes shows that it is really a free-energy "degeneracy" that is required. It is quite clear that were this not a requirement, the "products" activated complex could have higher energy than the "reactants" activated complex, so that in principle the return of products to reactants could be coupled to defeat the first law of thermodynamics. Similarly, the second law of thermodynamics restricts the entropy change during the electron jump to zero or positive values, since to a major extent the activated complex is an adiabatic system. The activated complex does communicate with the electrons of the environment through electron–electron polarization such that these electrons do follow the jump process, but this polarization inter-action is not an efficient means for moving energy into or out of the activated complex. Hence, in effect, the activated complex is adiabatically isolated in the thermodynamic sense. It is also apparent that if the "reactants" activated complex has a different spin state than that of the "products" activated complex (and here we cannot count the spin of the migrating electron), it is necessary that momentum move into or out of the spin system of the activated complex in times of the order of 10^{-15} sec, exactly as is required in photoexcitation processes of the intersystem type. Hence there is a very similar selection rule on the electron-spin quantum number.

The production of the degenerate activated complex is the activation process for the elementary step of electron transfer in any electron-transfer process. By thermal fluctuations the reactants must find themselves in a state of nuclear geometry (including solvent and solvent ions) such that the con-ditions for this special case of "free-energy degeneracy" exist. Furthermore, they must do so when their distance of separation is sufficiently short to provide the necessary electronic coupling through overlapping orbitals. The required ligand and solvent distortions away from the most common equilibrium geometry of reactants or products or both may be large and expensive in terms of activation enthalpy and entropy. Roughly speaking, if the overall electron-transfer reaction has a large negative (standard) free-energy change, the activated complex in outer-sphere reactions resembles in nuclear geometry the reactants; if the free-energy change is large and positive, the activated-complex geometry is similar to that of the products. If the free-energy change is small, the activated complex resembles neither reactants nor products. Thus, in a rough sort of way, highly "irreversible" electron-transfer processes will have small free energies of activation and be fast, but processes deviating only slightly from reversibility in the sense that the overall standard free energy is small will have significantly large free energies of activation, other things being equal. This consideration is not an incidental one, since, as a result of differences in the work required to form the activated complex, the range of rate constants for electron-transfer reactions covers at least 15 orders of magnitude. Furthermore, most biological oxidation–reduction reactions have been evolutionarily selected to involve small overall free-energy changes, so that protein control over these reactions can be very definite and very delicate. Significant nuclear reorganization, as is required to produce the activated complexes for electron transfer, is obviously subject to geometric control of just the type for which proteins are suited. The protein conformation can adjust the reaction part toward the activated complex in some desirable process or away from that of an undesirable process. It controls geometry changes and thus determines the probability that a given activated complex will occur. In hemoglobin, either because of a spin-selection rule between oxyhemoglobin, which is low-spin, and oxidized hemoglobin, which is high-spin, or simply because the protein restricts the ligand distortions of the heme group necessary to form the activated complex, electron transfer out of the ferrous iron to oxygen is slow. On the other hand, the physiological as well as some nonphysiological electron transfer reactions of cytochrome are fast (rate constants of the order of 10^6 sec^{-1} or faster). Hence, the heme group is held by the protein in a nuclear configuration similar to that of the activated complexes, so that the fluctuations that take it into this complex are inexpensive and thus frequent. We see that it is just those restrictions on geometric changes in the reaction part in which the protein conformation excels that control electron-transfer reactions, regardless of the specific protein groups or coenzymes that act as electron donors and acceptors. It seems reasonably certain that the delicacy of balance in physiological oxidation–reduction processes involving coenzymes and substrates requires that the protein be much more than a passive surface on which to bind the coenzyme. These binding processes must be as delicately constructed and as fully coupled to the dynamic changes of the protein as those of any multiply bonded substrate. It is important to realize that this control over the ease or difficulty of forming the activated complex for electron-transfer reactions involving coenzymes cannot be based on passive association of coenzyme and protein. The coenzyme must be distorted toward or away from the activated complex in binding or it must be distorted in a subsequent process if the protein is to play much more of a role in the oxidation–reduction process than to provide propinquity between electron donor and acceptor.

It is not always realized that electrons need not jump one at a time in electron-transfer reactions. A number of examples of simultaneous two-electron transfers in a single activated complex is known, and insofar as the probability of the electron-transfer act is concerned, two-electron-transfer probabilities can be and probably are very frequently of the same magnitude as one-electron-transfer probabilities. However, the chemical difference between an oxidant–reductant pair before transfer of two electrons and after this transfer is often considerably larger than in a one-electron case, so that, other things being equal, the activation free energy will be larger in the two-electron reaction than in the one-electron reaction. When one-electron intermediates in an overall two-electron process are very unstable, as it often the case for organic reactants, the two-electron jump may be preferred.

In conclusion we note first that the restrictions placed by the Franck–Condon requirement provide remarkable opportunities for protein conformation to control oxidation–reduction reactions through the control of substrate and coenzyme geometry. Second, we note that in a cytochrome c the electron trap is at least the total heme group and may include some of the protein fabric. Since a large part of the porphyrin ring is not deeply buried and much is actually exposed to solvent, we expect the cross section for oxidation or reduction of the heme group to be rather large. The cross section for reduction of cytochrome c by hydrogen atoms in water solution has been reported to be more than half of the surface (181).

D. A Word about Indole

Thus far we have little knowledge of any special reactions of indole side chains in proteins that are important in "normal" physiological processes. However, it is worth pointing out in connection with the special role that indole plays in the photoexcitation processes of proteins that indole compounds have an exotic photochemistry. These exotic properties should soon make the response of indole groups in proteins to exciting light a very powerful source of information about both static and dynamic behavior of proteins. In the first place it should be noted that indole excited to its lowest excited singlet state via $^{1}L_{b} \leftarrow {}^{1}A$, $^{1}L_{a} \leftarrow {}^{1}A$, or $^{1}B_{b} \leftarrow {}^{1}A$ transitions customarily forms an excited-state complex with any C—O or O—H dipolar group in its environment, provided that this group can be brought into proper alignment with the indole nucleus during the lifetime of the excitation of the latter. These complexes, called exciplexes (102, 169), are formed in $1:1$ or $1:2$ stoichiometry between indole and dipolar groups, depending on the nature of the group. The emission from exciplex states is considerably shifted toward the red and is in large part responsible for the appearance of fluorescence with maximum emission well above 300 nm. In nonpolar solvents noncomplexed indole groups fluoresce with a maximum near 305 nm in a well-structured band. The structure is lost in exciplex emission.

Relatively little is yet known about the ability of buried indole groups to form exciplexes.* Kronman, Holmes, and Robbins (*193*) propose that shrinking processes of the α-lactalbumin conformation reduce internal motion of indoles so as to prevent exciplex formation. Experiments on such proteins as human serum albumen (*194*), which has only one indole group and that thought to be buried, do suggest that exciplexes can be formed at internal points, but much more detailed investigation of this matter is necessary.

It should be pointed out that in proteins the overlap of emission spectra from those indole groups unable to form exciplexes with the absorption spectrum of unexcited indole groups is probably sufficient to facilitate resonance transfer. On the other hand, the emission spectrum of an exciplexed indole group does not overlap the absorption spectrum of an unexcited indole. As a consequence migration of electronic quanta among indoles must cease once the quantum has been transferred to an indole group able to form an exciplex, provided only that the exciplex forms rapidly relative to the lifetime of the quantum in the indole group. In small models exciplex formation is very rapid. This may not often be the case when indole groups are buried in proteins, though we really know very little about the temporal behavior of the motions of such groups. Indole groups freely exposed to water can form exciplexes very rapidly (10^{-10} sec), and we may generally expect that electronic quanta originally introduced into indole groups of proteins by photoexcitation will migrate with high probability to any indole groups in direct contact with water. Since conditions for quenching by ammonium, imidazole, phenol, and carboxylate groups at the protein surface are apparently favorable to quenching, complex quantitative behavior of indole fluorescence is to be expected. With sufficient study, fluorescence quenching should become a valuable source of information about proteins.

The second matter of interest in connection with the photoexcitation of indole groups of proteins is that the exciplexes formed after such excitation can decay through electron ejection (cf. Section V) (A). When electrons are the product of spontaneous intramolecular fluorescence quenching of indole exciplexes, rather drastic chemical consequences can be expected. The most

---

* Longworth has recently shown that the emission from the single tryptophyl residue of ribonuclease T is that expected for an indole group unable to form an exciplex during the lifetime of its excited singlet state (*189*). This important result demonstrates that not all indole groups of proteins are able to form exciplexes. This condition is probably due to the absence of polar groups in its immediate vicinity or the rigidity of the protein structure around the indole group. Studies of the spectral characteristics of this indole group should provide information via the dielectric shift about the "effective local dielectric constant" in this protein.

obvious of these is the ejection of the electron, not into water, but directly into some acceptor group of the protein to initiate chemical reactions with or without the participation of oxygen. We suspect that this pathway and not the triplet pathway is the common pathway for photodegradation of proteins following indole or phenol excitation. Konev and Volotovskii (195) do indeed find that the triplet state of indole is much less efficient than the first excited singlet state in photoinactivation of trypsin. It is of course well known (196, 197) that ultraviolet light of shorter wavelengths than that absorbed by indole groups can produce broken disulfide bridges and very possibly other free-radical products. There is, however, sufficient energy in an excited indole group to produce radicals at disulfide bonds, and electrons ejected from indole groups of proteins may migrate to disulfide groups, where they disrupt that linkage and become trapped (206).

The destruction of excited indole groups by direct attack of oxygen molecules is very efficient (198). Hence studies of the photodegradation of indole groups in the presence of varying concentrations of oxygen should also yield useful information about the exposure of indole groups to solvent and about the migration of electronic quanta from one indole group to another.

E. Determination of Protein Conformation

The original proposal (20) that polypeptides spontaneously fold into their conformational states of lowest free energy, that is, that thermodynamic rather than dynamic criteria determine the folded state of a protein, was based on a voluminous literature of spontaneous reversible refolding following unfolding produced by heat, urea, detergents, and so on. The proposal seemed obvious at the time, even insofar as spontaneous reductive disulfide exchange was required to position disulfide groups in thermodynamically most favorable positions, since Huggins's crossing (207), as this interchange was then known, had been shown to be rapid. Subsequently Anfinsen and co-workers (230), using ribonuclease, demonstrated that disulfide bonds could be correctly positioned by spontaneous processes. [See also Ref. (123).] This thermodynamic mechanism for the folding of polypeptides into proteins is entirely consistent with the bulk of precise information about such processes. The high degree of improbability associated with the random search for the thermodynamically stable conformation is adequately demonstrated in the large negative entropies for these folding processes (cf. Tables 4, 6, and accompanying text). It is undoubtedly possible that in rare instances, especially favored processes producing a folded protein in a state that is not its state of lowest free energy do occur. We have elsewhere (2) discussed the stability problem of α-chymotrypsin and γ-chymotrypsin, two sisters which

·

have the same amino-acid composition but which now appear to be conformational isomers with large free energies of activation restricting their
interchange. Nevertheless, one of these species is the stable form in one pH
range and the other the stable form in another pH range, and the desired
species is obtained by exposure to the appropriate pH. Extensive studies of
the refolding of unfolded chymotrypsinogen, α-chymotrypsin, γ-chymotrypsin, δ-chymotrypsin, and a large number of chemical derivatives of these
proteins, including some that have been heavily succinylated, some with
oxidized methionines that fold only partially, and some that are acyl
derivatives of the hydroxyl group of serine 195 or histidine 57 carried out in
plain-water solvents, urea, and guanidine–HCl solutions at various pH
values, salt concentrations (45–48), and temperatures all show a completely
reversible refolding to produce species that are very closely related as regards
the thermodynamic properties or, when NMR studies have been made (28, 29)
have very similar NMR patterns that are interpretable in terms of total or
partial unfolding in A and B chains on a fixed folded foundation that appears
to include most of the C chain. Furthermore, it should be noted that the
physical properties such as ORD patterns and absorption spectra indicate
close similarities. The minimal stability of these proteins, examples of which
are shown in Tables 4–6, for a broad sampling does not appear to be any
more consistent with the determination of folded form by accidental details
of the kinetics of the folding process than does the invariant behavior of the
several proteins and their derivatives in different solution conditions. Finally
we note that Biltonen was able to predict the temperature of maximum
stability and the value of the low-temperature transition temperature for
α-chymotrypsin from Brandts's transition-I data for chymotrypsinogen, given
only the transition-I heat-capacity change for α-chymotrypsin. Chymotrypsinogen and α-chymotrypsin are significantly different, and it is very unlikely
that data from one could be used to make predictions about the other unless
the folding in both were determined by free-energy considerations. Benson
(199), some years ago, showed that all the exchangeable protons of myoglobin
could be exchanged with water at pH 8 and 20° in about one hour. Rosenberg
and Enberg (31) have shown that all exchangeable protons of chymotrypsinogen A exchange, even at conditions of maximum stability for the folded
species. These observations appear to demonstrate that even the most deeply
buried protons are exposed to water in periods that are short compared to the
probable lifetime of the proteins. It hardly seems possible that in general a
folded structure initially determined by kinetic considerations rather than
thermodynamic considerations could persist indefinitely under these circumstances. The aggregate of these observations, though restricted primarily to
proteins of the chymotrypsinogen family, leaves little room for doubt that
thermodynamic determination of folded structure is a fact for at least some

globular proteins. Experiments that are constructed to demonstrate the alternative will have to be at least as convincing as those carried out with these proteins, ribonuclease (*63*), and lysozyme (*71, 200*).

ACKNOWLEDGMENTS

I am grateful for the support provided by the Office of Naval Research [No. 710(55)], The Air Force Office of Aerospace Research (1222-67), and the Atomic Energy Commission [No. AT(11-1)-894] not only for assistance in the preparation of this chapter but also for much of the experimental work discussed herein. I wish to give special thanks to Dr. Shyamala Rajender and for her assistance and criticism in the preparation of the chapter, and to Dr. Rodney Biltonen for suggestions and criticisms as well as for prepublication information about the NMR analysis of unfolding in the chymotrypsinogen family of proteins. I am indebted to Drs. J. L. Hoard, B. Chance, and B. Schoenborn for prepublication information about myoglobin X-ray information and its interpretation.

Addendum (February 1970)

This chapter was written in the winter and spring of 1968. It was revised in March 1969. Additional revision has not been possible. As a result the reader will find that the list of references is somewhat out of date. Some of the research results on which the concepts and discussion are based have proved to be in error. However, relatively little modification in the chapter is necessary because of these changes, and the most important of these modifications are discussed in this addendum. The principal criticism of the earlier version is the naivete with which many ideas have been presented. During the past year it has become apparent that proteins are much more complicated than has been indicated in this document. By and large the topics are treated correctly at the first level of approximation, but our level of understanding is entirely too incomplete to justify any attempt to describe proteins at the second level of approximation. Chief among current complications is the appearance of what we have called the "Vaslow–Doherty pattern of enthalpy–entropy compensation" in a variety of protein reactions (*205*). There is good evidence for this pattern in reactions of ferri-hemoglobin, ferri-myoglobin, chymotrypsin, and acetylcholinesterase. There is fragmentary evidence for its presence in reactions of ferro-hemoglobin, ferromyoglobin, cytochrome *c*, ribonuclease A, carboxypeptidase, α-amylase, and fumarase. The pattern of linear compensation of enthalpy by entropy now appears to be the universal thermodynamic manifestation of "structure making" and "structure breaking" in liquid water. The latter two terms are operationally defined by changes in Raman and infrared spectroscopy, relaxation parameters obtained from ultrasound absorption and dispersion, dielectric dispersion, NMR relaxation and neutron diffraction, etc. (*233*), although the quantitative measures of changes in "structure" do not always agree. In another place Lumry and Rajender (*205*) have discussed these matters in some detail, both in connection with processes of small solutes and processes of proteins. It appears that most processes in water have been analyzed and treated theoretically in incomplete ways that do not take into account the ubiquitous participation of the property of water responsible for the compensation pattern. From the point of view of the enzymologist and protein chemist the major puzzles are those now associated with the mechanism whereby the functional processes of proteins are linked to bulk water and perhaps to surface water, and the specific role played by water in processes of proteins. At present the best guess is that the protein expands and contracts in the elementary steps of its catalytic reactions, ligand binding, etc. Short discussions of the water phenomenon and its meaning are available in two other references (*214, 215*). The treatment given in this chapter is highly incomplete, and the major reference (*205*) should be consulted.

Specific Remarks

In several cases developments that have come to our attention in the period from April 1969 through January 1970 have made additions or modifications of material in the text necessary. At the points where this is necessary the symbol "(A)" has been inserted in the text to show that the item is discussed in this addendum. The page numbers index the cases in the addendum. Time limitations preclude consideration of any of these in adequate detail and in addition greatly limit the topics that can be treated. It has not been possible to bring the reference list completely up to date.

PAGE 7

The estimate of the enthalpy and entropy contributions from the protein–water system of hemoglobin given in this section have been improved. See Ref. (*236*).

PAGES 4, 7, AND 73

The X-ray diffraction evidence for and against changes in protein conformation accompanying ligand binding to metmyoglobin does not appear to be in useful shape. The changes on binding of CN^-, N_3^-, OH^- were detected with the two-dimensional Fourier difference method and have not been verified by results obtained with the three-dimensional Fourier difference method (*8*). With ferromyoglobin the behavior of heme iron in a low-spin complex is a matter of considerable uncertainty. Hoard *et al.* suggested that the iron ion would be expected to move into the porphyrin plane in the conversion of high-spin to low-spin forms (*137*). Recently Countryman *et al.* (*216*) have shown that this does occur in crystals of bis-imidazole, $\alpha,\beta,\delta,\gamma$ tetraphenylporphinato iron(III) chloride. This complex ion appears to be a reasonable model for the proteins. X-ray structures of low-spin derivatives of ferro-myoglobin such as oxy-myoglobin have not been reported. On the basis of studies of low-spin derivatives of metmyoglobin it is assumed that the iron ion remains well displaced out of the porphyrin plane, in contrast to the behavior of this ion in the small low-spin model. This conclusion, if correct, may not carry over into ferrous-iron forms. Padlan and Love (*217*) find in the hemoglobin from the marine worm *Glycera dibranchiata* that the iron ion remains substantially unchanged in position on oxygenation, but the porphyrin plane moves over the iron, producing easily detected changes in the protein conformation.

In more recent X-ray diffraction studies, Schoernborn (*244*) has found the second xenon-binding site in myoglobin.

PAGE 10

Calorimetric determinations of the enthalpy of oxidation have now been made (*218*). The results are in good agreement with those of Sullivan given in the text.

PAGE 48

It is no longer clear that information based on the behavior of small hydrophobic molecules can be translated directly into information about the solvation of nonpolar groups of unfolded sections of proteins. Data for small molecules may not all be correct and in any event have been obtained with dilute solutions of solute, a situation that may not correspond to the state of water about the nonpolar side chains of netlike polypeptides. There are serious problems in this respect based on the small volume changes in unfolding processes of proteins. The problem is discussed elsewhere (*205*).

PAGES 14 AND 35

The conclusion of Shiao, Lumry, and Fahey (*46*) that the heat capacity of unfolding in at least some unfolding processes of proteins is temperature independent has recently been questioned by Brandts (*219*) on the basis of direct calorimetric determinations of the absolute heat capacity of the folded and unfolded states of chymotrypsinogen A. Brandts finds that the heat capacity of the folded state increases with temperature, but that of the unfolded state is independent of temperature. The overall free-energy change thus decreases with increasing temperature, in contrast to the original analyses of unfolding transitions by Brandts and by Brandts and Hunt (see text).

PAGES 16 AND 74

As newer studies accumulate, the charge–charge interactions in folding processes of proteins and also in physiological processes seem more and more to require participation of many charged groups. Similarly, co-operation among large regions of the protein is suggested by studies of interactions of inhibitors and substrates with enzymic proteins. A typical example is the study by Parker (52) and Kim (58) of the spectral changes produced in phenol and indole chromophores of α-chymotrypsin when different inhibitors are bound. Recent X-ray studies by Steitz *et al.* (232) show that neither type of chromophore is near the slot of this protein in which the side chain of inhibitors is bound yet each inhibitor studied by these workers produced a significantly different change in spectrum. These findings demonstrate that the binding process induces changes at points some distance away and that a continuously variable set of changes is possible with different inhibitor side chains.

PAGE 17

Volini and Tobias (221) have recently reported C.D. bands at 202 nm and 228 nm in α-chymotrypsin solutions. Their results confirm those of Biltonen *et al.* (51). In addition they found C.D. bands in the red end of the indole bands. Blow and co-workers (222) have shown that there are extensive regions of disordered antiparallel β structure in this protein. It is not yet clear whether the latter type of structure is related to the C.D. bands, but it seems probable that there is an important connection.

PAGE 23

The approach to the interaction between protein and water through the internal-pressure concept has proved to be naive. See Ref. (205).

PAGE 27

The disagreement between the values for the enthalpy change in transition I of chymotrypsinogen appears to be resolved by the finding by Bolen and

Biltonen (*224*) and by Shiao (*225*) that this protein has a large heat of titration between pH 2 and 4 in its folded state. Jackson and Brandts (*226*), using heat-capacity calorimetry, obtained enthalpy changes in excellent agreement with Brandts's van't Hoff values and the van't Hoff values of Shiao, Lumry, and Fahey. Schwartz, Wadsö, and Biltonen (*76*) produced transition I by adjustment of pH from 4 to 2 at an elevated temperature and thus included the heat of titration. Calorimetric determinations of the heat of titration have made it possible to correct the enthalpy change in transition I. This correction brings the results of Schwartz *et al.* into agreement with the others. The case resembles the model described in the footnote, since chymotrypsinogen has two substates of its folded state and both substates have the same value of the observable used for the van't Hoff studies, absorbance at 293 nm. According to Freer and co-workers (*227*), chymotrypsinogen A has an ion pair formed between the carboxylate group of ASP 194 and the imidazolium group of HIS 40. It is not unlikely that the disruption of this ion pair between pH 4 and 2 is responsible for the extra step of about 35 kcal/mole found in this pH change. We are grateful to Drs. J. Brandts, T. Schwartz, and J. Kraut for making manuscripts available to us prior to publication.

PAGE 29

Proton-exchange studies by Rosenberg and Woodward (*223*) indicate that there is a considerable difference between the states B of α-chymotrypsin and chymotrypsinogen A. It becomes apparent that the A and B states of proteins can be adequately characterized only with the exhaustive application of physical and chemical methods.

PAGE 47

Proton-exchange studies demonstrate that the number of protein protons exchanging with water protons via the unfolding transition, transition I, is about 15 percent fewer in chymotrypsinogen A that in α-chymotrypsin, in good agreement with the results of heat capacity and NMR (*223*).

PAGE 58

A variety of new examples of Vaslow–Doherty compensation is given in Ref. (*205*). Most of these were obtained at high precision levels. The ubiquitous appearance of the Vaslow–Doherty compensation pattern in small-solute processes establishes its dependence on liquid water and its importance. Most, if not all, types of processes in water solution appear to require reconsideration in light of this property. Through mechanisms that are not understood the pattern is appearing in an increasing number of protein systems to indicate that even the physiologically functional processes of proteins are directly coupled to liquid water. When the compensation pattern shows that water is involved in a protein reaction, a totally different analysis of temperature, pH behavior, solvent-effects, and so on is required.

PAGE 62

We failed to mention in the early version that although only a small part of the catalytic efficiency of chymotrypsin can be attributed to the Vaslow–Doherty phenomenon and thus presumably to water, side-chain specificity for inhibitors and substrates is very closely tied to the phenomenon. Recently Belleau and Lavoie and Belleau and diTullio (210) have shown the same thing for acetylcholinesterase.

PAGES 74 AND 76

The point of view that the weight of the evidence favors the idea that conformation changes of the protein provide the vehicle for heme–heme interaction, the Bohr effect, etc. is still valid, but the evidence is far from satisfactory. Increased resolution in X-ray-diffraction studies of hemo-globin (228) has failed to demonstrate more than a few displacements in the comparison of low-spin and high-spin derivatives, so that conformation changes, if they occur in the polypeptide and are extensive, apparently must be below the resolution of about 2.8 Å (212, 228). This is of course entirely reasonable. However, recent work by Shulman and co-workers (229), using NMR spectra, place a severe limitation on sequential relationship between ligand binding and possible conformation changes. In addition the work of Alben et al. (152), using infrared studies of the C—O stretching frequency of carbon monoxide bound to heme iron, casts considerable doubt on the existence of any important coupling between the electronic state of the heme group and the protein conformation. Conformation changes are claimed to be the basis of changes in the values of a number of observables, but these changes may involve only water and surface groups of the protein. The possibility of direct involvement of bulk water as indicated by the appearance of the Vaslow–Doherty compensation pattern is raised by plotting the enthalpy and entropy data obtained by Roughton, Lyster, and Otis (6) for the single-step constants (215). A reasonable compensation plot is obtained, but the demands on the data are such that despite their high precision, the validity of this test result is not confirmed. The entire matter of conformation changes is thus still quite obscure and probably will remain so until the several possible roles for participation of water are better understood.

PAGES 59 AND 97

Hopkins and Lumry (*208*) have studied electron ejection from a variety of indole compounds, using scavenger methods. The electron is ejected not from the first excited singlet of the exciplex but from a singlet species formed from the first excited singlet via a process with a large activation energy. This species is probably an example of a type of separated-charge states proposed by Stein *et al.* (*231*). The formation of the activated complex for production of this state from the excited singlet state probably also involves charge rearrangement, so that this process, like the hydrolysis of *t*-butyl chloride or the ionization of acetic acid, reflects the effects of solvation changes which occur with a change in charge separation. Recent analyses of the alcohol-perturbed behavior of the solvolysis rate process of *t*-butyl chloride by Arnett *et al.* (*111*) show that the effects of alcohol additions to water solvent are larger for the ground state than for the more polar activated complex. The remarks in the text about the volume required for dissolving one mole of electrons have thus become irrelevant.

Note added in proof. The bibliography of this article, already inadequate in 1969, has not improved with time. Many topics require reconsideration in detail, though fortunately only rarely in spirit. Such reconsideration is not possible, but a few interesting or important points should be discussed. The first of these has to do with the use of the heat-capacity change in protein-unfolding transitions to estimate the change in exposure of nonpolar groups to water. Jackson and Brandts (*238*) have continued their work on the heat capacities of folded and unfolded states of proteins and on small model substances. The temperature dependence of the over-all heat-capacity change in unfolded states is still not established, but the heat capacity of state A appears to decrease with increasing temperature and that of state B to be nearly temperature independent. These workers have shown that the amide (peptide) bond also produces a significant increase in heat capacity of the protein solution when it becomes exposed to water. The contribution from the latter source is a significant fraction of the total heat-capacity increase due to unfolding so that quantitative comparisons between the amount of "free-moving" polypeptide measured by NMR band areas and by heat-capacity changes on unfolding becomes even more hazardous than has been pointed out in the text (cf. p. 48), despite the good agreement which has been obtained thus far. Neither the Brandts nor the Shiao–Lumry–Fahey expressions for the heat capacity of transfer and the enthalpy, entropy, and free-energy changes in an unfolding transition are quantitatively consistent with the temperature dependencies of the heat capacity of proteins in states A and B. However, over the limited experimental temperature range which has thus far been used in unfolding studies, the thermodynamic analysis is not very sensitive to the assumed form of the heat-capacity change and the differences among the average heat-capacity changes obtained with the different mathematical models are smaller than the experimental errors involved in their determination by the van't Hoff method.

Perutz and co-workers (*241*) have carried out a number of three-dimensional difference-Fourier studies of hemoglobins bearing different ligands or chemically modified. In these most recent studies, a wealth of changes in protein conformation associated with ligand binding are beginning to emerge. The data analyzed in terms of molecular models have suggested to the workers a heme-heme linkage mechanism generated by the change in the position of the iron ion relative to the porphyrin plane consistent with the proposal of Hoard *et al.* (*137*) and with X-ray-diffraction data on single-subunit hemoglobins obtained by Love (*240*) and by Hüber (*239*) and their respective co-workers. Changes in prophyrin geometry are proposed, but the major system of conformation change is thought to pass through the F helix and the C-terminal section of the H helix of all four subunits. Although this locus of change and type of change, "subtle change," is the rack model, the mechanism may be significantly different from the original rack mechanism as suggested by the following argument:

The X-ray data indicate that the sixth position of the heme groups of the β chains may be blocked by side chains, so that oxygen cannot be bound without displacement of the side chains, which requires full reorganization of the four chains. This situation is consistent with the NMR results of Wüthrich, Shulman, and co-workers (*242*) and the infrared studies of Alben *et al.* (*152*), all of which appear to establish that there are only two electronic states of the heme group, the unliganded and the liganded states, and no stable states associated with intermediate degrees of oxygenation, at least in normal behavior of hemoglobin. Hence heme-heme interaction does not require a direct linkage path stretching from heme group to heme group. Rather, it is a consequence of free-energy considerations which cannot be assessed from structural information. The simpler aspects of rack mechanisms are nevertheless still apparent in the latest data, since the electronic states of the heme groups undergo large changes when the subunits are dissociated from each other, a situation which requires direct transmission of information about the state of subunit association to the heme groups.

Since the mechanism given on p. 72 to explain the first-order independence of heme-heme interaction and the Bohr effect in hemoglobin is still the correct explanation for the way such phenomenological independence is obtained, it is important to note that, despite the new findings by Wüthrich *et al.* using NMR and the older findings of Alben and co-workers using infrared spectra, the electronic states of the heme groups are not altered in heme-heme interaction; that is, there is only a single nonliganded and a single liganded state of the heme group in normal function, and the free energies (and undoubtedly, even more, the enthalpies and entropies) of the nonliganded and the liganded protein species are not invariant and change with presence or absence of Bohr protons, 2,3-diphosphoglycerate, etc., just as described in the text.

REFERENCES

1. J. C. Kendrew, G. Bodo, H. M. Dintzis, R. G. Parrish, and H. Wyckoff, *Nature*, **181**, 662 (1958).
2. R. Lumry and R. Biltonen, in *Structure and Stability of Biological Macromolecules* (S. Timasheff and G. Fasman, eds.), Marcel Dekker, New York, 1969, p. 65.
3. H. Muirhead and M. Perutz, *Nature*, **199**, 633 (1963); H. Muirhead, J. Cox, L. Mazzarella, and M. Perutz, *J. Mol. Biol.*, **28**, 117 (1967).
4. S. Ogawa and H. McConnell, *Proc. Natl. Acad. Sci. U.S.*, **58**, 19 (1967); J. Boeyens and H. McConnell, *ibid.*, **56**, 22 (1966); S. Ohnish, J. Boeyens, and H. McConnell, *ibid.*, *idem.*, 809.
5. M. Reichlin, personal communication, 1968.
6. F. J. W. Roughton, *Biochem. J.*, **30**, 2117 (1936); *J. Physiol.* (*London*), **126**, 259 (1954); F. J. W. Roughton, A. B. Otis and R. L. T. Lyster, *Proc. Roy. Soc.* (*London*), *Ser. B.*, **144**, 29 (1955); R. L. J. Lyster, Dissertation, Cambridge Univ., 1955.
7. H. Watson and B. Chance, in *Hemes and Hemoproteins* (B. Chance, R. Estabrook and T. Yonetoni, eds.) Academic Press, New York, 1966, p. 151.
8. B. Chance, personal communications, 1970.
9. P. George in *Currents in Biochemical Research* (D. Green, ed.), Wiley-Interscience, New York, 1956, p. 338.
10. M. Keyes and R. Lumry, *Federation Proc.*, **27**, 895 (1968).
11. M. Keyes, Dissertation, Univ. Minnesota, 1968.
12. H. Theorell, *Biochem. Z.*, **268**, 73 (1934).
13. C. L. Nobbs, H. C. Watson, and J. C. Kendrew, *Nature*, **209**, 339 (1966); J. C. Kendrew, in *Biological Structure and Function* (T. W. Goodwin and O. Lindberg, eds.), Academic Press, London, 1961, Vol. 1, p. 5.
14. B. P. Schoenborn, *Federation Proc.*, **27(3)**, 888 (1968).
15. L. J. Banaszak, H. C. Watson, and J. C. Kendrew, *J. Mol. Biol.*, **12**, 130 (1965).
16. A. Chanutin and R. Curnish, *Arch. Biochem. Biophys.*, **106**, 433 (1964); **121**, 96 (1967); **123**, 163 (1968); *Proc. Soc. Exp. Biol. Med.*, **120**, 291 (1965).
17. J. Monod, J. Wyman, and J. Changeux, *J. Mol. Biol.*, **12**, 88 (1965).
18. R. Lumry and S. Takashima, *Symp. Enzyme Chem., Tokyo and Kyoto, Japan, 1957* (Publ. 1958), p. 215; R. Lumry, *Biophys.* (*Japan*), **1**, 1 (1961).
19. L. Pauling and C. D. Coryell, *Proc. Natl. Acad. Sci. U.S.*, **22**, 159 (1936); *ibid.*, *idem.*, 210 (1936).
20. R. Lumry and H. Eyring, *J. Phys. Chem.*, **58**, 110 (1954).
21. R. Lumry, A. Solbakken, J. Sullivan, and L. Reyerson, *J. Am. Chem. Soc.*, **84**, 142 (1967).
22. K. Okunuki, *Advan. Enzymol.*, **23**, 29 (1961).

23. E. Stellwagen, *Biochem.*, **3**, 919 (1964); *J. Biol. Chem.*, **242**, 602 (1967).
24. J. F. Sullivan, Dissertation, Univ. Minnesota, 1966.
25. R. H. Kretsinger, H. C. Watson, and J. C. Kendrew, *J. Mol. Biol.*, **31(2)**, 305 (1968).
26. C. Tanford, K. Kawahara, and S. Lapanje, *J. Am. Chem. Soc.*, **89**, 729 (1967); Y. Nozaki and C. Tanford, *ibid.*, *idem.*, 742; C. Tanford, K. Kawahara, S. Lapanje, T. Hooker, A. Salshuddin, K. Aune, and T. Takage, *ibid.*, *idem*, 5023.
27. J. F. Brandts and R. Lumry, *J. Phys. Chem.*, **67**, 1484 (1963).
28. D. P. Hollis, G. McDonald, and R. L. Biltonen, *Proc. Natl. Acad. Sci. U.S.*, **58**, 758 (1967).
29. R. L. Biltonen, G. McDonald and D. P. Hollis, to be submitted.
30. K. Chakravarti and A. Rosenberg, *J. Biol. Chem.*, **243**, 3193 (1968).
31. A. Rosenberg and J. Enberg, *J. Biol. Chem.*, **244**, 6153 (1969).
32. H. Frank and M. Evans, *J. Chem. Phys.*, **13**, 507 (1945).
33. D. Eley, *Trans. Faraday Soc.*, **35**, 1281, 1421 (1939); D. Eley and M. Evans, *ibid.*, **34**, 1093 (1938).
34. J. Brandts, *J. Am. Chem. Soc.*, **86**, 4302 (1964).
35. J. Brandts, in *Structure and Stability of Biological Macromolecules* (S. Timasheff and G. Fasman, eds.), Marcel Dekker, New York, 1969, p. 213.
36. P. Flory and W. Miller, *J. Mol. Biol.*, **15**, 284 (1966).
37. W. Miller, D. Brant and P. Flory, *J. Mol. Biol.*, **23**, 67 (1967).
38. D. A. Brant, *Abstr.*, *Natl. Meeting*, *Am. Chem. Soc.*, *155th, San Francisco, 1968.*, Abstr. No. T079.
39. K. Linderstrøm-Lang and S. O. Nielsen, in *Electrophoresis* (M. Bier, ed.), Academic Press, New York, 1959, pp. 35–89.
40. J. Steinhardt and S. Beychok, *The Proteins*, Academic Press, New York, 2nd ed., 1964, Vol. 2, p. 140.
41. J. Kendrew and H. Watson, personal communication, 1967.
42. B. Matthews, P. Sigler, R. Henderson, and D. Blow, *Nature*, **214**, 652 (1967).
43. H. W. Wyckoff, K. D. Hardman, N. M. Allewell, T. Inagami, L. N. Johnson, and F. M. Richards, *J. Biol. Chem.*, **242**, 3984 (1967).
44. M. Eisenberg and G. Schwert, *J. Gen. Physiol.*, **34**, 583 (1951).
45. J. Brandts, *J. Am. Chem. Soc.*, **86**, 4291 (1964).
46. D. Shiao, Dissertation, Univ. Minnesota, 1968; D. Shiao, R. Lumry, and J. Fahey, *J. Am. Chem. Soc.*, **93** (1971), in press.
47. R. Biltonen, Dissertation, Univ. Minnesota, 1965.
48. R. Biltonen and R. Lumry, *J. Am. Chem. Soc.*, **91**, 4251, 4256 (1969); **92** (1970).
49. H. Parker and R. Lumry, *J. Am. Chem. Soc.*, **85**, 483 (1962).
50. H. Havsteen and G. P. Hess, *J. Am. Chem. Soc.*, **85**, 791 (1963).
51. R. Biltonen, R. Lumry, V. Madison, and H. Parker, *Proc. Natl. Acad. Sci. U.S.*, **54**, 1018, 1412 (1965).
52. H. Parker, Dissertation, Univ. Minnesota, 1967.
53. M. A. Marini and C. Wunsch, *Biochemistry*, **2**, 1454 (1967).
54. A. Rosenberg, *J. Biol. Chem.*, **241(21)**, 5119 5126 (1966).
55. S. Rajender, unpublished observations from this Laboratory.
56. F. Roughton and L. Rossi-Bernardi, *Proc. Roy. Soc.* (*London*), **B164**, 381 (1966); N. Meldrum and F. Roughton, *J. Physiol.*, **80**, 143 (1933); H. Constantine, L. Rossi-Bernardi, and F. Roughton, *ibid.*, **169**, 21P (1965).
57. J. Chipperfield, *Proc. Roy. Soc.* (*London*), **B164**, 401 (1966).
58. Y. Kim, Dissertation, Univ. Minnesota, 1968; Y. Kim and R. Lumry, *J. Am. Chem. Soc.*, **93**, 1003 (1971).

59. W. Kauzmann, *Advan. Protein Chem.*, **14**, 1 (1959).
60. L. Josefsson, *Abstr. Intern. Biochem. Congr.*, *5th, Moscow, 1961;* L. Josefsson, *Biochem. Biophys. Acta*, **74**, 774 (1963); J. Josefsson and P. Edman, *ibid.*, **25**, 614 (1957).
61. R. Lumry, R. Biltonen, and J. Brandts, *Biopolymers*, **4**, 917 (1966).
62. I. Wadsö, communicated by R. Biltonen.
63. J. F. Brandts, *J. Am. Chem. Soc.*, **87**, 2759 (1965); J. F. Brandts and L. Hunt, *J. Am. Chem. Soc.*, **89**, 4826 (1967).
64. R. Danforth, H. Krakower, and J. M. Sturtevant, *Rev. Sci. Instr.*, **38**, 484 (1967).
65. R. Steiner and H. Edelhoch, *Biochim. Biophys. Acta*, **66**, 352 (1963).
66. Y. Kim and R. Lumry, *J. Am. Chem. Soc.*, **93** (1971), in press.
67. C. C. Bigelow, *J. Mol. Biol.*, **8(5)**, 696 (1964); C C. Bigelow, *Compt. Rend. Trav. Lab. Carlsberg, Ser. Chemie*, **31**, 305 (1960); *J. Biol. Chem.*, **236**, 1706 (1961); and I. I. Geschnind, *Compt. Rend. Trav. Lab. Carlsberg, Ser. Chemie*, **31**, 283 (1960); *Ibid.*, **32**, 89 (1961); C. C. Bigelow and M. Sonnenberg, *Biochemistry*, **1**, 197 (1962); C. C. Bigelow and T. A. Krenitsky, *5th Intern. Cong. Biochem., Moscow, 1961*, Abstr. Comm. 25.
68. W. A. Klee, *Biochemistry*, **6(12)**, 3736 (1967).
69. J. Brandts, personal communication, 1968.
70. K. Linderstrøm-Lang, *Stanford Univ. Publ. Med.*, **6**, 52 (1952).
71. C. Martin, personal communication, 1967.
72. D. Holcomb and K. van Holde, *J. Phys. Chem.*, **66**, 1999 (1963).
73. M. Lazdunski and M. Delaage, *Biochim Biophys. Acta*, **140**, 417 (1967).
74. B. F. Erlanger and W. Cohen, *J. Am. Chem. Soc.*, **85**, 348 (1963).
75. J. A. Schellman, *Compt. Rend. Trav. Lab. Carlsberg, Ser. Chim.*, **29**, 223 (1955).
76. T. Schwartz, I. Wadsö and R. Biltonen, submitted, to *Biochemistry* (1971).
77. M. Bender, G. Schonbaum, and B. Zerner, *J. Am. Chem. Soc.*, **84**, 2540 (1962); M. Caplow and W. Jencks, *J. Biol. Chem.*, **238**, 1907 (1963); T. Spencer and J. Sturtevant, *J. Am. Chem. Soc.*, **81**, 1874 (1959); H. Guttfreund and J. Sturtevant, *Proc. Natl. Acad. Sci. U.S.*, **42**, 71 (1956); B. Zerner, R. Bond, and M. Bender, *J. Am. Chem. Soc.*, **86**, 3674 (1964); H. Gutfreund and J. Sturtevant, *Biochem. J.*, **63**, 656 (1956); R. Epand and I. Wilson, *J. Biol. Chem.*, **238**, 1718 (1963); *Ibid.*, **239**, 4138 (1964); G. Dixon and H. Neurath, *J. Biol. Chem.*, **225**, 1049 (1957).
78. H. Weiner, C. W. Bartt, and D. E. Koshland, *J. Biol. Chem.*, **241**, 3687 (1966).
79. J. Edsall, *J. Am. Chem. Soc.*, **57**, 1506 (1935); J. Edsall, in *Proteins, Amino Acids and Peptides* (E. Cohn and J. Edsall, eds.), Reinhold, New York, 1943, p. 165.
80. C. Tanford, *J. Am. Chem. Soc.*, **84**, 4240 (1962).
81. D. Wetlaufer, S. Malik, L. Stoller, and R. Coffin, *J. Am. Chem. Soc.*, **86**, 508 (1964).
82. G. Krescheck and L Benjamin, *J. Phys. Chem.*, **68**, 2476 (1964).
83. P. Desneulle, *The Enzymes*, 2nd. ed., Academic Press, New York, 1960, Vol. 4. p. 93.
84. R. Biltonen, personal communication.
85. B. S. Hartley, *Nature*, **201**, 1284 (1964); B. S. Hartley, *Atlas of Protein Sequence and Structure 1967–69*, Natl. Biomed. Res. Found., Maryland, p. 206; L. B. Smillie and B. S. Hartley, *J. Mol. Biol.*, **12**, 933 (1965); D. G. Smyth, W. H. Stein, and S. Moore, *J. Biol. Chem.*, **238**, 227 (1963).
86. M. A. Lauffer, *Biochemistry*, **3**, 731 (1964); M. A. Lauffer, A. T. Ansevin, T. R. Cartwright, and C. T. Brinton, *Nature*, **181**, 1338 (1958); M. A. Lauffer, *Biochemistry*, **5**, 1952, 2440 (1966).
87. W. Drost-Hansen, *Advan. Chem.* **67**, 70 (1967) and references therein; W. Drost-Hansen, *Intern. Symp. Water Desalination, Washington, D.C., Oct., 1965.*

88. G. Kortum and K. A. Steiner, *Angew. Chem.*, **6**, 1087 (1967).

89. A. Wishnia and T. W. Pinder, Jr., *Biochemistry*, **8**, 5064 (1969).

90. M. Keyes and R. Lumry, in *Probes for Mitochondrial Structure and Function* (B. Chance, ed.), Academic Press, New York, 1970, in press.

91. J. Leffler and E. Grunwald, *Rates and Equilibria of Organic Reactions*, Wiley, New York, 1963, Chap. 9.

92. L. G. Hepler and W. F. O'Hara, *J. Phys. Chem.*, **65**, 811 (1961); L. G. Hepler, *J. Am. Chem. Soc.*, **85**, 3089 (1963); G. L. Bertrand, F. J. Millero, Ching-hsieu Wu, and L. G. Hepler, *J. Phys. Chem.*, **70**, 699 (1966); R. N. Goldberg, R. G. Riddell, M. R. Wingard, H. P. Hopkins, C. A. Wulff, and L. G. Hepler, *J. Phys. Chem.*, **70**, 706 (1966); B. J. Hales, G. L. Bertrand, and L. G. Hepler, *J. Phys. Chem.*, **70**, 3970 (1966).

93. D. Doherty and F. Vaslow, *J. Am. Chem. Soc*, **74**, 931 (1952); F. Vaslow and D. Doherty, *ibid.*, **75**, 928 (1953).

94. A. Yapel, Dissertation, Univ. Minnesota, 1967.

95. A. Yapel and R. Lumry, *J. Am. Chem. Soc.*, **86**, 4499 (1964); A. Yapel, M. Han, R. Lumry, A. Rosenberg, and D. F. Shiao, *ibid.*, **88**, 2573 (1966).

96. S. Rajender, M. H. Han and R. Lumry, *J. Am. Chem. Soc.* **92**, 1378 (1970).

97. B. Hartley and B. Kilby, *Biochem. J.*, **56**, 288 (1954).

98. M. L. Bender, *J. Am. Chem. Soc.*, **84**, 2582 (1962); M. L. Bender and F. J. Kezdy, *ibid.*, **86**, 3704 (1964); J. Mercouroff and G. P. Hess, *Biochem. Biophys. Res. Commun.*, **11**, 293 (1963); M. Capton and W. P. Jencks, *Biochemistry*, **1**, 883 (1962); R. M. Anderson, E. H. Cordes, and W. P. Jencks, *J. Biol. Chem.*, **236**, 455 (1961); M. L. Bender, *Chem. Rev.*, **60**, 53 (1960); D. H. Staumeyer, W. N. White and D. E. Koshland, Jr., *Proc. Natl. Acad. Sci. U.S.*, **50**, 931 (1963); R. A. Oosterbaan and M. E. Van Andrichem, *Biochem. Biophys. Acta*, **27**, 423 (1958); H. Weiner and D. E. Koshland, Jr., *J. Biol. Chem.*, **240**, 2764 (1965).

99. G. L. Likhtenshtein, *Biofizika*, **11(1)**, 24 (1966).

100. H. Wu and E. F. Yang, *Chinese J. Physiol.*, **6**, 51 (1932); T. Huang and H. Wu, *ibid.*, **4**, 221 (1930).

101. N. Gralen, *Biochem. J.*, **33**, 1907 (1939); J. Roche, A. Roche, G. S. Adair, and M. E. Adair, *Biochem. J.*, **26**, 1811 (1932).

102. M. Walker, T. Bednar, and R. Lumry, in *Molecular Luminescence* (E. C. Lim, ed.), W. A. Benjamin, Inc., New York, 1969.

103. L. Grossweiner and H. Joschek, *Advan. Chem.*, *Ser.* **50**, 279 (1965).

104. J. K. Thomas, in *Radiation Research* (G. Silini, ed.), North-Holland, Amsterdam, 1967; J. Jortner and S. A. Rice, *Advan. Chem.*, **50**, (1965).

105. D. Ives and P. Marsden, *J. Chem. Soc.*, **1965**, 649 (1965).

106. S. Winstein and A. H. Fainberg, *J. Am. Chem. Soc.*, **79**, 5937 (1957).

107. H. S. Golinkin and J. B. Hyne, *Can. J. Chem.*, **46**, 125 (1968).

108. E. M. Arnett, in *Physico-Chemico Processes in Mixed Aqueous Solvents* (F. Franks, ed.), Heineman Educational Books, London, 1967.

109. M. Smith and M. C. R. Symons, *Discussions Faraday Soc.*, **24**, 206 (1957); M. Smith and M. C. R. Symons, *Trans. Faraday Soc.*, **54**, 338, 346 (1958).

110. F. Franks and D. Ives, *Quart. Rev.*, **20**, 1 (1966).

111. E. M. Arnett, in *Hydrogen Bonded Solvent Systems* (A. Covington and P. Jones, eds.), Taylor and Francis, Ltd., London, 1967.

112. I. Klotz, in *Horizons in Biochemistry* (B. Pullman and M. Kacha, eds.), Academic Press, New York, 1962.

113. J. Leffler, *J. Org. Chem.*, **20**, 1202 (1955); **31**, 533 (1968).

114. M. S. Jhon, J. Grosh, T. Ree, and H. Eyring, *J. Chem. Phys.*, **44**, 1465 (1966).

115. R. A. Horne and R. P. Young, *J. Phys. Chem.*, **71**, 3824 (1967), and references therein; R. A. Horne and D. S. Johnson, *Nature*, **209**, 82 (1966).

116. R. Lumry, in *Jena, 1963, Abhandl. Deut. Akad. Wiss. Berlin Kt. Chem. Geol. Biol.*, **1964** (6), 125.

117. J. Wyman, *J. Am. Chem, Soc.*, **89**, 2202 (1967).

118. R. Benesch and R. E. Benesch, *Federation Abstr. 696, Federation Proc.*, **27**, 339 (1968).

119. J. Changeux, *Cold Spring Harbor Symposia on Quantitative Biology*, Long Island Biol. Assn., Cold Spring Harbor, N.Y., 1963, p. 497; J. C. Gerhart and A. B. Pardee, *ibid.*, p. 491.

120. D. E. Atkinson and G. M. Watson, *J. Biol. Chem.*, **240**, 757 (1965).

121. E. R. Stadtman, *Advan. Enzymol.*, **28**, 49 (1966) and references therein.

122. W. Jencks, in *Current Aspects of Biochemical Energetics: Fritz Lipman Dedicatory Volume* (N. Kaplan and E. Kennedy, eds.), Academic Press, New York, 1966, p. 273.

123. T. Isemura, T. Tokazi, and H. Toda, *J. Biochem. (Japan)*, **52**, 16 (1962).

124. R. Lumry, in *Photosynthesis Mechanisms in Green Plants*, Publication 1145, Natl. Acad. Sci. and Natl. Res. Council, 1963, p. 625.

125. W. Reynolds and R. Lumry, *Mechanisms of Electron-Transfer Reactions*, Ronald Press, New York, 1966, Chap. 5, p. 83.

126. S. Maricic, J. S. Leigh and D. E. Sunko, *Nature*, **214**, 462 (1967).

127. L. Casola, P. E. Brumberg, and V. Massey, *J. Biol. Chem.*, **241**, 4977 (1966); L. Casola and V. Massey, *ibid.*, *idem.*, 4985 (1966); D. V. DerVartanian, W. H. Orme-Johnson, R. E. Hansen, H. Beinert, R. L. Tsai, J. C. M. Tsibris, R. C. Bartholomaus, and I. C. Gunsalus, *Biochem. Biophys. Res. Comm.*, **26**, 569 (1967); V. Aleman-Aleman, K. V. Rajagopalan, P. Handler, H. Beinert, and G. Palmer, *Proc. Symp., Amherst, Mass., 1964*, pp. 380–416 (1965);

128. H. A. Harbury, J. R. Cronin, M. W. Fanger, T. P. Hettinger, A. J. Murphy, Y. P. Myer and S. N. Vinogrodov, *Proc. Natl. Acad. Sci. U.S.*, **54(6)**, 1658 (1965).

129. A. Lein and L. Pauling, *Proc. Natl. Acad. Sci. U.S.*, **42**, 51 (1956).

130. P. George, S. C. Glauser, and A. Schejter, *J. Biol. Chem.*, **242**, 1690 (1967); P. George and R. L. J. Lyster, *Proc. Natl. Acad. Sci. U.S.*, **44**, 1013 (1958); *Natl. Acad. Sci., Natl. Res. Council, Publ. No.* **557**, 53 (1958).

131. E. Antonini, Physiol. Rev., **45**, 123 (1965).

132. E. Fleischer, *J. Am. Chem. Soc.*, **85**, 146 (1963); E. Fleischer, C. Miller, and L. Webb, *J. Am. Chem. Soc.*, **86**, 2342 (1964).

133. H. Eyring, R. Lumry, and J. Spikes, in *Mechanisms of Enzyme Action* (B. Glass and W. McElroy, eds.), Johns Hopkins Press, Baltimore, 1954.

134. L. Vaska, *Science*, **140**, 809 (1963).

135. J. McGinnety, R. Doedens, and J. Ibers, *Science*, **155**, 709 (1967).

136. R. J. P. Williams, in *The Enzymes*, 2nd ed. (Boyer, Lardy, and Myrback, eds.), Academic Press, New York, 1959, Vol. 1.; *Chem. Rev*, **56**, 299 (1956).

137. J. L. Hoard, M. Harrow, T. Harrow, and W. Caughey, *J. Am. Chem. Soc.*, **87**, 2312 (1965); J. L. Hoard, in *Hemes and Hemoproteins* (B. Chance, R. Estabrook, and T. Yonetani, eds.), Academic Press, New York, 1966, p. 9.

138. J. L. Hoard, in *Structural Chemistry and Molecular Biology* (A. Rich and N. Davidson, eds.), Freeman, San Francisco, 1968, p. 573.

139. L. J. Parkhurst and Q. H. Gibson, *Federation Abstr. 1732, Federation Proc.*, **27**, (1968); Q. H. Gibson, L. Parkhurst, and G. Geraci, *J. Biol. Chem.*, **244**, 4668 (1969).

140. W. M. Clark, *Oxidation–Reduction Potentials in Organic Systems*, Williams and Wilkins, Baltimore, Md., 1960.
141. J. Wyman, Jr. and D. Allen, *J. Polymer Sci.*, **7**, 499 (1951).
142. K. Winterhalter and D. Deranleau, *Biochemistry*, **6**, 3136 (1967).
143. K. Javeherian and S. Beychok, *J. Mol. Biol.*, **37**, 1 (1968).
144. E. Antonini, J. Wyman, A. Rossi-Fanelli, and A. Caputo, *J. Biol. Chem.*, **237**, 2773 (1962); A. Rossi-Fanelli and E. Antonini, *Arch. Biochem. Biophys.*, **80**, 299 (1959); A. Rossi-Fanelli, E. Antonini, and A. Caputo, *ibid.*, **85**, 37 (1959); E. Antonini, M. Brunori, E. Chiancone, A. Caputo, A. Rossi-Fanelli, and J. Wyman, *Biochem. Biophys. Acta.*, **79**, 284 (1964); Q. H. Gibson and E. Antonini, *J. Biol. Chem.*, **238**, 1384 (1963); A. Rossi-Fanelli and E. Antonini, *ibid.*, **235**, PC4 (1967).
145. D. W. Urry, *J. Biol. Chem.*, **242**, 4441 (1967).
146. S. Beychok, I. Tyuma, R. E. Benesch, and R. Benesch, *J. Biol. Chem.*, **242**, 2460 (1967).
147. H. Mizukami and R. Lumry, *Arch. Biochem. Biophys.*, **118**, 434 (1967).
148. E. Antonini, E. Bucci, L. Fronticelli, J. Wyman, and A. Rossi-Fanneli, *J. Mol. Biol.*, **12**, 375 (1965); M. Brunori, R. Noble, E. Antonini, and J. Wyman, *J. Biol. Chem.*, **241**, 5238 (1966).
149. M. Brunori, J. Wyman, and S. R. Anderson, *J. Mol. Biol.*, **34**, 352 (1968).
150. J. Barcroft, *The Respiratory Function of the Blood*, Part I, Cambridge Univ. Press, London, 1928; R. Lemburg and J. W. Legge, *Hematin Compounds and Bile Pigments*, Wiley-Interscience, New York, 1949, p. 290.
151. T. Shigu, K. Hwang, and I. Tyuma, *Arch. Biochem. Biophys.*, **123**, 203 (1968).
152. J. O. Alben and W. S. Caughey, in *Hemes and Hemoproteins* (B. Chance, R. W. Estabrook, and T. Yonetani, eds.), Academic Press, New York, 1966, p. 139; *Biochemistry*, **7**, 175 (1968).
153. W. Junge, E. Reinwald, B. Rumberg, U. Siggel and H. T. Witt, *Naturwiss.*, **55**, 1 (1968); W. Junge and H. T. Witt, *Z. Naturforsch.*, **236**, 244 (1968); B. Rumberg, E. Reinwald, H. Schroder and U. Siggel, *Naturwiss.*, **55**, 77 (1968).
154. J. E. Coleman and B. L. Vallee, *Federation Proc.*, **20**, 220 (1961); *J. Biol. Chem.*, **236**, 2244 (1961); B. L. Vallee, T. L. Coombs, and F. L. Hoch, *ibid.*, **235**, PC45 (1960); *ibid., idem.*, 64 (1960).
155. T. C. Bruice and S. J. Berkovic, *Bioorganic Mechanisms*, Benjamin, New York, 1966, pp. 236–242; G. H. Dixon, W. J. Dreyer, and H. Neurath, *J. Am. Chem. Soc.*, **78**, 4810 (1956); R. M. Anderson, E. H. Cordes, and W. P. Jencks, *J. Biol. Chem.*, **236**, 455 (1961).
156. L. A. E. Sluyterman, *Biochim. Biophys. Acta*, **151(1)**, 178 (1968).
157. E. L. Smith, *Proc. Natl. Acad. Sci. U.S.*, **35**, 80 (1959).
158. L. Cunningham, *Science*, **125**, 1145 (1957); L. Cunningham and C. Brown, *J. Biol. Chem.*, **221**, 287 (1956).
159. L. Parker and J. H. Wang, *Abstr. Natl. Meeting, Am. Chem. Soc.*, *154th, Chicago, 1967*, Abstr. CO31.
160. S. Bernhard, *The Structure and Function of Enzymes*, Benjamin, New York, 1968.
161. F. Karush, *J. Am. Chem. Soc.*, **72**, 2705 (1950).
162. D. Blears and S. Danyluk, *Biochem. Biophys. Acta*, **154**, 17 (1968).
163. S. Mizushima and T. Shimanouchi, *Advan. Enzymol.*, **23**, 1 (1961); *The Nature of the Peptide Bond*, Univ. Tokyo, 1967, pp. 305–17; S. Mizushima, T. Shimanouchi, M. Tsubon, and T. Arakawa, *J. Am. Chem. Soc.*, **79**, 5357 (1957).
164. D. Koshland, Jr., *Proc. Natl. Acad. Sci. U.S.*, **44**, 98 (1958); *Cold Spring Harbor Symp. Quant. Biol.*, **28**, 473 (1963).

165. W. D. Britt and D. Keilin, *Proc. Roy. Soc.* (*London*), **B156,** 429 (1962).

166. E. Margoliash, *Advan. Protein Chem.*, **21,** 113 (1966); E. L. Smith and E. Margoliash, *Federation Proc.*, **23,** 1243 (1964); *Evolving Genes and Proteins* (C. Bryson and H. J. Vogel, eds.), Academic Press, New York, 1965, pp. 221–242.

167. R. Benesch and R. E. Benesch, *Biochem. Biophys. Res. Commun.*, **26,** 162 (1967); R. E. Benesch, R. Benesch, and G. Macduff, *Proc. Natl. Acad. Sci. U.S.*, **54,** 535 (1965).

168. R. Yue, unpublished observations from this Laboratory, 1964.

169. M. Walker, T. Bednar, and R. Lumry, *J. Chem. Phys.*, **47,** 1020 (1967).

170. J. Wyman, *Advan. Protein Chem.*, **4** 407 (1948).

171. D. E. Koshland, Jr., G. Nemethy, and D. Filmer, *Biochemistry*, **5,** 365 (1966).

172. J. Wyman, *Advan. Protein Chem.*, **19,** 223 (1964).

173. J. Wyman, *J. Mol. Biol.*, **11,** 631 (1965).

174. J. P. Changeaux, J. C. Gerhart and H. K. Schachman, *Biochemistry*, **7(2),** 531 (1968). J. C. Gerhart and H. K. Schachman, *ibid.*, 538 (1968); J. P. Changeaux and M. M. Rubin, *ibid.*, 553 (1968).

175. K. Kirschner, M. Eigen, R. Bittman, and B. Voigt, *Proc. Natl. Acad. Sci. U.S.*, **56(7),** 1661 (1966).

176. M. Eigen, *Abstr. 154th Ann. Meeting, Am. Chem. Soc., Pittsburgh, Sept. 1967;* M. Brunori, M. Eigen, T. Schuster, in *Probes for Mitrochondrial Structure and Function* (B. Chance, ed.), Academic Press, New York, 1970, in press.

177. F. J. W. Roughton and R. L. J. Lyster, *Hvalradets, Skrifter No. 48*, 185 (1961).

178. J. Rifkind, unpublished observations from this Laboratory.

179. R. Lumry, in *The Enzymes*, 2nd ed. (Boyer, Lardy, and Myrback, eds.), Academic Press, New York, 1959, Vol. 1, p. 157.

180. D. D. Eley and R. B. Leslie, *Advan. Chem. Phys.*, **7,** 238 (1964).

181. G. Czapski, N. Frohwith, and G. Stein, *Nature*, **207,** 1191 (1965).

182. R. K. Clayton, *Molecular Physics in Photosynthesis*, (R. K. Clayton ed.), Blaisdell, Inc., New York, 1965.

183. T. Förster, *Fluoreszenz Organischer Verbindungen*, Vanderhoeck und Ruprecht, Göttingen, 1951.

184. J. Perrin, *2nd Conseil de Chimie Solvay*, Grauthier and Viller, Paris, 1925, p. 322; J. Perrin, C. R. Hebd, *Seances Acad. Sci.*, **184,** 1097 (1927); F. Perrin, *J. Phys.*, **VI(7),** 390 (1926); *Ann. Phys.*, **X(12),** 169 (1929); *J. Phys.*, **VII(5),** 497 (1934); *Ann. Phys.*, **17,** 283 (1932).

185. H. Kallman and F. London, *Z. Phys. Chem., Ser. B.*, **2,** 207 (1928).

186. R. W. Noble, M. Brunori, J. Wyman, and E. Antonini, *Biochem.*, **6(4),** 1216 (1967); M. Brunori, R. Noble, E. Antonini and J. Wyman, *J. Biol. Chem.*, **241,** 5238 (1966); Q. H. Gibson and S. Ainsworth, *Nature*, **180,** 1416 (1957); T. Bucher and J. Kaspers, *Biochim. Biophys. Acta*, **1,** 21 (1947); *Naturwiss.*, **33,** 93 (1946); T. Bucher, *Advan. Enzymol.*, **14,** 1 (1953); T. Bucher, *Angew. Chem.*, **63,** 256 (1950).

187. Y. Elkaner, *J. Phys. Chem.*, **72,** 3654 (1968).

188. R. F. Steiner and H. Edelhoch, *Chem. Rev.*, **62(5),** 457 (1962); G. M. Edelman and W. O. McClure, *Accounts Chem. Res.*, **1(3),** 65 (1968).

189. J. Longworth, *Photochem. Photobiol.*, **7,** 587 (1968).

190. R. A Harris, J. T. Penniston, J. Asai, and D. E. Green, *Proc. Natl. Acad. Sci. U.S.*, **59,** 830 (1968); J. T. Penniston, R. A. Harris, J. Asai, and D. E. Green, *ibid., idem.*, 624 (1968).

191. L. Packer, P. A. Siegenthaler, and P. S. Nobel, *J. Cell. Biol.*, **26,** 593 (1965); P. S. Nobel, *Biochim. Biophys. Acta*, **153,** 170 (1968).

192. W. Reynolds and R. Lumry, *Mechanisms of Electron-Transfer Reactions*, Ronald Press, New York, 1966.

193. M. J. Kronman, *Biochim. Biophys. Acta*, **133**, 19 (1967); M. J. Kronman, L. G. Holmes, and F. M. Robbins, *ibid.* **133**, 46 (1967).

194. F. W. J. Teale, *Biochem. J.*, **76**, 381 (1960).

195. S. V. Konev and I. D. Volotovskii, *Biofizika*, **11**, 791 (1966).

196. L. G. Augenstein and C. A. Ghiron, *Proc. Natl. Acad. Sci. U.S.*, **47**, 1530 (1961).

197. S. V. Konev, *Fluorescence and Phosphorescence of Proteins and Nucleic Acids*, S. Udenfriend (English ed.), Plenum Press, New York, 1967.

198. M. Walker, unpublished observations from this Laboratory, 1967.

199. E. Benson, *Compt. Rend. Trav. Lab. Carlsberg.*, *Ser. Chem.*, **31**, 235 (1959).

200. T. Takage and C. Tanford, *Federation Abstr. 695, Fed. Proc.*, **27**(2)(1968); C. Tanford, in *Subunits in Biological Systems* (S. Timasheff and G. Fasman, eds.), Marcel Dekker, New York, 1971.

201. J. McConn, G. Czerlinski and G. P. Hess, *Federation Proc.*, **27**, 456 (1968).

202. C. Jorgensen, *Progr. Inorg. Chem.*, **4**, 73 (1962).

203. R. M. Noyes, *J. Am. Chem. Soc.*, **84**, 513 (1962).

204. A. Hayashi, T. Suzuki, A. Shimizu, H. Morimoto, and H. Watari, *Biochim. Biophys. Acta*, **147**, 407 (1967).

205. R. Lumry and S. Rajender, *Biopolymer*, **9**, 1125 (1970).

206. J. Koudelka and L. Augenstein, *Photochem. Photobiol.*, **7**, 613 (1968).

207. C. Huggins, D. F. Tapley, and E. V. Jensen, *Nature*, **167**, 592 (1951).

208. T. R. Hopkins and R. Lumry, *Abstr. 5th Intern. Conf. Photobiol.*, *Hanover, New Hampshire, 1968*, Abstr. EF-6.

209. J. G. Beetlestone and D. H. Irvine, *J. Chem. Soc. (A)*, 951 (1968); A. C. Anusiem, J. G. Beetlestone, and D. H. Irvine, *ibid.*, 960, 1337 (1968); J. G. Beetlestone and D. H. Irvine, *ibid.*, 1340 (1968); J. G. Beetlestone, A. A. Epega, and D. H. Irvine, *ibid.*, 1346 (1968); J. E. Bailey, J. G. Beetlestone, and D. H. Irvine, *ibid.*, 2778 (1968); J. E. Bailey, J. G. Beetlestone, and D. H. Irvine, *ibid.*, 2913 (1969); J. E. Bailey, J. G. Beetlestone, and D. H. Irvine, *ibid.*, 241 (1969).

210. B. Belleau and J. Lavoie, *Can. J. Biochem.*, **46**, 1397 (1968); B. Belleau and V. di Tullio, *Can. J. Biochem.* **92**, 6320 (1970).

211. M. Keyes and R. Lumry, submitted to *Biochemistry*, 1970.

212. M. F. Perutz, H. Muirhead, J. M. Cox, and L. C. G. Coarman, *Nature*, **219**, 131 (1968).

213. G. Pool and D. F. Shiao, unpublished results from this Laboratory.

214. R. Lumry, in *Probes for Mitrochondrial Structure and Function*, (B. Chance, ed.), Academic Press, New York, 1970, in press.

215. R. Lumry, *Eying Festschrift*, (D. Henderson and J. Hirschfelder, eds.), Wiley-Interscience, New York, in press.

216. R. Countryman, P. Collins and J. L. Hoard, *J. Am. Chem. Soc.*, **91**, 18 (1969).

217. E. Padlan and W. Love, *Nature*, **220**, 376 (1968); personal communication, 1970.

218. P. George, W. A. Eaton, and M. Trachtman, *Federation Proc.*, **27**, 526 (1968).

219. J. F. Brandts, personal communication New York, 1969.

220. M. Lund and R. Lumry, unpublished observations from this laboratory.

221. M. Volini and P. Tobias, *J. Biol. Chem.*, **244**, 5105 (1969).

222. D. Blow, *Biochem. J.*, **112**, 261 (1969).

223. A. Rosenberg and C. K. Woodward, *J. Biol. Chem.*, **245**, 4677 (1970).

224. W. Bolen and R. Biltonen, personal communication 1970.

225. D. F. Shiao, personal communication 1970.

226. W. M. Jackson and J. F. Brandts, *Biochemistry*, **9**, 2294 (1970).
227. S. T. Freer, J. Kraut, J. D. Robertus, H. T. Wright and Ng. H. Xuong, *Biochemistry*, **9**, 1997 (1970).
228. M. Perutz, H. Muirhead, J. Cox, L. C. G. Goaman, F. Mathews, E. McGandry, and L. Webb, *Nature*, **219**, 29 (1968).
229. R. Shulman, S. Ogawa, K. Wüthrich, T. Yamane, J. Peisach, and W. Blumberg, *Science*, **165**, 251 (1969).
230. C. Anfinsen, E. Haber, M. Sela, and F. White, Jr., *Proc. Natl. Acad. Sci. U.S.*, **47**, 1309 (1961); M. Sela, C. Anfinsen, and W. Harrington, *Biochem. Biophys. Acta*, **26**, 502 (1957); **24**, 229 (1957); C. Epstein, and C. Anfinsen, *J. Biol. Chem.*, **237**, 2175 (1962).
231. J. Jortner, M. Ottobenghil and G. Stein, *J. Phys. Chem.*, **68**, 247 (1964).
232. T. A. Steitz, R. Henderson, and D. M. Blow, *J. Mol. Biol.*, **46**, 337 (1969).
233. G. E. Walrafen, in *Hydrogen-Bonded Solvent Systems* (A. Covington and P. Jones, eds.), Taylor & Francis, London, 1968, pp. 9–29.
234. C. D. Ritchie and W. Sager, *Progr. Phys. Org. Chem.*, **2**, 323 (1964).
235. L. Stryer, J. Kendrew, and H. Watson, *J. Mol. Biol.*, **8**, 96 (1964).
236. M. Keyes and R. Lumry, *J. Am. Chem. Soc.*, **93** (1971).
237. A. Rossi-Fanelli and E. Antonini, *Nature*, **186**, 895 (1960); *Arch. Biochem. Biophys.*, **77**, 478 (1958).
238. W. Jackson and J. F. Brandts, personal communication, 1970.
239. R. Huber, *Abstracts, 8th Intern. Cong. Biochem.*, Interlaken, Switzerland, 1970, p. 2.
240. W. E. Love, *Abstracts, 8th Intern. Cong. Biochem.*, Interlaken, Switzerland, 1970, p. 2.
241. M. F. Perutz, *Abstracts, 8th Inter. Cong. Biochem.*, Interlaken, Switzerland, 1970, p. 5.
242. K. Wüthrich and R. G. Shulman, *Phys. Today*, **23**, 43 (1970).
243. H. L. Oppenheimer, B. Labouesse, and G. P. Hess, *J. Biol. Chem.*, **241**, 2720 (1966).
244. B. P. Schoenborn, *J. Mol. Biol.*, **45**, 279 (1969).

CHAPTER 2

MODEL OXIDATIVE PHOSPHORYLATION SYSTEMS

Jui H. Wang

KLINE CHEMISTRY LABORATORY
YALE UNIVERSITY
NEW HAVEN, CONNECTICUT

Investigations of the respiratory chain oxidative phosphorylation during the past two decades have yielded a wealth of information on the subject, but the molecular mechanism by which phosphorylation is coupled to electron transfer has not yet been elucidated. While intensive work on the biological system is being continued, concurrent studies of simple model systems could lead to more definite conclusions that may be biologically relevant. Although the chemistry of model oxidative phosphorylation systems deserves careful study in its own right, the scientific significance will be greater if it can also illuminate the biological mechanism of free-energy conversion. Therefore both the molecular mechanisms in model systems and their biological relevance will be discussed in this chapter. However, in order to limit the scope, only those model reactions that have been found experimentally to couple phosphorylation with electron transfer will be included in the present discussion.

I. Quinol Phosphate

Clark, Hutchinson, and Todd (*1*) reported in 1961 that the oxidation of quinol phosphates by bromine in dry *N,N*-dimethylformamide (DMF)

$$\tag{1}$$

solution containing mono-(tetra-*n*-butylammonium) salt of adenosine-5'-monophosphate (AMP) produces adenosine-5'-diphosphate (ADP): This very interesting observation stimulated the speculation that respiratory chain oxidative phosphorylation might also take place through a similar mechanism, since quinones abound in mitochondria.

However, because of their structural resemblance to enol phosphates the quinol phosphates may already be "energy rich" before oxidation to quinones. Consequently a satisfactory oxidative phosphorylation mechanism that involve quinol phosphates as intermediates must also account for their formation. In their model reaction, the quinol phosphates was synthesized by Clark *et al.* through the base-catalyzed reaction between the substituted naphthaquinone and dibenzyl phosphite, followed by debenzylation with sodium thiocyanate. Since phosphites are absent in biological systems,

Vilkas and Lederer (2) suggested the biological mechanism shown in Fig. 1 for coupling phosphorylation to oxidation through quinol phosphates as

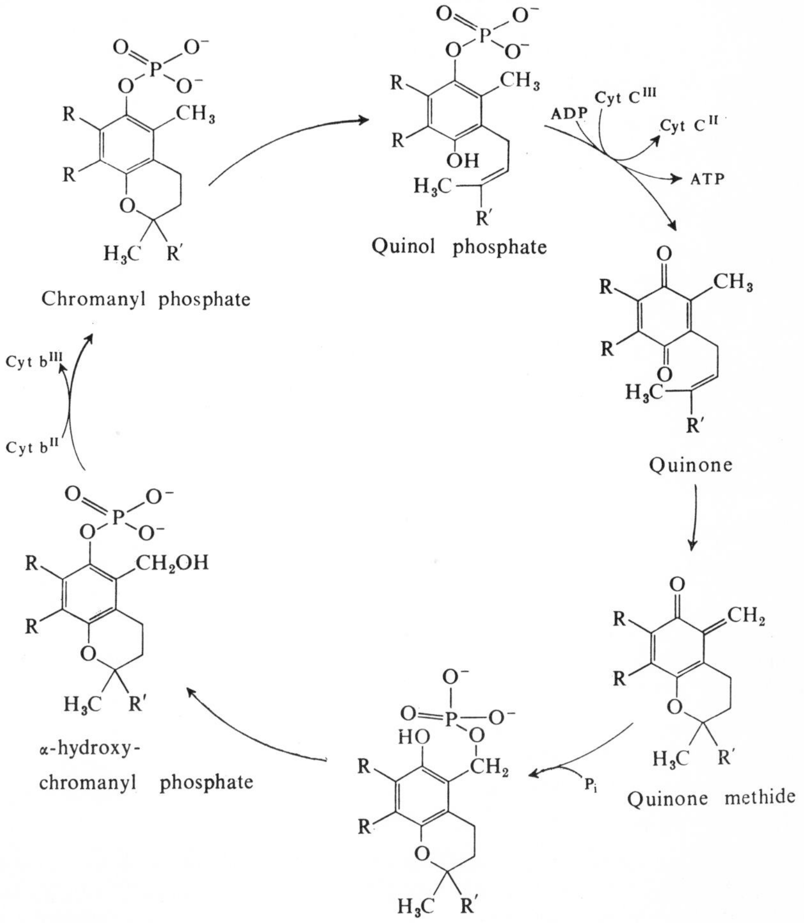

Fig. 1. The quinol phosphate mechanism for coupling phosphorylation to oxidation according to Vilkas and Lederer.

intermediates. This mechanism predicts hydrogen exchange between the substituted quinone and water.

Although earlier studies indicate tritium incorporation into the quinone from tritium-labeled water in the medium (3), later investigation by Rapoport *et al.* (4) of possible tritium incorporation from labeled water to the quinone in the cell-free extract from *Mycobacterium phlei* during oxidative phosphorylation gave unambiguously negative results. Both intact extracts (native quinone) and vitamin K_1-reconstituted, light-inactivated systems were used in the experiments with cell-free extracts that display phosphate fixation coupled to oxidation with either malate or pyruvate as the substrate. The data of Rapoport *et al.* indicate the lack of any significant tritium incorporation into the quinones.

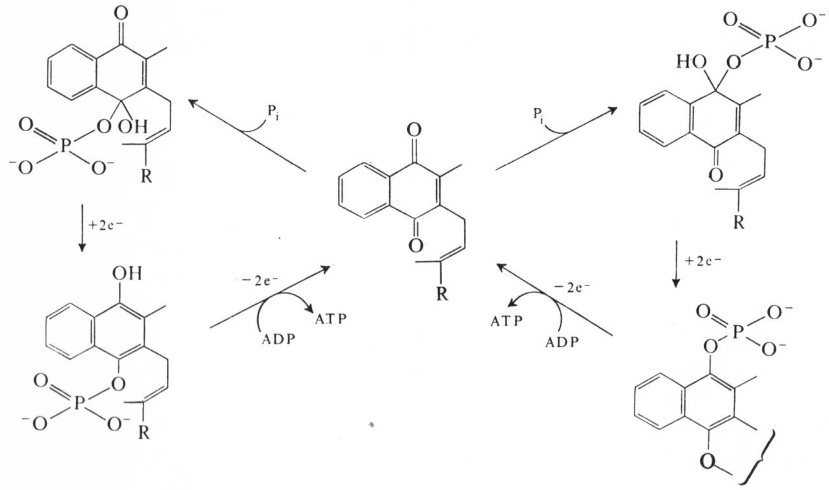

Fig. 2. The quinol phosphate mechanism for coupling phosphorylation to oxidation according to Snyder and Rapoport.

An analogous study by Gutnick and Brodie (5) using mammalian mitochondria and native ubiquinone, and by Parson and Rudney (6) using endogenous ubiquinone in *R. rubrum* also failed to detect significant tritium incorporation. Thus one must conclude that quinone involvement in oxidative phosphorylation cannot proceed through the mechanism in Fig. 1.

An alternative quinone mechanism, proposed by Brodie (7), involves the addition of phosphate to the carbonyl group of quinone, followed by reduction to a quinol phosphate with or without chromanol formation as illustrated in Fig. 2. To test the mechanism suggested in Fig. 2, Rapoport *et al.* (8) reconstituted oxidative phosphorylation in cell-free, light inactivated extracts from *Mycobacterium phlei* with both uniformly and specifically ^{18}O-labeled vitamin K_1 and determined the retention of the ^{18}O label in the quinone after the oxidative phosphorylation experiments. The recovered vitamin K_1 showed complete retention of ^{18}O except for a slow loss roughly proportional to the exposure time of the quinone to the medium, which probably occurred via simple exchange between the carbonyl group with water. That such an exchange is not related to oxidative phosphorylation is established by the parallel exchange observed for the KCN-treated system, where no consumption of molecular oxygen occurred. These results indicate that in contrast to the suggested mechanism in Fig. 2, quinone involvement in oxidative phosphorylation must proceed with the original C—O bond remaining intact.

II. Imidazole-Hemochrome

Brinigar and Wang (9) observed in 1964 that oxidation of chlorocruoro-heme dimethyl ester in a N,N-dimethylacetamide (DMAC) solution containing imidazole, orthophosphate (P_i), and AMP resulted in the production of appreciable amounts of ADP and ATP. Their earlier mechanistic interpretation in terms of the chemical participation of the formyl group of chlorocruorohemin was incorrect, because later more careful study (10) showed that molecular oxygen is required for these phosphorylation reactions and that "chlorocruorohemin" works only when an appreciable fraction of it is in the Fe(II) state. It was found that the phosphorylation of AMP and/or ADP takes place only when there is net oxidation of ferro- to ferri-heme by molecular oxygen in the presence of imidazole and that the formyl group is neither oxidized nor hydrated in the oxidation process, because the infrared spectrum of chlorocruorohemin dimethyl ester reisolated from the reaction mixture appears identical with that of the starting material with the carbonyl stretching frequency of the formyl group remaining unchanged at $1660 \; cm^{-1}$. In fact it is possible to produce phosphorylation through the aerobic oxidation of ferrohemes without formyl side chain (10).

The initial choice of the imidazolium salts of AMP and P_i was made because of their suitable solubility in polar organic solvents such as DMAC. However, when other salts (1-methylimidazolium, pyridinium, tetrabutyl-ammonium, etc.) were used to replace the imidazolium salts, either no or very little phosphorylation occurred. Imidazole was therefore implicated as an important component of this model oxidative phosphorylation system.

In order to explore the possible existence of an intermediate capable of transphosphorylation reactions leading to the formation of ATP in the above model reactions, protohemochrome dimethyl ester was oxidized by air in two separate experiments: (a) The diimidazole-ferrohemochrome solution in DMAC was added to a DMAC solution of the diimidazolium salts of HPO_4^{2-} and ADP that was preequilibrated with air; (b) the ferrohemochrome was added to an air-equilibrated DMAC solution containing only di-imidazolium hydrogen phosphate, and then ADP was added 1.5 hr later when the oxidation was already complete. The results are illustrated in Fig. 3.

The data in Fig. 3 clearly demonstrate that when ferrohemochrome is oxidized by air in a DMAC solution containing imidazole and P_i, an intermediate is formed that can subsequently react with ADP to form ATP.

If the nucleotides are omitted and the P_i is labeled with ^{32}P, a radioactive "energy-rich" intermediate product can be detected that has hydrolytic and

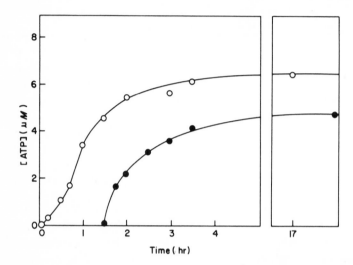

Fig. 3. Phosphorylation of ADP caused by the oxidation of ferrohemochrome. Open circles, ADP present prior to the addition of ferrohemochrome; solid circles, ADP added 1.5 hr after the ferrochrome. Final concentrations: ADP (0.13 mM), diimidazolium hydrogen phosphate (1.3 mM), and diimidazole-ferrohemochrome (0.4 mM). The experiments were carried out at 25°C.

paper chromatographic behavior indistinguishable from that of synthetic 1-phosphoimidazole (*10*). Since 1-phosphoimidazole is known to phosphorylate AMP and ADP in polar organic solvents (*11*), these observations identify it as the "energy-rich" intermediate in the above model oxidative phosphorylation reactions.

But how is 1-phosphoimidazole itself formed through the oxidation of heme by O_2? Studies with substituted imidazoles (*12*) suggest that molecular oxygen first extract two electrons from the ferrohemochrome to produce a complex of ferriheme and the reactive imidazolyl radical ($C_3H_3N_2$):

$$HN \diagup\!\!\!\diagdown N - Fe^{II} - N \diagdown\!\!\!\diagup NH \xrightarrow{O_2} HN \diagup\!\!\!\diagdown N - Fe^{III} - N \diagdown\!\!\!\diagup N \cdot + H^+ \qquad (2)$$

This radical can then react with P_i to form an unstable phosphoimidazolyl radical, which is subsequently reduced by another ferrohemochrome

$$(3)$$

molecule either to eliminate H_2O to produce 1-phosphoimidazole, or to eliminate imidazole and regenerate P_i:

$$(4)$$

$$(5)$$

The formation of the above trigonal–bipyramidal intermediate compound 1-orthophosphoimidazole ($C_3H_3N_2PO_4^{2-}$) by the usual nucleophilic attack at the phosphorus atom is a very slow process. But since radical reactions generally require a much lower activation free energy, the above trigonal–bipyramidal phosphoimidazolyl radical ($C_3H_3N_2PO_4^{-}$) can be formed much more rapidly. In the subsequent step driven by the oxidation–reduction free energy, this phosphoimidazolyl radical is reduced to the unstable 1-ortho-phosphoimidazole, which can spontaneously eliminate water to form 1-phosphoimidazole. In this way, oxidation can be coupled to phosphorylation.

III. Oxidation of Thioesters and Thioethers

The oxidation of thioesters (13, 14) and thioethers (15, 16) by iodine or bromine has been shown to lead to the formation of the pyrophosphate bond. According to Higuchi and Gensch (15), the mechanism of this model reaction is as follows:

$$(6)$$

If the resulting sulfoxide could be reduced back to the thioether form under physiological conditions, the latter could react again, and consequently we would have a functioning catalytic cycle. Further studies of this attractive reaction may provide us with valuable information relevant to oxidative phosphorylation.

IV. A Molecular Mechanism of Oxidative Phosphorylation Suggested by Model Reactions

One apparent advantage of the imidazolyl radical mechanism for coupling electron transport to phosphorylation is that the ferrohemochrome can be

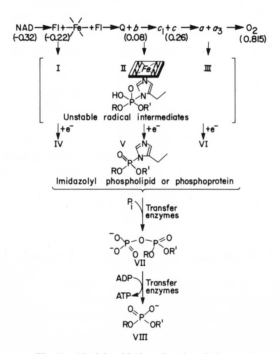

Fig. 4.   Model oxidative phosphorylation systems.

rapidly regenerated through electron transport to complete the catalytic cycle. Recent experiments show that even in aqueous solution, imidazole can be rapidly phosphorylated by the radical mechanism (17). A molecular mechanism of oxidative phosphorylation based on this particular model reaction has been proposed (18, 19). It was suggested that a similar radical mechanism may be operative in the inner mitochondrial membrane, with the important difference that the corresponding imidazolyl radical cannot diffuse away to cause biological damage but is covalently bonded to the protein of cytochromes as part of the histidine side chain. Furthermore, if the responsible cytochrome is bound to the membrane in such a manner that the substituted imidazole is adjacent to the phosphoester group of a phospholipid molecule, $(RO)(R'O)PO_2^-$, or phosphoprotein molecule, the imidazolyl radical may react rapidly with the latter to form a phosphoimidazolyl radical, which can then be reduced to the "energy-rich" intermediate **IV, V,** or **VI** as illustrated in Fig. 4. In this figure, the direction of electron transport under normal conditions is represented by the heavy arrows, and the normal *in vitro* midpoint potentials in volts at pH 7 for different electron carriers are indicated by the numbers in parentheses. The "energy-rich" intermediates **V** or **VI** can presumably react with P_i in the presence of a specific enzyme or coupling factor to give compound **VII**, which can in turn react with ADP in the presence of another enzyme or coupling factor to give ATP and regenerate the phospholipid or phosphoprotein **VIII** for the next round of oxidative phosphorylation. Since the phosphoryl group in **IV, V,** or **VI** is not from P_i, each of these "energy-rich" compounds should behave as a "nonphosphorylated" intermediate according to the conventional criterion (*20*).

In addition to the indirect evidences from the model system, the proposed mechanism in Fig. 4 is also consistent with many other well-known observations on mitochondrial oxidative phosphorylation. These will be discussed in the following.

A. Coupling between Electron Transport and Phosphorylation

Although the midpoint reduction potentials of the respiratory electron carriers *in vivo* are not precisely known, it is generally believed that the value increases steadily from NAD to O_2 for efficient electron transport along the respiratory chain (Fig. 4). That electron transport is necessary for phosphorylation is expected not only for thermodynamic reasons but also from a mechanistic point of view, because it is the transfer of electrons that gives rise to the radical intermediates and then changes them to the "energy-rich" intermediates **IV, V,** and **VI**. The more puzzling requirement of phosphorylation for electron transport becomes understandable if we recall that unless ADP and P_i are present in sufficient concentration to discharge the intermediates **IV, V,** and **VI** continually, the "electron carriers" will be left in these "energy-rich" states, each with a considerably lower or higher midpoint reduction potential than both of its neighboring carriers in the respiratory chain. Such a modified sequence will prevent efficient electron transport in the normal direction. In aqueous solution ferriheme has a midpoint reduction potential of about -0.12 V at pH 7. By complexing with strong coordinating ligands such as pyridine, its midpoint reduction potential can be raised by as much as 0.3 or 0.4 V (21). Conversely, since the phosphorylated imidazole group is a much weaker coordinating ligand for the Fe(II) state than the original imidazole group, one may expect the midpoint reduction potential of a (or a_3) in compound VI to be <0.26 V. Likewise the midpoint reduction potentials of compounds IV and V could for structural reasons be very different from their normal values. Consequently, unless the concentrations of ADP and P_i are sufficiently high and that of ATP sufficiently low so that compounds IV, V, and VI are discharged continually by the mitochondrial enzyme system, normal electron transport cannot continue. Thus regulation of the ratio $[ADP][P_i]/[ATP]$ controls the rate of respiration.

B. P:O Ratio

Inspection of the midpoint reduction potentials of oxygen at pH 7 as indicated in the following diagram shows that O_2 serves more efficiently as a 2- or 4-electron than 1- or 3-electron acceptor:

$$O_2 \xrightarrow{-0.7 \text{ V}} O_2^- \xrightarrow{1.3 \text{ V}} H_2O_2 \xrightarrow{0.4 \text{ V}} HO + H_2O \xrightarrow{2.3 \text{ V}} 2H_2O$$

$$\underbrace{\qquad\qquad}_{0.27 \text{ V}} \qquad \underbrace{\qquad\qquad}_{1.35 \text{ V}}$$

$$\underbrace{\qquad\qquad\qquad\qquad\qquad\qquad}_{0.815 \text{ V}} \qquad (7)$$

In the respiratory chain, once the bound O_2 has extracted an electron from the Fe(II) of heme a (or a_3), it has a strong tendency also to extract a second electron from the coordinating imidazole group, thereby converting the latter to the highly reactive imidazolyl radical, which could immediately attack the neighboring phospholipid to form the phospholipid–imidazolyl radical. In the subsequent reduction by cytochrome $c_1 + c$, the phospholipid–imidazolyl radical is first reduced to compound VI with elimination of H_2O, followed by the formation of compound VII and the reduction of the Fe(III) of heme back to Fe(II.) Consequently one ATP is formed at heme a (or a_3) for every pair of electrons transferred. The same P:O ratio is expected at the heme b (or c_1) phosphorylation site where compound V is formed. At the nonheme iron or at the quinone plus flavin site, the phosphorylation mechanism proposed in Fig. 4 is similar and hence also produces one ATP per two electrons transferred. Consequently the overall P:O ratio expected from the proposed mechanism is 3. The present interpretation also gives a plausible account for using heme derivatives, which are one-electron carriers, to transport electrons from the two-electron donor NADH to the four-electron acceptor O_2.

C. Crossover Points

If ATP is formed according to the scheme in Fig. 4 from ADP, P_i, and compounds **IV**, **V**, and **VI** through enzyme-controlled transphosphorylation reactions, then depletion of ADP will cause accumulation of compounds **IV**, **V**, **VI**. Since the midpoint reduction potentials of these "energy-rich" compounds should be quite different from their normal values they could fail to oxidize or reduce their respective neighboring carriers in the respiratory chain. Consequently the difference spectrum of ADP-depleted systems may show crossover points at the coupling sites as observed in the ingenious pioneering work of Chance and Williams (22).

D. Coupling Factors

Because the triesters of phosphoric acid are generally much more reactive than the monoesters (23) and because 1-phosphoimidazole rapidly hydrolyzes in even very weakly acidic solutions (24), it may be inferred by analogy that "energy-rich" intermediates such as compound VI should be much more susceptible to hydrolysis than even 1-phosphoimidazole and hence unsuitable to function as a free-energy transducer at an exposed site. Protecting this reactive group by a second lipid layer is hardly satisfactory, since the latter would also hinder the elimination of water and prevent the subsequent reaction with P_i. The only conceivable protection that satisfies all the necessary requirements is to combine the reactive part of compound **VI** with a specific enzyme or coupling factor that catalyzes the transphosphorylation of P_i by compound **VI** in preference to the hydrolysis of the latter. Similarly the enzyme that catalyzes the transphosphorylation of compound **VII** by ADP may also be regarded as a dual-function coupling factor.

If any of these endogenous coupling factors is leached out during the preparation of the so-called electron-transport particles (ETP) from mitochondria (25), the corresponding "energy-rich" intermediate will be susceptible to rapid hydrolysis. Consequently compounds **IV**, **V**, and **VI** will be spontaneously discharged to restore the midpoint reduction potentials of the respective electron carriers to their normal values for efficient electron transport without phosphorylation. By complexing the electron-transport particles with coupling factors discovered by Racker (26), the phosphorylating properties of the former can partially be restored.

These interpretations are entirely consistent with the mechanism proposed in Fig. 4. Moreover, according to the proposed mechanism, each of the transfer enzymes in Fig. 4 should exhibit ATP-ase activity when it forms a complex with a partial complement of the phospholipid.

E. Uncouplers

When the phosphoimidazolyl radical in our model system is reduced, it could either eliminate water to form 1-phosphoimidazole or eliminate imidazole to regenerate P_i. The success of the model oxidative phosphorylation experiments shows that the elimination of water is quite favorable under the experimental conditions. In respiratory chain oxidative phosphorylation, water is presumably preferentially eliminated from the mitochondrial membrane to form compound **IV**, **V**, or **VI**.

Molecules such as 2,4-dinitrophenol and dicoumarol may, because of their lipid solubility, penetrate the inner mitochondrial membrane and uncouple the oxidative phosphorylation there, either by reducing the radical intermediates **I**, **II**, or **III** and regenerating the imidazole group, or by displacing the imidazole group from compound **IV**, **V**, or **VI** to form the phenolic ester of the corresponding phospholipid. In either case, the midpoint reduction potential of the cytochrome will have been restored to its normal value for further efficient electron transport without phosphorylation having taken place. In other words, the mechanism is uncoupled. These molecules may also cause ion leakage across the mitochondrial membrane and consequently dissipate ATP.

It is also of interest to note that according to this interpretation, P_i is not necessary for the uncoupling of respiratory chain by dinitrophenol, an inference that is in agreement with the work of Slater (27).

F. Inhibitors

In terms of the proposed mechanism, inhibitors of electron transport could be molecules that complex with the electron carriers and thereby prevent electron transport either by increasing the energy barrier for electron transfer between neighboring carriers or by drastically changing the reduction potential of the carriers or both. On the other hand, inhibitors of the specific transfer reactions in Fig. 4 should behave as the inhibitors of phosphorylating oxidation reported in the literature, e.g., oligomycin and aurovertin.

G. Phosphohistidine

The isolation of ^{32}P-labeled phosphohistidine by Boyer et al. (28) stimulated many interesting speculations on its role in oxidative phosphorylation (29, 30). Recently Bieber and Boyer concluded that phosphohistidine probably does not serve as an intermediate in the formation of ATP by oxidative phosphorylation in mitochondrial particles, because the observed rate of labeling of phosphohistidine from $^{32}P_i$ during oxidative phosphorylation did not reach its maximum specific radioactivity when or before the rate of ATP labeling was maximal (31). Slater et al. had previously reached a similar conclusion (32).

However, according to the mechanism in Fig. 4, the "energy-rich" intermediates **IV**, **V**, and **VI** are formed through reactions of a bound radical species with the phosphodiester groups of the respective phospholipids, not with the P_i that was initially labeled in both the experiment of Slater et al. and the experiment of Bieber and Boyer. Consequently any ^{32}P-labeled phosphohistidine isolated by their procedure has to originate from subsequent transphosphorylation reactions, which need not bear strict sequential relationships to the labeling of ATP from $^{32}P_i$ during oxidative phosphorylation.

In addition, if oligomycin, aurovertin, etc. are indeed specific inhibitors of the transfer enzymes (coupling factors) in Fig. 4, they would be expected to inhibit the labeling of ATP from $^{32}P_i$ but not the labeling of phosphohistidine from $^{32}P_i$, as was also observed by Bieber and Boyer.

H. Reversal of Electron Transport by ATP

Since catalysts do not alter chemical equilibrium, the same enzymes that catalyze the transformations from **IV**, **V**, and **VI** to **VII** and **VIII** respectively in Fig. 4 must also catalyze the reverse transformations. Thus in the presence of an excess of ATP at high concentration, the following reverse reaction may take place:

$$\text{ATP} + \text{VII} \underset{\longleftarrow}{\overset{K_1}{\longrightarrow}} \text{ADP} + \text{VI}, \tag{8}$$

$$\text{VI} + a^{III} \underset{\longleftarrow}{\overset{K_2}{\longrightarrow}} P_i + a^{III*} \tag{9}$$

where a^{III} represents the Fe(III) state of cytochrome a (or a_3), and a^{III*} its "energy-rich" form, i.e., compound **VI**, and K_1 and K_2 are the equilibrium constants.

Combining equilibria (8) and (9) we get

$$\frac{[\text{ADP})[P_i][a^{III*}]}{[\text{ATP})[\text{VII}][a^{III}]} = K_1 K_2 \tag{10}$$

The presumably much lower midpoint reduction potential of a^{III*}, relative to a^{III}, could cause electrons to flow in the NAD direction, provided that

ATP or other "energy-rich" intermediates are present in sufficient concentration and that the O_2 side of the respiratory chain is intercepted, e.g., by cyanide.

Making the simplifying approximation that practically all the reduced cytochrome are in the unphosphorylated form a^{II}, we obtain from Eq. (10) the relationship

$$\frac{[a^{III}] + [a^{III*}]}{[a^{II}]} = B \left\{ 1 + K_1 K_2 [\text{VII}] \left(\frac{[\text{ATP})}{[\text{ADP})[\text{P}_i)} \right) \right\} \qquad (11)$$

where B represents the concentration ratio $[a^{III}]/[a^{II}]$.

The last equation states that the ratio of the equilibrium concentrations of oxidized to reduced cytochrome a (or a_3) increases with the ratio [ATP]/ ([ADP][P_i]). Both uncouplers and inhibitors are expected from the proposed mechanism to interrupt this type of reverse electron transport through the relevant path.

I. "Energy-Rich" Intermediate

The most obvious weakness of any chemical coupling mechanism of respiratory chain oxidative phosphorylation is that, in spite of the intensive efforts during the last twenty-five years, no "energy-rich" intermediate has ever been isolated from mitochondria. The failure to isolate such an intermediate could either be due to the fact that the intermediate does not exist or that it cannot be isolated by conventional methods. The mechanism in Fig. 4 illustrates the second possibility, because the "energy-rich" N—P bond in compound IV, V, or VI is between a hydrophilic protein molecule and a phospholipid molecule with bulky hydrophobic side chains. Consequently, compound IV, V, or VI should be insoluble in most solvents, and any successful attempt to separate the hydrophilic and hydrophobic moieties of such an intermediate would invariably rupture the crucial N—P bond. Recent experimental results (33) with oligomycin and aurovertin on rat liver mitochondria support the existence of "energy-rich" intermediates.

J. Conformational and Morphological Changes

The newly formed N—P bond between the cytochrome and the phospholipid could trigger conformational change in either macromolecular subunit. The cumulative effect of such conformational rearrangements could result in morphological changes of mitochondrial as well as chloroplastic membranes in a manner similar to the oxygenation of hemoglobin, which is known to cause tertiary structural changes at the molecular level (34) and morphological changes at the cellular level (35).

Inasmuch as the formation of such a new N—P bond and the conformational changes are thermodynamically linked processes (36), they both affect the free energy of the "energy-rich" state. Consequently, competing interpretations based on the "energy-rich" intermediate and on the conformationally (37) or morphologically (38) energized state are not really mutually exclusive, but may reflect different features of the same molecular process.

K. Concentration Gradients

The free energy of oxidation may also be stored in the form of concentration gradients, because the electron-transport chain may take up protons by reduction and release protons by oxidation at different locations in space. Application of Gibbs–Donnan equilibrium to membranes of selective permeability allows the calculation of the concentration gradients of other ions from this proton gradient (39). Moreover, inasmuch as the formation of each mole of **IV**, **V**, or **VI** at neutral pH involves the uptake of a mole of proton, and the regeneration of each mole of **VIII** from **VII** involves the release of $\frac{1}{2}$ mole of proton, these reactions may cause changes in the net membrane charge as well as in local cation concentrations. Ionic concentration gradients are also produced by active transport driven by the hydrolysis of ATP.

In general, differences in local concentration lower the entropy and hence also raise the free energy of the system. Since this generation of concentration gradients and ATP hydrolysis are also linked processes, their reciprocal thermodynamic relationship allows the formation of ATP by externally imposed sudden concentration changes (40).

A proposal has indeed been made (41) to consider such concentration gradients as the primary form of converted free energy with which ATP is formed from ADP and P_i. Hypothetical models based on this premise are in general less attractive because they are electrochemically all equivalent to the inefficient concentration cells. For example, at pH 7 the net reaction

$$H^+ + NADH + \tfrac{1}{2}O_2 \rightarrow NAD^+ + H_2O$$

has a midpoint potential of 1.14 V. To build an equivalent concentration cell we would need a concentration (or activity) ratio of $e^{1.14/(RT)} \approx 10^{19}$. In other words, if the reacting solute concentration is 1 M on one side of the membrane, the concentration of this solute on the other side would have to be $10^{-19}\,M$ or less. With such a low concentration of the reacting solute responsible for one side of the membrane, it would be impossible for the system to attain the steady-state metabolic rates of terrestrial life. Attempts to remedy this chemiosmotic hypothesis by introducing loops in the electron-transport path (42) are also unsatisfactory, because at steady state the net production or uptake of protons as a primary reaction can occur only at the source or sink of electrons, but not at any other point or junction along a fixed electron-transport chain. The net uptake of protons can, of course, occur as the result of a chemical reaction, e.g., the formation of a non-phosphorylated "energy-rich" intermediate. But such an occurrence would be inconsistent to the basic postulate of producing proton gradient as a primary energy storage process.

On the other hand, in view of the wide range of midpoint potentials of oxidation–reduction couples in the chemical world, it would seem much more promising to use net chemical reactions as a primary concept and build upon it a theory of reaction-induced conformational and concentrational changes to account for cellular regulatory processes (43).

ACKNOWLEDGMENT

The experimental work discussed in Section II of this article was supported in part by research grants from the National Institute of Health (GM 04483) and the National Science Foundation (GB 3631).

REFERENCES

1. W. M. Clark, D. W. Hutchinson, and A. Todd, *J. Chem. Soc.*, 722 (1961).
2. M. Vilkas and E. Lederer, *Experimentia*, **18**, 546 (1962).
3. D. L. Gutnick and A. F. Brodie, *J. Biol. Chem.*, **240**, PC3698 (1965).
4. C. D. Snyder, S. J. DiMari, and H. Rapoport, *J. Am. Chem. Soc.*, **88**, 3868 (1966).
5. D. L. Gutnick and A. F. Brodie, *J. Biol. Chem.*, **241**, 255 (1966).
6. W. W. Parson and H. Rudney, *Biochemistry*, **5**, 1013 (1966).
7. A. F. Brodie, in *Biochemistry of Quinones* (R. A. Morton, ed.), Academic Press, New York, 1965, p. 355.
8. C. D. Snyder and H. Rapoport, *J. Am. Chem. Soc.*, **89**, 1269 (1967).
9. W. S. Brinigar and J. H. Wang, *Proc. Natl. Acad. Sci. U.S.*, **52**, 699 (1964).
10. W. S. Brinigar, D. B. Knaff, and J. H. Wang, *Biochemistry*, **6**, 36 (1967).
11. W. S. Brinigar and J. H. Wang, *Proc. Intern. Congr. Biochem., 6th, New York*, **32**, 263 (1964).
12. T. A. Cooper, W. S. Brinigar, and J. H. Wang, *J. Biol. Chem.*, **243**, 5854 (1968).
13. A. B. Falcone, *Proc. Natl. Acad. Sci. U.S.*, **56**, 1043 (1966).
14. T. Wieland and E. Bäuerlein, *Chem. Ber.*, **100**, 3869 (1967).
15. T. Higuchi and K.-H. Gensch, *J. Am. Chem. Soc.*, **88**, 3874; 5486 (1966).
16. D. O. Lambeth and H. A. Lardy, *Biochemistry*, **8**, 3395 (1969).
17. S. I. Tu and J. H. Wang, *Biochemistry*, **9**, 4505 (1970).
18. J. H. Wang, *Proc. Natl. Acad. Sci. U.S.*, **58**, 37 (1967).

19. J. H. Wang, *Science*, **167**, 25 (1970).

20. C. P. Lee and L. Ernster, *European J. Biochem.*, **3**, 385 (1968).

21. J. H. Wang, in *Oxygenases* (O. Hayaishi, ed.), Academic Press, New York, 1962, p. 505.

22. B. Chance and G. R. Williams, *Advan. Enzymol.*, **17**, 65 (1956).

23. J. R. Cox and O. B. Ramsey, *Chem. Rev.*, **64**, 317 (1964).

24. D. E. Hullquist, R. W. Moyer, and P. D. Boyer, *Biochemistry*, **5**, 322 (1966).

25. D. E. Green, *Advan. Enzymol.*, **21**, 73 (1959).

26. E. Racker, *Proc. Natl. Acad. Sci. U.S.*, **48**, 1659 (1962).

27. E. C. Slater, *Proc. Intern. Congr. Biochem.*, *5th, Moscow*, **V**, 325 (1961).

28. P. D. Boyer, C. H. Suelter, M. DeLuca, J. B. Peter, and M. E. Dempsey, *Proc. Intern. Congr. Biochem.*, *5th, Moscow*, **V**, 274 (1961).

29. P. D. Boyer, *Science*, **141**, 1147 (1963).

30. T. I. Bieber, *Biochem. Biophys. Res. Comm.*, **16**, 501 (1964).

31. L. L. Bieber and P. D. Boyer, *J. Biol. Chem.*, **241**, 5375 (1966).

32. E. C. Slater, A. Kemp, and J. M. Tager, *Nature*, **201**, 781 (1964).

33. R. L. Cross, B. A. Cross, and J. H. Wang, *Biochem. Biophys. Res. Comm.*, **40**, 1155 (1970).

34. M. F. Perutz, H. Muirhead, J. M. Cox, and L. C. G. Goaman, *Nature*, **219**, 131 (1968).

35. L. Pauling, H. A. Itano, S. J. Singer, and I. C. Wells, *Science*, **110**, 543 (1949).

36. J. Wyman, *Advan. Protein Chem.*, **4**, 436 (1948).

37. B. Chance, C. P. Lee, and L. Mela, *Federation Proc.*, **26**, 1341 (1967).

38. R. A. Harris, J. T. Penniston, J. Asai, and D. E. Green, *Proc. Natl. Acad. Sci. U.S.*, **59**, 830 (1968).

39. F. G. Donnan, *Z. Elektrochem.*, **17**, 572 (1911).

40. A. T. Jagendorf, *Federation Proc.*, **26**, 1361 (1967).

41. P. Mitchell, *Nature*, **191**, 144 (1961).

42. P. Mitchell, in *Regulation of Metabolic Processes in Mitochondria* (J. M. Tager, S. Papa, E. Quagliariello, and E. C. Slater, eds.), Elsevier, Amsterdam, 1966, p. 65.

43. H. A. Lardy, S. N. Graven, S. Estrada-O., *Federation Proc.*, **26**, 1355 (1967).

CHAPTER 3

DPNH DEHYDROGENASES OF MITOCHONDRIAL ORIGIN

F. M. Huennekens

DEPARTMENT OF BIOCHEMISTRY
SCRIPPS CLINIC AND RESEARCH FOUNDATION
LA JOLLA, CALIFORNIA

B. Mackler

DEPARTMENT OF PEDIATRICS
UNIVERSITY OF WASHINGTON MEDICAL SCHOOL
SEATTLE, WASHINGTON

I. Introduction

Oxidation of DPNH by various electron acceptors, such as dyes, ferricyanide, quinones, etc., can be accomplished by a number of isolated flavoproteins.* In the cell, however, DPNH oxidation is carried out mainly

---

* This is usually referred to as "diaphorase" activity after the classical Straub diaphorase (DPNH → dye → O_2) (*1*), which was later shown to be lipoyl dehydrogenase (*2*).

by the mitochondrial electron transport system. The latter is an array of oxido-reduction components (Fig. 1), which serve to link DPNH to molecular oxygen. This formulation rests upon data obtained from two different approaches: (a) spectrophotometric or EPR* techniques to monitor the oxidation and reduction of the components in the presence of various inhibitors (3); and (b) resolution of the overall system into subordinate structural and functional units (4).

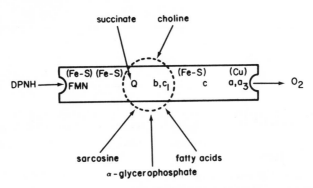

Fig. 1.   Mitochondrial electron transport systems. A component in parentheses is one whose precise position in the sequence is not yet established. Fe–S represents a nonheme iron-labile sulfide complex.

The high degree of efficiency with which electrons flow through this system is due to the fact that the components are enclosed in a matrix of lipid and structural protein (5). While this arrangement is undoubtedly advantageous for the interaction of each redox component with its neighbor, rather than with extraneous oxidants, it creates obvious problems in studying the detailed mechanisms of the electron-transfer processes. The following questions illustrate some of the uncertainties that still prevail in the mito-chondrial electron-transport system: (a) Is the sequence linear, as shown in Fig. 1, or are some of the components on side pathways? (b) Have all of the redox components in the chain been discovered, especially those that may lack prominent spectra or EPR signals? (c) Do all components pass through an individual oxido-reduction cycle or are some paired as complexes? and (d) Are discrete enzymes responsible for each transfer of reducing power between components?

* The following abbreviations are used: EPR, electron paramagnetic resonance; ETP, electron transport particle; Q, Q_1, coenzyme Q, coenzyme Q_1, etc.; Men, menadione; DCIP, 2,6-dichlorophenolindophenol; c or cyt. c, cytochrome c, etc.; PCMS, p-chloro-mercurisulfonate; PCMB, p-chloromercuribenzoate; DTNB, 5,5'-dithiobis-(2-nitro-benzoic acid).

Mitochondria can be comminuted into a smaller "electron-transport particle" (ETP), which catalyzes the oxidation of both DPNH and succinate by oxygen (6). Reducing power from the latter substrate is transferred, via a separate sequence of carriers, into the DPNH → O_2 chain at the level of coenzyme Q. Choline, sarcosine, α-glycerophosphate, fatty acids, and TPNH are also oxidized by mitochondrial electron-transfer systems but in these instances the exact site of attachment to the main chain (Fig. 1) has not yet been elucidated. Nevertheless, it seems reasonable to assume that the segment from coenzyme Q to oxygen is the actual "core" of the mitochondrial electron-transport system and that various substrates, including DPNH, are linked to this core by individual "dehydrogenases" containing a flavin, usually accompanied by additional electron carriers. The flow of reducing power (hydride ions and/or electrons) is usually thought of as originating from the reduced substrate and terminating in oxygen. The integration of these dehydrogenase chains with the core, however, makes it possible for the reduced form of one substrate to be linked with the oxidized form of another, provided that the equilibrium for the overall reaction is favorable. Thus, in ETP preparations, DPNH reduces fumarate and, in the presence of ATP to overcome the thermodynamic (and perhaps mechanistic) barrier, succinate will reduce DPN (7).

II. DPNH Dehydrogenase Preparations

Although DPNH dehydrogenases of mitochondrial origin have been studied intensively by a number of investigators, there are still many areas of experimental and theoretical disagreement (8). In part, this stems from the various types of "DPNH dehydrogenase" that can be derived from the linear sequence of carriers between DPNH and O_2. For example, if three oxido-reduction carriers (A, B, and C) were interposed between DPNH and coenzyme Q, the system conceivably could be fragmented in different ways, leading to one protein containing only A, another containing A and B, and a third containing A, B, and C. Each of these derived enzymes would be equally valid as a DPNH dehydrogenase, although, to be rigorous, a given dehydrogenase (e.g., the AB-containing fragment) should retain the capacity to utilize the next carrier (in this case, C) as an external electron acceptor. In practice, though, artificial electron acceptors are commonly employed for this purpose, and this does not invalidate mechanistic studies provided that the acceptor interacts only with the terminal redox component bound to the enzyme.

For reasons cited above, we shall consider the DPNH dehydrogenase region to extend only from DPNH to coenzyme Q. Thus, the "maximum" DPNH dehydrogenase would consist of all carriers between DPNH and coenzyme Q. This arbitrary definition has its counterpart in fact, since, as discovered by Hatefi and his colleagues (9), the mitochondrial electron-transport system can be resolved into four subordinate, particulate complexes, each of which represents one segment of the overall chain (Fig. 2). These

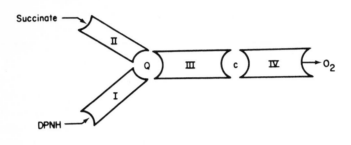

Complex I   DPNH-Coenzyme Q reductase
       II   Succinate-Coenzyme Q reductase
      III   Reduced Coenzyme Q-cytochrome c reductase
       IV   Reduced cytochrome c oxidase

Fig. 2.   Resolution of the mitochondrial electron transport system into complexes I, II, III, and IV [from Hatefi (10)].

complexes can be recombined physically to regenerate the entire operational sequence. Complex I, a DPNH–coenzyme Q reductase,* contains FMN, nonheme iron, labile sulfide, lipid, and structural protein in the proportions shown in Table 1; the catalytic activities of this preparation are also given in the table. Although there is some uncertainty in assigning a molecular weight to a particulate complex, the value of 570,000 (calculated from the flavin content of 1.5 μmole/mg protein) can be taken as an approximation. Since this enzyme has been treated in detail elsewhere (10), we shall not consider it further in this review. For comparison, Table I also includes data on two other high-molecular-weight DPNH dehydrogenases: (1) The preparation of Singer and his colleagues [see Ref. (11) for a comprehensive review]; and (2) the particulate "PD dehydrogenase" described by King's laboratory (12).

Although the particulate DPNH–Q reductase (10) is very likely the most "physiological" (i.e., in the sense of being the least damaged) of the DPNH

* DPNH dehydrogenase preparations are often designated by the name of the investigator, and it will be necessary, unfortunately, to continue this practice in the present review. Where possible, however, enzymatic activities will be referred to in terms of the acceptor used (e.g., DPNH–Q reductase or DPNH–FeCN reductase).

TABLE 1
HIGH-MOLECULAR-WEIGHT DPNH DEHYDROGENASES

Investigator and preparation	Molecular weight	Composition [mμmoles (or mμatoms)/mg protein]				Activity (μmoles DPNH ox./min/mg protein)			Temperature	Ref.
		FMN	Nonheme iron	Labile sulfide	Lipid (%)	FeCN	Quinones	DCIP or cyt. c		
Hatefi DPNH-Q reductase (Complex I)	570,000	1.4–1.5	26	25 [a]	22% [b] 4.2–4.5 Q	100–105	$\left(\begin{array}{l}13.5\text{-}Q_1\\1.5\text{-Men.}\end{array}\right)$ [c]	4-cyt. c [d]	38°	(10)
Singer DPNH dehydrogenase	550,000 [e]	1.12 0.15–FAD	19.6	31.1	0	200	~0	~0	30°	(11)
King PD dehydrogenase	—	0.08–FMN	5.3	Present	45%	6–8	0	~0	23°	(12)

[a] Unpublished data of Y. Hatefi; a value of 1.0 for the ratio of nonheme iron:labile sulfide has been reported for other preparations of complex I (13).

[b] Complex I also contains 4.2–4.5 mμmoles of coenzyme Q.

[c] Q_1 (but not menadione) reductase activity is completely sensitive to Amytal.

[d] This activity is probably due to a slight contamination by complex III (9), since it is inhibited by Antimycin A.

[e] Estimated value of Cremona and Kearney (14).

dehydrogenases prepared to date, it is not well suited for mechanistic studies involving electron transfer between individual components. To overcome this difficulty, investigators have utilized a number of precursors, reviewed in detail elsewhere (12), for deriving *soluble* DPNH dehydrogenases from mitochondria or related particulate preparations. Most of the preparative variations have centered about the nature of the starting material and the solubilization procedure.

The former point may be illustrated by Fig. 3, which shows the operational relationship between particulate precursors and various soluble dehydrogenases. In general, it is desirable to effect the maximum reduction in particle size and complexity before applying the solubilization procedure. This reduces the possibility of contamination of the extracted dehydrogenase by extraneous proteins and it also allows the extraction time to be shortened, thereby minimizing damage to the dehydrogenase. On the other hand, the mechanical and chemical manipulations involved in the comminution of particles may possibly inactivate or alter the emergent dehydrogenase.

Empirical methods that have been used successfully for solubilization of DPNH dehydrogenases fall into three categories: (a) treatment of the particles at an elevated temperature (40–45°) with ca. 10% ethanol at pH values ranging from 4.5 to 5.4; (b) treatment of the particles at 30 to 37° with snake venom (*Naja naja*); and (c) treatment of the particles at 37° with thiourea or urea. These diverse agents are able to extract essentially the same DPNH dehydrogenase from particulate precursors, and it would be interesting to see, via the electron microscope, whether they have a common mode of attack upon the structure of the particles. Urea, thiourea, and acid–ethanol evidently exert their effects without breaking any covalent bonds, and even the lipase component of snake venom probably acts on the lipid sheath of the particles rather than on the dehydrogenase itself. Thus, it appears as if the DPNH dehydrogenase preexists as a discrete entity, either trapped mechanically within the particle or held by noncovalent forces (hydrogen, ionic or hydrophobic bonds).* If "solubilization" consists mainly of disrupting the surrounding matrix or loosening the attachment to other enzymes, it is understandable why there is considerable reproducibility in each preparation and why only a small additional purification is required to bring these

* It is of interest that succinic dehydrogenase also appears to be bound noncovalently to the electron-transport system, although it is not released by any of the above three methods. It is readily extracted, however, by dilute alkaline solution to the reconstitutively inactive (15) and active (15a) forms, a procedure that fails to liberate the DPNH dehydrogenase. Thus, noncovalent bonds not only hold together small-molecular-weight protein subunits but also appear to be instrumental in linking individual protein components within complexes I, II, III, and IV (Fig. 2) and in keeping the complexes together as the electron-transport system.

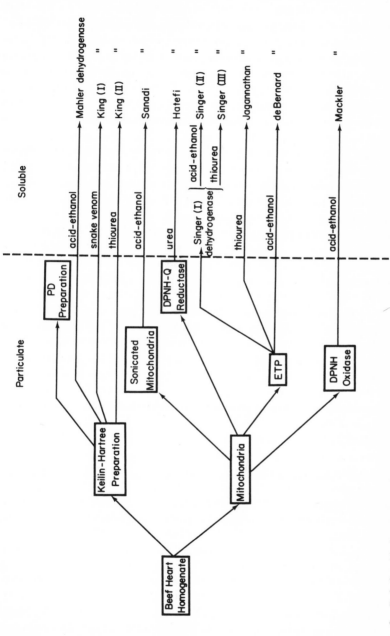

Fig. 3. Relationship between particulate precursors (enclosed in boxes) and the soluble, derived DPNH dehydrogenases (identified by the name of the principal investigator; roman numerals are used to distinguish between several preparations from the same laboratory).

enzymes to a high degree of homogeneity. The soluble DPNH dehydrogen-ases should be considered, therefore, as having been "released" from a particulate precursor, rather than having been produced by "degradation" [in a prejudicial sense (16)] of the precursor.

The acid–ethanol treatment, adapted from an earlier procedure (1) for the isolation of "Straub diaphorase," was first utilized by Mahler and his colleagues (17) in 1952 to prepare a soluble DPNH dehydrogenase from pig heart. After further purification by ammonium sulfate fractionation, this flavoprotein was shown to be reasonably homogeneous when examined by electrophoresis and ultracentrifugation. The prosthetic group did not appear to be either FAD or FMN, although later work by Rao et al. (18) on the anomalous behavior of flavin nucleotides during paper chromatography suggests that the earlier preparation probably contained FMN. The Mahler enzyme catalyzed the dehydrogenation of DPNH at the expense of indo-phenol or cytochrome c (17, 19), and, although Mahler (20) did not regard the latter activity as physiological (since it was insensitive to antimycin a and amytal), it was rather commonly assumed that the entire DPNH → c segment of the respiratory chain had been isolated as a unit. The view was strengthened by the fact that a comparable TPNH–cytochrome c reductase, containing FMN as its prosthetic group, had been obtained earlier in highly purified form from yeast (21). The subsequent discovery (22) that the Mahler enzyme contained four atoms of nonheme iron per mole of flavin formed the basis for an interesting hypothesis that electron transport from DPNH to one-electron acceptors (e.g., cytochrome c) involved both the flavin and iron, whereas two-electron acceptors (indophenol) interacted directly with the flavin. Although later evidence indicated that the electron-transfer network is more complex, this formulation was useful in focusing attention upon the need for "crossover" mechanisms to link one- and two-electron carriers.

DeBernard (23) extended this work by preparing a similar enzyme (except for an iron:flavin ratio of 2) from beef heart ETP. In addition to its ability to reduce cytochrome c, the DeBernard enzyme was able to utilize ferri-cyanide as an acceptor. Later, Mackler utilized DPNH oxidase as the starting material and obtained still another acid–ethanol preparation (24, 25) whose antigenic properties are similar to those of the parent DPNH oxidase (25). This enzyme was originally characterized (24) with respect to electrophoretic homogeneity, absorption spectra of the oxidized and reduced forms, com-ponents (flavin, nonheme iron, and thiol groups), and catalytic activities (viz. 40, 3, 3, and 0.2 μmoles DPNH oxidized/min/mg protein when indophenol, cytochrome c, FAD, and O_2 were used as acceptors). More recent preparations of the enzyme (25) have about twice the above specific activities with various acceptors. Kumar et al. (26) have further purified the

Mackler-type enzyme by fractionation with ammonium sulfate and filtration through Sephadex G-25 and have shown that the protein migrates as a single band when subjected to electrophoresis on starch gel or polyacrylamide. Based upon a molecular weight of 70,000 [determined by filtration through Sephadex G-200 according to the Andrews's modification (27) of Whitaker's method (28)], each molecule of protein contains almost exactly 1 FMN, 2 nonheme iron atoms, 1 labile sulfide group, and 7 thiol groups.

Recently, a soluble DPNH dehydrogenase has been prepared from yeast (*Saccharomyces cerevisiae*) ETP by a modification of the acid–ethanol method (29). This enzyme has a molecular weight of about 55,000 and catalyzes the oxidation of DPNH with the usual artificial acceptors. It differs, however, from the heart muscle dehydrogenases in several respects: (a) FAD, rather than FMN, is the prosthetic group. (b) Upon removal of the FAD, all catalytic activities, including that with ferricyanide, are lost— incubation of the apoenzyme with FAD completely restores these activities. (c) The nonheme iron can be removed without loss of catalytic activity. (d) No labile sulfide is present. (e) Preincubation of the enzyme with DPNH protects against inhibition by mercurials. It is also of interest that the yeast dehydrogenase is not inhibited by the specific antiserum prepared against heart muscle dehydrogenase (25).

Sanadi and his colleagues (30–33) applied the acid–ethanol technique to sonicated mitochondrial particles and obtained a soluble dehydrogenase that showed good activity with coenzyme Q homologues and menadione. The total units of activity and ratio of the Q_6: menadione activities varied when different particulate precursors were used. When snake venom, rather than acid–ethanol, was the solubilizing agent, there was a selective loss of DPNH-Q reductase activity relative to that with menadione.

This preparation could be resolved further into two components (each containing 1 FMN and 4 nonheme irons) by electrophoresis on cellulose acetate or by ultracentrifugation. Rotenone, an inhibitor that blocks the DPNH-Q segment in particulate preparations (34–36), has an interesting multiphasic effect upon the soluble DPNH-Q_6 reductase, viz. partial inhibition at low levels, activation at intermediate levels, and inhibition again at high levels (31). This has been interpreted in terms of multiple binding sites for rotenone. Singer's group has utilized ^{14}C-labeled rotenone to inhibit DPNH-Q activity in particles and has shown that the rotenone remains in the insoluble fraction after the soluble DPNH-Q_6 reductase has been extracted (36). The rotenone (and amytal) sensitivity of the soluble enzyme, like the DPNH-Q_6 activity itself, is believed to be different from that preexisting in particulate precursors (35–37). Possible reasons for this difference are discussed by Sanadi elsewhere in this volume (38). Phospholipase A destroys the DPNH-Q activity and the rotenone sensitivity of particulate preparations, but both effects can be partially restored by the addition of phospholipids (*35, 39*).

Solvents other than ethanol have also been used to release DPNH dehydrogenase from particles. Thus, Cunningham *et al.* (*40*), while searching for the solvent-sensitive site in the DPNH $\rightarrow$ O_2 chain, discovered that diethyl ether causes the progressive release from heart mitochondria of a soluble DPNH–menadione reductase activity. The specific activity and total yield of this enzyme are considerably lower than that obtained by the acid–ethanol procedure.

Another commonly used procedure for solubilizing DPNH dehydrogenase consists of treating the particulate preparation with snake venom. Although venom is a rich source of various hydrolytic enzymes, its action in releasing DPNH dehydrogenase is referable to the endogenous lipases. In the early 1960s, Singer's laboratory reported (*41*) that treatment of beef heart ETP at 30° with snake venom (or with a phospholipase A obtained from the venom) produced a soluble* DPNH dehydrogenase whose properties were believed to correspond closely to those of some portion of the mitochondrial DPNH chain (Fig. 1). That is, it was implied that a dehydrogenase, representing the carrier between DPNH and some unspecified component in the chain, had been lifted from the matrix in essentially an undamaged form (*42*). However, the relatively high molecular weight of the Singer preparation, its iron:flavin ratio of about 16, and its high specific activity with ferricyanide suggest that it is very closely related to the DPNH–Q reductase except for the loss of lipid (cf. Table 1).

In contrast to these results, King and Howard in 1960 reported (*43*) that treatment of a Keilin–Hartree preparation with snake venom at 37° gave rise to a soluble DPNH dehydrogenase whose properties (*12, 44*) were generally similar to those of the acid–ethanol preparations. The nonheme iron:flavin ratio, for example, was about 4. The action of snake venom thus appears to be progressive: (a) release from the particle of an entity (Singer preparation) which has the general composition of the DPNH–Q segment, but is more soluble than Complex I owing to a loss of lipid and perhaps also structural protein; and (b) further disruption or resolution of this quasi-soluble preparation to yield a smaller-molecular-weight flavoprotein (King preparation).

---

* When first released from the particles, this enzyme behaves like a soluble protein in the following respects: (a) It does not sediment when centrifuged at 144,000 $\times$ G for 1 hr. (b) It does not precipitate following extensive dialysis or repeated freezing and thawing. (c) It can be fractionated by conventional techniques (ammonium sulfate precipitation, chromatography on hydroxylapatite, or filtration through Sephadex) used for soluble proteins. However, the purified enzyme tends to precipitate out of solution unless a detergent, such as Triton, is present (*42*).

The claim that the Singer preparation is "the respiratory chain dehydrogenase" rests, to a large measure, upon the assumption that essentially all of
the DPNH–ferricyanide activity demonstrable in the particulate precursor
has been recovered in the soluble preparation (45). However, if either the
precursor or the soluble dehydrogenase has more than one site at which
ferricyanide can interact, this agreement may be entirely fortuitous. But
even if there is only *one* site (i.e., a specific oxido-reduction component
between DPNH and coenzyme Q) for ferricyanide, it follows that, in the
particle, reducing power from DPNH is being diverted from that component
to a nonphysiological acceptor outside of the chain in preference to con-
tinuance through the chain. In the isolated enzyme this competition may or
may not exist, depending upon whether the ferricyanide-linked component is
terminal. However, if the soluble enzyme happened to contain any oxido–
reduction components on the oxygen side of the ferricyanide site, these
would be nonfunctional in the ferricyanide assay and would have to be
considered, therefore, as impurities.

Singer's DPNH dehydrogenase has been further purified by conventional
means of protein fractionation (14) and its properties [kinetics (46), labile
sulfide content (47), EPR signal (48), and transformation into the smaller-
molecular-weight dehydrogenases (49)] have been documented extensively.
This preparation, however, has several shortcomings: (a) It is not easily
reproduced in the hands of other investigators (12) (although this is also true
for other DPNH dehydrogenases), and it tends to aggregate (14) in the
absence of detergents (42); (b) although the enzyme appears to be nearly
homogeneous (42, 50) upon electrophoretic or ultracentrifuge examination
(an S_{20} value of 12.6 was reported in the latter determination), the molecular
weight has been revised from the original value of greater than one million
(42) to its present figure of about 550,000 (14); and (c) earlier preparations
contained a mixture of FAD, FMN, and AMP (50), a result that delayed the
acceptance by Singer's laboratory (14) of the fact that FMN is the only flavin
(18, 51–53) in the DPNH → O_2 chain in mitochondria.

Present evidence indicates that the Singer preparation is quite similar in
size, components, and catalytic activity to the DPNH–Q reductase (Table 1),
except that it has lost lipid and no longer interacts with coenzyme Q via the
amytal- and rotenone-sensitive pathway (55, 56). It may also contain
additional components [e.g., FAD, AMP (50)] and activities [e.g., transhydrogenase (45)] not found in complex I. The implication in the phrase
"*the* respiratory chain DPNH dehydrogenase" and the criticism that other
dehydrogenase preparations are unphysiological (11, 16, 57–62) would
therefore appear to be without foundation.*

* Despite the recent claim (54) that two flavoproteins are involved in this chain, none
other than the DPNH dehydrogenase has yet been isolated.

Perhaps the most effective procedure yet devised for solubilization of DPNH dehydrogenases involves the use of chaotropic agents such as thiourea or urea. Thiourea was first employed by Jagannathan and Chapman (*63*) to extract a low-molecular-weight dehydrogenase from particles, and subsequently, similar enzymes have been obtained by King *et al.* (*12*) and by Cremona *et al.* (*64*) using this method. More recently, Hatefi and Stempel (*65*) have demonstrated that the DPNH–Q reductase can be resolved by

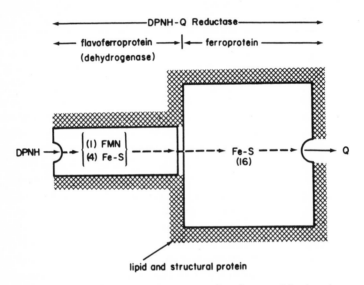

Fig. 4. Components of the DPNH–coenzyme Q reductase. Numbers in parentheses indicate approximate composition in moles (or atoms) per mole of protein.

treatment with urea into: (a) a soluble flavoprotein (actually a "flavoferroprotein") (containing FMN, nonheme iron and labile sulfide in the mole ratio of 1:4:4); (b) a ferroprotein (containing nonheme iron and labile sulfide in a ratio of 1:1); and (c) an insoluble material similar to mitochondrial structural protein (*5*). The flavoprotein has the highest catalytic activity (ca. 200 μmoles DPNH/min/mg protein with ferricyanide as acceptor) of any of the small-molecular-weight DPNH dehydrogenases; activities with other acceptors are also quite high.

The important finding that the soluble ferroprotein can serve as the oxidant for the DPNH-reduced flavoprotein suggests, in agreement with earlier predictions (*10, 31, 66*), that the DPNH $\rightarrow$ Q sequence in mitochondria may be formulated as follows: DPNH $\rightarrow$ flavoferroprotein $\rightarrow$ ferroprotein $\rightarrow$ Q. Figure 4 illustrates a possible structural and functional relationship of these components, from which conclusions may be drawn concerning the various other DPNH dehydrogenases. Thus, the high-molecular-weight (>300,000) particulate or quasisoluble dehydrogenases (Table 1) with a high ratio of nonheme iron:flavin appear to be closely related to the DPNH–Q reductase except for some loss of lipid, structural protein, and a small amount of nonheme iron. Alternatively, all of the low-molecular-weight (<100,000), soluble dehydrogenases having nonheme iron:flavin ratios of 2:1 or 4:1 are similar to the flavoprotein of Hatefi and Stempel (65). This interpretation is consistent with the demonstration by Singer's group that their high-molecular-weight DPNH dehydrogenase preparation can also be transformed by a variety of methods (treatment with proteolytic enzymes, urea, thiourea, or acid–ethanol) into smaller-molecular-weight, soluble dehydrogenases (49).

Table 2 summarizes the composition (in terms of FMN, nonheme iron, and labile sulfide) of the various low-molecular-weight dehydrogenases; all values are expressed as mμmoles (or mμatoms) of the component per mg of protein. In some instances, the data have been rounded off in order to make comparisons more evident. Ideally, these values should be expressed as moles (or atoms) per mole of protein, but in view of uncertainties in the molecular weights, this is generally not feasible. An exception is provided by one of the acid–ethanol preparations (26), whose molecular weight of 70,000 (determined by filtration through Sephadex) coincides very nearly with that calculated from analytical values for the oxido-reduction components. Most preparations, however, are best characterized by the mole ratio of various components. On this basis, all of the enzymes contain either 2 or 4 atoms of nonheme iron per mole of FMN.* On the other hand, the labile sulfide content of the DPNH dehydrogenase is much more variable, ranging from less than (26), equal to (25, 33, 65), or more than (31) that of the nonheme iron. In Hatefi's preparation (65), the ratio is almost exactly 1:1, as it is in various ferredoxins (68). The ease with which labile sulfide is destroyed by exposure to acidic pH values suggests that the acid–ethanol preparations will generally have low values for this component.

The molecular weights of the soluble dehydrogenases must be regarded as only approximate. None of the present preparations has yet been crystallized, and the possibility of impurities makes uncertain molecular weights based on physical methods or calculated from the content of some fixed component. FMN has often been used as the basis for such a calculation (assuming one mole of flavin per mole of enzyme), but the molecular weight is then apt to be high owing to the facile loss of flavin during purification. Variations in molecular weight (from 70,000 to 12,000) may also be caused by

---

* Soluble succinic dehydrogenases have also been isolated with two or four atoms of nonheme iron per mole of peptide-bound flavin (67).

TABLE 2

Low-Molecular-Weight, Soluble DPNH Dehydrogenases

Investigator and solubilization procedure	Molecular weight	Composition [mμmoles (or mμatoms)/mg protein]			Activity (μmoles DPNH ox./min/mg protein)				Temperature	References
		FMN	Nonheme iron	Labile sulfide	FeCN	Indophenol	Cyt. c	Quinones		
Mahler Acid–ethanol	75,000–80,000	13	52	—	—	8	15	Men–9	24°	(17)
de Bernard Acid–ethanol	—	7	14	—	30	—	10	—	38°	(23)
Mackler Acid–ethanol	80,000–90,000	10	20	—	40	40	8	—	30°	(24)
Mackler (II)[a] Acid–ethanol	80,000	11	30	23	95	80	6	—	30°	(25)
Kumar Acid–ethanol	70,000	14	29	13	35	36	3	Q_1–11	38°	(26)
Sanadi Acid–ethanol (fraction II)	100,000	10	42	50–53	25–80	—	15	Q_1–120 Men–150	30°	(32, 33)
King Snake venom	110,000–120,000	7	30	Present	20	10	20	Men–4	23°	(43)
Jagannathan Thiourea	—	5	—	—	38 35–40	3	3	Q_1–40 Men–40	25°	(63)
Hatefi Urea	—	15	63	60	215	100	43	Q_1–150 Men–160	38°	(65)
Mackler (III) Acid–ethanol (from yeast)	50,000–60,000	17 (FAD)	6	0	192 (pH 5.5)	30	18 (pH 5.5)	—	38°	(29)
Singer (II) Acid–ethanol	S_{20} = 5.6	—	34	—	—	—	—	—	30°	(49)
King (II) Thiourea	—	5	49	Present	92	28	11	Q_2–6 Men–13	23°	(12)
Singer (III) Thiourea	—	—	—	—	—	—	—	—	30°	(64)

[a] Roman numerals are used to distinguish preparations from the same laboratory.

the presence of different amounts of extraneous protein from other segments of the electron-transport chain that remains firmly attached to the dehydrogenase during the solubilization procedure.

Also shown in Table 2 is the catalytic activity of each enzyme (expressed as μmoles of DPNH oxidized per minute per mg of protein) with four of the commonly used artificial acceptors: ferricyanide, indophenol, cytochrome c, and coenzyme Q. Coenzyme Q activity has usually been tested with Q_1 or with menadione. The technical difficulties in assaying soluble DPNH dehydrogenases (or even DPNH–Q reductase) with the water-insoluble homologues (Q_2 to Q_{10}) have been encountered by Sanadi (32) and by Hatefi (10, 69).

It is difficult to compare activities of the various dehydrogenase preparations, since the assay conditions have not been identical. It may be assumed, though, that assays have been carried out at optimal pH values (ranging from 7.4 through 8.5 for the four activities) and that initial velocities have been measured in order to minimize the time-dependent decline in activity often seen with DPNH dehydrogenases. The temperature at which assays have been performed ranges from ambient (23–25°) to 37°. In some instances (45), experimentally determined activities have been extrapolated to an infinite concentration of acceptor in order to provide a fundamental (and maximal) catalytic constant for the enzyme.

Except for the Hatefi preparation (65), all of the dehydrogenases show activities ranging between 25 and 100 for ferricyanide, and about the same level for indophenol and various quinones. On the other hand, the reported cytochrome c activity is uniformly lower, i.e., about 10–25% of the above values, probably because this particular activity is not being assayed under optimal conditions.* The specific activity of the urea-extracted enzyme (65) with all acceptors is somewhat higher than that of the other low-molecular-weight, soluble dehydrogenases. A number of the derived dehydrogenases have been shown to be specific for the B form of DPNH, as are the particulate precursors (70).

Fragmentary data exist regarding the effects of various inhibitors on the multiple catalytic activities associated with the soluble dehydrogenases. Mercurials, such as PCMB, at low concentrations (10^{-6} M) inhibit all the catalytic activities of all the soluble enzymes (12) except that from yeast, which requires 10^{-4} M PCMS for inhibition (71). Preincubation of the enzyme with DPNH (in the absence of an acceptor) leads to an increased sensitivity toward mercurials (26). Phosphate was reported to inhibit the cytochrome c activity of the Mahler enzyme (17) and the Q_6 reductase

---

* Hatefi and Stempel (65) have shown that the activity of their preparation with cytochrome c is 43 when tested at the usual concentration of acceptor (0.15 mM) and 220 when the concentration of cytochrome c is raised to 0.6 mM.

activity of Sanadi's preparation (32), but the Mackler preparation is stimulated by this ion (24). Amytal and rotenone, both of which inhibit the
particulate DPNH–Q reductase at low concentrations (10), have a variable
and partial effect (discussed earlier) on the Q_6 and menadione reductase
activities of the derived dehydrogenases. Incubation of the Hatefi enzyme
with o-phenanthroline leads to inhibition of all of the activities (65) but not
the cytochrome c, menadione, or Q_6 activities of the acid–ethanol preparation
(26). Cytochrome c activity of the Hatefi preparation is depressed after
prolonged incubation with thenoyltrifluoroacetone and excess DPNH inhibits
the Q reductase activity of this enzyme (65).

III. Mechanisms of Electron Transfer in DPNH Dehydrogenase

As defined earlier, DPNH dehydrogenase may be considered in the
broadest sense as that segment of the electron-transport chain beginning with
DPNH and terminating with the carrier that interacts with the ferroprotein
(Fig. 4). Although there is considerable variation in the molecular weight
and composition of the dehydrogenases listed in Table 2, it is assumed that
all share a common mechanism for oxidizing DPNH at the expense of various
artificial electron acceptors. It is further assumed that, although these
enzymes may have suffered a certain amount of damage as a result of being
sheared from the lipid–structural protein matrix of mitochondria, there has
been no internal "scrambling" of the oxido–reduction components. The
mechanism of action for this type of enzyme may then be considered in terms
of the following questions: (a) Which of the potential oxido–reduction
groups are actually functional? (b) Are any of these groups paired, thereby
functioning as a unit? (c) Is the network of electron transport linear or
branched when the various acceptors are used? (d) Are there actually specific
binding sites on the protein for the various acceptors? (e) What is the
mechanism for each transfer of reducing power?
    The functional nature of an individual oxido–reduction component in the
dehydrogenase can usually be established by the same methods (see Section
I) that are used for the intact mitochondrial chain: (a) observation of an
oxidation–reduction cycle for the component by spectrophotometry or
EPR; (b) loss of catalytic activity following deletion or inhibition of the
component; and (c) fragmentation of the dehydrogenase into a smaller unit
containing the component.
    In the DPNH dehydrogenases much attention has been directed toward
the bound flavin, largely because of the ease with which its oxidation–
reduction states can be followed spectrophotometrically. Thus, when

DPNH, in the absence of an acceptor, is added to substrate amounts of the dehydrogenase, changes in absorption spectrum (24, 26, 33) indicate that the flavin has been fully reduced. Alternatively, recent work by Gibson et al. (72) has shown that when DPNH oxidation is linked to ferricyanide, the flavin exists largely as the semiquinone during the steady state; this raises the question whether the fully reduced state (which can be observed when the enzyme is treated with DPNH under anaerobic conditions) is catalytically significant. Another approach for determining the functionality of FMN is to delete this cofactor under conditions sufficiently mild to minimize damage to the protein or other redox groups. This can be accomplished by prolonged dialysis of the enzyme or passage through Florisil columns (18, 51). Concomitantly, dehydrogenase activity is lost when indophenol and cytochrome c are used as acceptors, but the ferricyanide activity is essentially unaffected (18). A holoenzyme, prepared by incubation of FMN with the apoenzyme, exhibits the original cytochrome c activity, but the indophenol activity is not restored under these conditions.

Depending upon the method of isolation, DPNH dehydrogenases contain two or four atoms of nonheme iron, probably in the Fe^{3+} state, per mole of protein. This component is largely responsible for the amber color (absorbency in the 400-mμ region) of the preparations. Nonheme iron is rather firmly attached to the protein and is not removed, even after reduction, by treatment with various chelating agents (24, 26, 31, 65) or by passage through Chelex resin. The functional nature of the iron, however, has not yet been demonstrated. Reduction of the enzyme nonenzymatically (e.g., with dithionite) is accompanied by a decreased absorbance in the 400-mμ region and by the appearance of an EPR signal at $g = 1.94$; these effects cannot be achieved, however, by the addition of DPNH under anaerobic conditions. The $g = 1.94$ signal seen in the DPNH–Q reductase (10) and in Singer's dehydrogenase (48) is due most likely to the ferroprotein portion (cf. Fig. 4) of these preparations, rather than to the iron in the flavoferroprotein.

Although there is frequently an equivalence between the amount of labile sulfide and the nonheme iron in oxido–reduction enzymes, the various dehydrogenases described in Table 2 do not necessarily conform to this generalization. Unlike most of the ferredoxins, DPNH dehydrogenases can apparently lose labile sulfide without an equivalent loss of iron.* In part because of the uncertainty about the structure of labile sulfide, it is not clear whether this component undergoes an oxidation–reduction cycle during operation of the DPNH dehydrogenase. There is, however, one peripheral piece of evidence that may bear on this point: The original Mackler enzyme

---

* Similar findings have been reported for alfalfa ferredoxin (73) and for both the particulate and supernatant fractions of Mycobacterium phlei (74).

(24, 26) contains only one labile sulfide per mole and has a specific activity of 40 with ferricyanide as acceptor. A more recent preparation from Mackler's laboratory (25) contains two labile sulfides per mole and has an activity of approximately 90. The corresponding values for the Hatefi enzyme (65) are four labile sulfides and an activity of 200. These data suggest a direct proportionality between the content of labile sulfide and the catalytic activity. The intact operational unit for the dehydrogenase, therefore, may be a cluster of four nonheme irons and four labile sulfides. The increased sensitivity toward mercurials when the enzyme is preincubated with DPNH (see below) would also be consistent with the idea that labile sulfide is a functional entity whose reduced form is the hydrosulfide anion* HS^-.

As determined by either the DTNB or the PCMB methods, the acid–ethanol dehydrogenase contains approximately seven thiol groups† per mole of protein (26). This value is in close agreement with the total number of cysteine residues determined by amino acid analysis. Although addition of mercurials at ca. $10^{-5}\ M$ leads to displacement of FMN (26, 71) and a concomitant loss of catalytic activity,‡ it is likely that thiol groups play a structural, rather than a functional, role in the enzyme. One or more thiol groups may be directly involved in binding FMN to the protein, or mercuration of these groups may cause conformational changes that weaken ionic and hydrogen bonds holding FMN to the protein. If thiol groups were directly involved as oxido–reduction components in catalysis, the mechanism would require reduction of a disulfide bridge to a vicinal dithiol or reduction of a thioketone to a thioalcohol. The failure of arsenite or Cd^{2+} to appreciably inhibit the dehydrogenase would appear to rule out the former possibility, while the latter, although intriguing, has few counterparts in enzymology (77).

We shall assume, then, that the dehydrogenase has two functional groups: FMN and a polynuclear complex§ composed of iron and labile sulfide. The latter would be expected to have an oxidation potential different from that of the ferredoxins or the iron–labile sulfide complexes that appear

---

* The pK$_a$ values for H_2S are approximately 10^{-7} ($H_2S \rightleftarrows H^+ + HS^-$) and 10^{-15} ($HS^- \rightleftarrows H^+ + S^=$) (75).

† Only two or three of these groups react immediately with the above reagents; the remainder are detected only in the presence of $4\ M$ urea. In the DPNH–Q reductase (10, 65) and in Singer's dehydrogenase (47, 76), the thiol groups appear to be inaccessible, since the ferricyanide activity of these enzymes is not inhibited by mercurials; the Q reductase activity, however, is inhibited.

‡ It is interesting that the various electron acceptors, all of which are capable of oxidizing thiol groups, do not appear to inhibit the enzyme or cause the FMN to be released.

§ The work of Saltman and his colleagues [see, for example, Ref. (78)] has emphasized that the ability of iron to form polynuclear complexes with anions may be a common feature of ferritin and nonheme iron proteins of the type discussed here.

elsewhere in the overall chain [i.e., in the ferroprotein and in the region of cytochrome $b-c_1-c$ (79)]. Each of these iron–sulfur complexes may have a common structure, possibly a square planar complex in which a cysteine residue in a peptide chain and a hydrosulfide anion are among the groups coordinated with the iron. The remaining two ligands (above and below the plane) would then confer individuality (e.g., oxidation potential) upon each type of complex. Inhibition of all the catalytic activities by low levels of mercurials ($<10^{-5}$ M) and the increased sensitivity to the inhibitor when the enzyme is preincubated with DPNH (26) in the absence of an acceptor [see

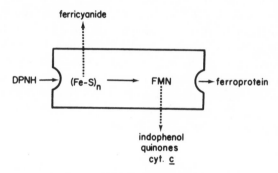

Fig. 5. Proposed mechanism for DPNH dehydrogenase. $(Fe–S)_n$ indicates an iron-labile sulfide polymer. Dashed arrows indicate reactions with artificial acceptors.

also Ref. (80) for a similar effect on DPNH oxidation in heart-muscle particles] would then be referable to an attack on the fully reduced form of the sulfur (HS^-)* in the iron–labile sulfide complex, rather than upon any cysteine residue, in the protein.

In line with the above assumptions, a possible mechanism for electron transfer within DPNH dehydrogenase is that shown in Fig. 5. The two functional oxido-reduction groups, FMN and the nonheme iron–labile sulfide (Fe–S) complex, probably occur in their fully oxidized states as the enzyme is isolated. During the catalytic cycle, the hydride ion from DPNH is assumed to interact first with the iron–sulfur complex. Reducing power is then transferred to the ferroprotein via the bound FMN. In the steady state, both redox groups may shuttle between the fully oxidized and semiquinone states†; the fully reduced states of these groups would appear only when DPNH is added in the absence of an acceptor. In the isolated dehydrogenase,

---

* This assumes that the sulfur, rather than the iron, is undergoing oxido-reduction. Recent evidence by DerVartanian et al. (81) demonstrates that sulfur may well contribute to EPR signals in the $g = 1.94$ region.

† Nonenzymatic model studies have shown that one-electron oxidants oxidize DPNH much faster than do two-electron oxidants (82).

and perhaps in the intact DPNH $\rightarrow$ O$_2$ chain as well, ferricyanide is assumed to react with (Fe–S), while the other acceptors interact with FMN via the site formerly occupied by the ferroprotein.

The above scheme is intended only as a guide for future experimentation. It is evident that any proposed mechanism must fit the data on the relative rates of DPNH oxidation by various acceptors, although, as noted previously, existing data on rates may not have been obtained under optimal conditions. Experiments in which DPNH oxidation is linked simultaneously to more than one acceptor should establish whether the electron transfer pathway in the dehydrogenase is linear or branched. Attention should also be given to exploring these pathways by means of reversed electron flow (reduced acceptor $\rightarrow$ DPN) or by means of coupled electron flow involving the oxidized and reduced forms of two acceptors in the absence of DPNH or DPN. A better understanding of the detailed mechanism of the DPNH dehydrogenase would also be facilitated by information obtained by the deletion of functional groups without damaging the remainder of the protein, or by further cleavage of the enzyme into a "minimal DPNH dehydrogenase" (i.e., protein plus the first functional group).

ACKNOWLEDGMENTS

The authors are indebted to Drs. Y. Hatefi and D. R. Sanadi for their helpful comments on this problem. Support of the experimental work from the authors' laboratories by the following grants is gratefully acknowledged: American Heart Association (66-796), National Cancer Institute, National Institutes of Health (CA 6522), and National Heart Institute, National Institutes of Health (H 5457).

REFERENCES

1. H. S. Corran, D. E. Green, and F. B. Straub, *Biochem. J.*, **33**, 793 (1939).
2. V. Massey, *Biochim. Biophys. Acta*, **30**, 205 (1958).
3. B. Chance and G. R. Williams, *Advan. Enzymol.*, **17**, 65 (1956).
4. D. E. Green, *Advan. Enzymol.*, **21**, 73 (1959).
5. R. S. Criddle, R. M. Bock, D. E. Green, and H. D. Tisdale, *Biochemistry*, **1**, 287 (1962).
6. F. L. Crane, J. L. Glenn, and D. E. Green, *Biochim. Biophys. Acta*, **22**, 475 (1956).
7. B. Chance and G. Hollunger, *J. Biol. Chem.*, **236**, 1534 (1961).
8. T. E. King, T. P. Singer, and B. Mackler, Discussion, in *Flavins and Flavoproteins* (E. C. Slater, ed.), Elsevier, Amsterdam, 1966, pp. 482–495.
9. Y. Hatefi, A. G. Haavik, and D. E. Griffiths, *Biochem. Biophys. Res. Commun.*, **4**, 441 (1961).
10. Y. Hatefi, in *Comprehensive Biochemistry*, Vol. 14 (M. Florkin and E. Stotz, eds.), Elsevier, Amsterdam, 1966, p. 199.

11. T. P. Singer, in *Nonheme Iron Proteins: Role in Energy Conversion* (A. San Pietro, ed.), Antioch Press, Yellow Springs, Ohio, 1965, p. 349.

12. T. E. King, R.L. Howard, J. Kettman, B. M. Hedgekar, M. Kuboyama, K. S. Nickel, and E. A. Possehl, in *Flavins and Flavoproteins* (E. C. Slater, ed.), Elsevier, Amsterdam, 1966, p. 441.

13. C. J. Lusty, J. M. Machinist, and T. P. Singer, *J Biol. Chem.*, **240**, 1804 (1965).

14. T. Cremona and E. B. Kearney, *J. Biol. Chem.*, **239**, 2328 (1964).

15. T. P. Singer, E. B. Kearney, and P. Bernath, *J. Biol. Chem.*, **223**, 599 (1956).

15a. T. E. King, *Biochim. Biophys. Acta*, **58**, 375 (1962).

16. T. Cremona, E. B. Kearney, M. Villavicencio, and T. P. Singer, *Biochem. Z.*, **338**, 407 (1963).

17. H. R. Mahler, N. K. Sarkar, L. P. Vernon, and R. A. Alberty, *J. Biol. Chem.*, **199**, 585 (1952).

18. N. A. Rao, S. P. Felton, F. M. Huennekens, and B. Mackler, *J. Biol. Chem.*, **238**, 449 (1963).

19. L. P. Vernon, H. R. Mahler, and N. K. Sarkar, *J. Biol. Chem.*, **199**, 599 (1952).

20. H. R. Mahler, and J. L. Glenn, in *Inorganic Nitrogen Metabolism* (W. D. McElroy and B. Glass, eds.), Johns Hopkins Press, Baltimore, 1956, p. 575.

21. E. Haas, B. L. Horecker, and T. R. Hogness, *J. Biol. Chem.*, **136**, 747 (1940).

22. H. R. Mahler and D. G. Elowe, *J. Am. Chem. Soc.*, **75**, 5769 (1953).

23. B. DeBernard, *Biochim. Biophys. Acta*, **23**, 510 (1957).

24. B. Mackler, *Biochim. Biophys. Acta*, **50**, 141 (1961).

25. B. Mackler, R. J. Erickson, S. D. Davis, T. D. Mehl, C. Sharp, R. J. Wedgwood, G. Palmer, and T. E. King, *Arch. Biochem. Biophys.*, **125**, 40 (1968).

26. S. A. Kumar, N. A. Rao, S. P. Felton, F. M. Huennekens, and B. Mackler, *Arch. Biochem. Biophys.*, **125**, 436 (1968).

27. P. Andrews, *Biochem. J.*, **96**, 595 (1965).

28. J. R. Whitaker, *Anal. Chem.*, **35**, 1950 (1963).

29. H. M. Duncan and B. Mackler, *Biochemistry*, **5**, 45 (1966).

30. R. L. Pharo and D. R. Sanadi, *Biochim. Biophys. Acta*, **85**, 346 (1964).

31. D. R. Sanadi, R. Pharo, and L. Sordahl, in *Nonheme Iron Proteins: Role in Energy Conversion* (A. San Pietro, ed.), Antioch Press, Yellow Springs, Ohio, 1965, p. 429.

32. R. L. Pharo, L. A. Sordahl, S. R. Vyas, and D. R. Sanadi, *J. Biol. Chem.*, **241**, 4771 (1966).

33. R. L. Pharo, L. A. Sordahl, H. Edelhoch, and D. R. Sanadi, *Arch. Biochem. Biophys.*, **125**, 416 (1968).

34. L. Ernster, M. Ljunggren, and L. Danieldson, *Biochem. Biophys. Res. Commun.*, 2, 88 (1960).

35. J. M. Machinist and T. P. Singer, *Proc. Natl. Acad. Sci. U.S.*, **53**, 467 (1965).

36. D. J. Horgan and T. P. Singer, *Biochem. Biophys. Res. Commun.*, **27**, 356 (1967).

37. J. Salach, T. P. Singer, and P. Bader, *Biochem. J.*, **104**, 22C (1967).

38. D. R. Sanadi, this volume, Chap. 4.

39. S. Fleischer, A. Casu, and B. Fleischer, *Federation Proc.*, **23**, 486 (1964).

40. W. P. Cunningham, F. L. Crane, and G. L. Sottocasa, *Biochim. Biophys. Acta*, **110**, 265 (1965).

41. R. Ringler, S. Minakami, and T. P. Singer, *Biochem. Biophys. Res. Commun.*, **3**, 417 (1960).

42. R. L. Ringler, S. Minakami, and T. P. Singer, *J. Biol. Chem.*, **238**, 801 (1963).

43. T. E. King and R. L. Howard, *Biochem. Biophys. Acta*, **37**, 557 (1960).

44. T. E. King and R. L. Howard, *J. Biol. Chem.*, **237**, 1686 (1962).

45. S. Minakami, R. L. Ringler, and T. P. Singer, *J. Biol. Chem.*, **237**, 569 (1962).

46. S. Minakami, T. Cremona, R. Ringler, and T. P. Singer, *J. Biol. Chem.*, **238**, 1529 (1963).
47. T. Cremona and E. B. Kearney, *J. Biol. Chem.*, **240**, 3645 (1965).
48. H. Beinert, G. Palmer, T. Cremona, and T. P. Singer, *J. Biol. Chem.*, **249**, 475 (1965).
49. H. Watari, E. B. Kearney, and T. P. Singer, *J. Biol. Chem.*, **238**, 4063 (1963).
50. S. Minakami, R. L. Ringler, and T. P. Singer, *Biochim. Biophys. Acta*, **50**, 610 610 (1961).
51. F. M. Huennekens, S. P. Felton, N. A. Rao, and B. Mackler, *J. Biol. Chem.*, **236**, PC57 (1961).
52. T. E. King, *Proc. Intern. Biochem. Congr.*, *5th, Moscow, 1961*, **5**, 207 (1961).
53. A. Merola, R. Coleman, and R. Hansen, *Biochim. Biophys. Acta*, **73**, 638 (1963).
54. B. Chance, L. Ernster, P. B. Garland, C. P. Lee, P. A. Light, T. Ohniski, C. I. Ragan, and D. Wong, *Proc. Natl. Acad. Sci. U.S.*, **57**, 1498 (1967).
55. J. Salach, T. P. Singer, and P. Bader, *J. Biol. Chem.*, **242**, 4555 (1967).
56. D. R. Biggs, J. Hauber, and T. P. Singer, *J. Biol. Chem.*, **242**, 4563 (1967).
57. T. P. Singer and T. Cremona, in *Oxygen in the Animal Organism* (F. Dickens and E. Neil, eds.), Pergamon, London, 1964, p. 179.
58. T. P. Singer, S. Minakami, and R. L. Ringler, *Proc. Intern. Congr. Biochem.*, *5th, Moscow, 1961*, **5**, 174 (1961).
59. T. P. Singer, in *Biological Oxidations* (T. P. Singer, ed.), Wiley-Interscience, New York, 1967, p. 339.
60. T. P. Singer, in *Biological Structure and Function*, Vol. II (T. W. Goodwin and O. Lindberg, eds.), Academic Press, New York, 1961, p. 103.
61. T. P. Singer and E. B. Kearney, in *Redoxfunktionen Cytoplasmatischer Strukturen* (T. Bucher, ed.), 1962, p. 241. (Symposium of the German Society of Physiological Chemists and Austrian Biochemical Society, Vienna, Sept. 26–29, 1962).
62. T. P. Singer, E. Rocca, and E. B. Kearney, in *Flavins and Flavoproteins* (E. C. Slater, ed.), Elsevier Publishing Co., Amsterdam, 1966, p. 391.
63. A. G. Chapman and V. Jagannathan, Information Exchange Group No. 1, Memo No. 45.
64. T. Cremona, E. B. Kearney, and G. Valentine, *Nature*, **200**, 673 (1963).
65. Y. Hatefi and K. S. Stempel, *Biochem. Biophys. Res. Commun.*, **26**, 301 (1967).
66. E. R. Redfearn and T. E. King, *Nature*, **202**, 1313 (1964).
67. T. P. Singer, E. B. Kearney, and P. Bernath, *J. Biol. Chem.*, **223**, 599 (1956).
68. B. B. Buchanan, in *Structure and Bonding*, Vol. 1 (C. K. Jorgensen *et al.*, eds.), Springer-Verlag, New York, 1966, p. 109.
69. Y. Hatefi, unpublished results.
70. L. Ernster, H. D. Hoberman, R. L. Howard, T. E. King, C. P. Lee, B. Mackler, and G. Sottocasa, *Nature*, **207**, 940 (1965).
71. B. Mackler, in *Non-Heme Iron Proteins: Role in Energy Conversion* (A. San Pietro, ed.), Antioch Press, Yellow Springs, Ohio, 1965, p. 421.
72. Q. H. Gibson, B. Mackler, and T. E. King, unpublished results.
73. S. Keresztes-Nagy and E. Margoliash, *J. Biol. Chem.*, **241**, 5955 (1966).
74. C. K. R. Kurup and A. F. Brodie, *J. Biol. Chem.*, **242**, 2909 (1967).
75. W. M. Latimer, *Oxidation Potentials*, Prentice-Hall, New York, 1938, p. 65.
76. H. Mersmann, J. Luthy, and T. P. Singer, *Biochem. Biophys. Res. Commun.*, **25**, 43 (1966).
77. L. W. Mapson and F. A. Isherwood, *Biochem. J.*, **86**, 173 (1963).
78. L. Pape, J. S. Multani, C. Stitt, and P. Saltman, *Biochemistry*, **7**, 607 (1968).
79. J. S. Rieske, D. H. McLennan, and R. Coleman, *Biochem. Biophys. Res. Commun.*, **15**, 338 (1964).

80. D. D. Tyler, R. A. Butow, J. Gonze, and R. W. Estabrook, *Biochem. Biophys. Res. Commun.*, **19**, 551 (1965).

81. D. V. DerVartanian, W. H. Orme-Johnson, R. E. Hansen, H. Beinert, R. L. Tsai, J. C. M. Tsibris, R. C. Bartholomaus, and I. C. Gunsalus, *Biochem. Biophys. Res. Commun.*, **26**, 569 (1967).

82. K. A. Schellenberg and L. Hellerman, *J. Biol. Chem.*, **231**, 547 (1958).

CHAPTER 4

NADH-UBIQUINONE REDUCTASE*

D. R. Sanadi, Pih-kuei C. Huang,† and Richard L. Pharo

INSTITUTE OF BIOLOGICAL AND MEDICAL SCIENCES
RETINA FOUNDATION
BOSTON, MASSACHUSETTS

* The unpublished work included in this article and the preparation of the manuscript were supported by a Public Health Service Research Grant (GM13641) from the National Institutes of General Medical Sciences.
† Scholar in Cancer Research, American Cancer Society, Grant No. T-435A.

Much of the confusion regarding the NADH dehydrogenase* of mitochondria and the profusion of repetitious literature has arisen because the
identity of the respiratory carrier that oxidizes the reduced dehydrogenase has
not been unequivocally established. There is strong evidence that ubiquinone
(coenzyme Q) (1, 2), an iron protein (nonheme iron protein) (1, 3, 4) and
cytochrome b (5) are in close association with the dehydrogenase, but
their sequence is uncertain. Consideration has been given to the possibility
that one or more of these components is on a side path, possibly communicat-
ing with succinate dehydrogenase or with other respiratory-chain assemblies
(6). Since ubiquinone is a potential direct oxidant of the NADH dehydrogenase, considerable efforts have been made to isolate an NADH–ubiquinone
reductase from mitochondrial particles (7).

I. Assay of Ubiquinone Reductase Activity with Q_6† or Q_{10} as Acceptor

Early attempts to demonstrate the enzymatic oxidation of NADH by exogenous higher isoprenologs failed, presumably because the assay conditions
were unsatisfactory. Surface-active agents were used in the medium with the
aim of increasing the solubility of the ubiquinone, but it was later found that
they inhibited the reaction strongly (8). The inhibition was less apparent
when quinones with greater solubility in aqueous media (e.g., Q_1, menadione)
were used, indicating that the detergents per se did not affect the enzyme. The
inhibition was probably related to removal of the low amount of the dissolved
quinone from the aqueous medium and transfer into the detergent micelles,
so that the enzyme was essentially left without the substrate. The inhibition
of the reaction by proteins in general (7) (including hemoglobin, insulin, and
crystalline bovine serum albumin) has been demonstrated, and it may be
related similarly to nonspecific binding of the quinone with decreased
availability for the reductase. A second difficulty with the assay was the use
of phosphate buffer, which also inhibits the reaction (8). Using improved

* The term "NADH dehydrogenase" will refer to the mitochondrial enzymes that
catalyze the oxidation of NADH by artificial electron acceptors (ferricyanide or dyes).
When referring to specific preparations, the name of the principal investigator associated
with the isolation will be used in order to avoid confusion. The term "iron protein" will
be used for that flavin-free ferredoxinlike nonheme iron protein that is known to be involved
in an unidentified manner in this region of the respiratory chain. Some investigators
regard the iron-containing flavoprotein and the flavin-free iron protein as one complex
functional unit of the respiratory chain.

† The abbreviations used are Q_1, Q_2, etc. for ubiquinone-1, ubiquinone-2, and so on,
the subscript referring to the number of isoprenoid units in the side chain. The term
"ubiquinone" will be used when referring to them in general.

assay conditions, it was possible to show that submitochondrial particles that had not been exposed to drastic conditions would catalyze the oxidation of NADH by Q_6 and Q_{10} at appreciable rates. Under these assay conditions, the Q_6 and Q_{10} are present largely as precipitates, with only a limited amount of dissolved quinone present in the aqueous phase. Since the amount of NADH oxidized is stoichiometric with the quinone added, when the former is in excess, it seems likely that the dissolved quinone acts as a shuttle between the respiratory chain and the precipitated quinones. Similar assay conditions were satisfactory with the soluble extracts from the particles (see later). The reaction medium is thus biphasic and requires precise control for reproducible and meaningful assays. Since these problems were not appreciated fully, Machinist and Singer (9) encountered difficulties with the assay. However, in their more recent work (10), the assay has yielded results similar to those reported by Pharo et al. (7). The biphasic assay system also introduces difficulties in the studies with lipophilic inhibitors such as rotenone, which may be partly occluded by or dissolved in the precipitated quinone, leaving a lower amount available for interaction with the enzyme. Such an effect has been demonstrated experimentally with submitochondrial particles (7). In these experiments, Q_6 reduced the efficiency of rotenone inhibition of aerobic NADH oxidation. The mutual interference of quinone and lipophilic inhibitor could be even greater when the lipid-free soluble enzyme is assayed. In the case of particles, the bound lipid present near the site of reduction could at least compete more effectively with the micelles for substrate and inhibitor.

It is pertinent to note that in all of the enzymic studies carried out so far, Q_6 and Q_{10} behaved quite similarly, but Q_6 has been used routinely in view of its ready availability. However, Q_1 and menadione often respond differently from Q_6 and Q_{10}, e.g., in the experiments on rotenone and Amytal inhibition, stability of reductase activity, or sensitivity of reductase to venom extracts. These differences (7) will be discussed subsequently.

II. NADH–Ubiquinone Reductase Activity of Particulate Preparations

A. Complex I

Using surface-active agents to disperse proteins and salt fractionation to separate them, Hatefi and co-workers (1) purified a complex form of NADH–ubiquinone reductase (complex I). It catalyzed the oxidation of NADH by Q_1 and Q_2 and had lower activity with cytochrome c. Because of the effect of added bile salts, the preparation appeared soluble. Complex I, whose composition is shown in Table 1, had bound lipid and Q_{10}, but was virtually

TABLE 1

COMPOSITION AND PHYSICAL PROPERTIES OF NADH–UBIQUINONE
REDUCTASE AND SIMILAR PREPARATIONS

| | Complex I | Purified reductase | | | |
| | | Pharo et al. (11) | | | |
Component	Hatefi (1)	Fraction I	Fraction II	Hatefi (12)	Singer (10)
Flavin[a]	1.4–1.5	10.3	9.3	13.5–14.5	13.5
Ubiquinone[a]	4.2–4.5	—	—	—	—
Iron (nonheme)[a]	26	42	42	60–65	—
Sulfide (acid labile)[a]	—	50	53	58–60	—
Lipid (mg/mg protein)	0.22	—	—	—	—
Molecular weight:					
Yphantis method	—	>70,000	98,000	—	—
Flavin basis	—	98,000	108,000	71,000	74,000

[a] nmoles/mg protein.

free of cytochromes. The Q_1 reductase activity was inhibited over 90%
by 3 mM Amytal and by 6 μM p-chloromercuriphenyl sulfonate.

B. Submitochondrial Particles

Particulate preparations made from sonically disrupted mitochondria
catalyze the oxidation of NADH by Q_6 or Q_{10} at a rate of 0.2–0.4 μmoles/
min/mg protein, which compares favorably with the rate of NADH oxidation
by oxygen under similar conditions (7). This activity is considered substantial,
since the dissolved quinone concentration in the medium is extremely low and
probably well below the saturation level for the enzyme. The activity with
Q_0 or menadione was also approximately the same, but vitamin K_2 compounds
with two to five isoprenoid units in the side chain showed little or no activity.

It is by no means clear where exactly the added ubiquinone derivatives
intercept electron flow in the respiratory chain. Since there is quinone present
in the particles, possibly concentrated in the bound lipid, it is likely that the
external quinone cannot compete effectively for electrons at this site. It is
more likely that the added ubiquinone intercepts the respiratory chain via
the endogenous quinone or some subsequent carrier in the respiratory chain.
Indirect and circumstantial evidence for this hypothesis comes from the
experiments of Machinist and Singer (9), who showed that lipid-extracted
particles lose Q_1 reduction activity, which can be restored by phospholipids.*

* The activity restored by lipid is sensitive to rotenone. The data have been interpreted
as indicative of lipid requirement for rotenone effect (9). However, the conclusion is
unsupported, since activity is dependent on lipid addition and rotenone sensitivity cannot
be measured in the absence of lipid.

III. Soluble NADH–Ubiquinone Reductase Preparations

A. Soluble Extracts

The availability of an acceptable assay with Q_6 and Q_{10} made it possible to test for an extracted soluble NADH-ubiquinone reductase (7). A satisfactory extraction procedure consists of warming submitochondrial particles at 43° for 15 min at pH 5.3 in approximately 10% ethanol. This method is a minor modification of the procedure of Mahler et al. (13). The extracts of the mitochondrial particles had high activity with Q_6 (about 15 μmoles NADH oxidized/min/mg protein at 30°), Q_{10} (about 8 μmoles/min/mg protein), and menadione (70–80 μmoles/min/mg), and somewhat variable activity with ferricyanide and cytochrome c (7). Mahler's cytochrome c reductase (13), which is extracted similarly from hog-heart particles, and Mackler's NADH dehydrogenase (14), extracted from a partially purified NADH oxidase, showed no activity with Q_6 (7). ETP made by the original procedure of Crane et al. (15) gave extracts with low Q_6 reduction activity; however, ETP made by a modified procedure (16)* has yielded high Q_6 reductase activity (17). NADH dehydrogenase made in King's laboratory (17a) also had no detectable activity with Q_6. These observations suggest that alterations in the flavoprotein can occur, even when it is bound to the insoluble elements of mitochondrial membrane. At least partly as a result of this, the activity of the extracted flavoproteins with respect to different oxidants varies from one preparation to another. The Q_6 or Q_{10} reduction activity may be the most labile one, since many of the preparations have high activity with ferricyanide, cytochrome c, or dichlorophenolindophenol, but none with Q_6 or Q_{10}.

Crane and co-worker (18) have described another type of extraction procedure for releasing NADH–menadione reductase from submitochondrial particles. It involves brief shaking with diethyl ether, centrifugation, and purification on a DEAE–cellulose column. The enzyme has relatively high activity towards menadione, which is sensitive to rotenone, amytal, and piericidin. The preparation resembles the NADH–ubiquinone reductase with respect to low iron/flavin ratio (2.3:1) and inhibition by rotenone. However, its activities towards menadione and ubiquinones are less than

---

* The term "ETP" introduced in Green's laboratory, has been used loosely in many communications to refer to preparations that had similar electron transport activity but were often prepared by different procedures. This has led to considerable confusion, since not all properties of the preparations are identical. The extracts made from different types of ETP show differences in activities.

1% of that shown by the NADH–ubiquinone reductase (*11*) on the basis of protein.

B. Purified Reductase Preparations

Chromatography of NADH–ubiquinone reductase extracts on Bio-Gel HTP hydroxylapatite* yielded two active flavoprotein fractions, fraction I and fraction II, which after further purification, have the properties shown in Tables 1 and 2 (*11*). The two components can be separated also by electrophoresis. Rechromatography of the two fractions separately on hydroxylapatite gave single bands with unaltered adsorption characteristics, indicating that they are not interconvertible after extraction from the particles. In other respects the two fractions were nearly indistinguishable and hence will be discussed without distinction unless specified. It is conceivable that the two proteins originate from the same flavoprotein complex of the respiratory system by disruption at different points. This suggestion has some support from the observation that minor changes in the procedure sometimes change the relative amounts of the two proteins.

The isolation of a soluble ubiquinone reductase with activity toward the higher isoprenologs has been confirmed by Singer and co-workers (*17, 19*) and Hatefi and Stempel (*12*). The latter workers used an entirely different isolation procedure involving dissociation of complex I in urea and obtained a flavoprotein preparation with activities similar to that reported by Pharo *et al.* (*11*). Despite earlier disagreements (*9*), the presence of substantial Q_6 and Q_{10} reduction activity associated with the flavoprotein can now be regarded as established. The significance of this activity and its relation to similar activity in the particulate systems is under investigation in many laboratories, and will be discussed in a later section.

IV. Properties of NADH–Ubiquinone Reductase

The purified fractions I and II of ubiquinone reductase have FMN, iron (nonheme) and sulfide (acid labile) in the amounts of 1.0, 4.2, and 5.0 moles (or g atoms) respectively per 100,000 g (Table 1). The levels of these components are about 50% higher per mg protein in the reductase reported by Hatefi and Stempel (12). The difference is probably not due to contaminating proteins in the preparations of Pharo *et al.*, since they are over 95% pure,

---

* Bio-gel HTP supplied by Bio-Rad Laboratories, Richmond, California, was used for the separation. The reproducibility of the separation was poor with other preparations of hydroxylapatite, although the nonflavoprotein impurities were consistently removed.

TABLE 2

ACTIVITY OF DIFFERENT ENZYME PREPARATIONS IN QUINONE REDUCTION

| | | Purified ubiquinone reductase | | | |
| | | Pharo *et al.* (*11*) and unpublished | | | |
Substrate	Particles (SP$_{HL}$) (*7*)	Fraction I	Fraction II	Hatefi (*18a*)	Singer (*17*)
Assay temperature	30°	30°		38°	30°
Menadione	0.22[a]	150[b]	180[b]	160[c]	—
Ubiquinone-1	0.31[a]	119	125	150[c]	29.9[b]
Ubiquinone-2	—	—	—	35–40[c]	—
Ubiquinone-6	0.32[a]	26[b]	28[b]	30–33[c]	15.2[b]
Ubiquinone-10	0.27[a]	14	13	12–16[c]	12.2[b]
Ferricyanide	1.9[d]	—	—	215[e]	138[h]
Cytochrome *c*	8.8[f]	—	—	43[g]	53.6[i]

The activities are presented as μmoles of NADH oxidized/min/mg protein. It should be noted that the assay temperatures used in the different laboratories vary. Reference to the data is given after the enzyme preparation. SP$_{HL}$ stands for sonic particles from heavy and light mitochondria. Fractions I and II refer to the two reductase peaks separated on hydroxylapatite and purified further independently.

The assay conditions were as follows:

a. The medium consisted of Tris sulfate, pH 8.0, 50 mM, 0.17 mM menadione or 0.17 mM Q_1 or 0.10 mM Q_6 or 0.07 mM Q_{10} (added in 0.05 ml methanol), 1 mM KCN and 0.13 mM NADH.

b. As under *a*, except no KCN was present.

c. Tris sulfate, pH 8.0, 50 mM, 0.4 mM menadione (1 % ethanol in final volume) or 0.4 mM Q_1 or 0.15 mM Q_2 or 0.15 mM Q_6 or 0.7 mM Q_{10} (in 3 % final volume of methanol), and 0.15 mM NADH.

d. Tris sulfate, pH 7.5, 50 mM ferricyanide, and 0.13 mM NADH.

e. Phosphate buffer, pH 7.5, 40 mM, 1.3 mM ferricyanide, and 0.75 mM NADH.

f. Tris sulfate, pH 8.5, 50 mM, 0.1 mM cytochrome *c*, and 0.13 mM NADH.

g. Phosphate, pH 8.5, 20 mM, 0.15 mM cytochrome *c*, 0.75 mM NADH, and 0.1 % serum albumin.

h. Phosphate buffer, pH 7.4, 40 mM, 0.15 mM NADH, concentration of ferricyanide was varied in the range of 0.5 to 1.7 mM and V_{max} was calculated.

i. Diol(2-amino-2-methyl-1,3-propanediol) buffer, pH 8.5, 67 μM, 0.60 μM NADH, concentration of cytochrome *c* was varied and V_{max} was calculated.

based on polyacrylamide gel or cellulose acetate strip electrophoresis and on sedimentation analysis in the ultracentrifuge. More likely, the reductase fractions have been cleaved somewhat differently from the respiratory chain compared to Hatefi's reductase and represent a slightly larger segment. They appear closer in size to the NADH dehydrogenase of Mackler with a molecular weight of approximately 80,000 and the dehydrogenase of King and Howard with a molecular weight of approximately 110,000 (see Chapter 3, this volume). It is interesting to note that the Mackler preparation has FMN, Fe and sulfide in the approximate molar ratio of 1:2:2, which is different from that seen in ubiquinone reductase (Table 1).

On the basis of protein, the Hatefi preparation has slightly lower activity with menadione as acceptor, compared to the reductase of Pharo *et al.* (Table 2). If allowance is made for the difference in temperature of the assay used in the two laboratories, the activity is even lower. Since the

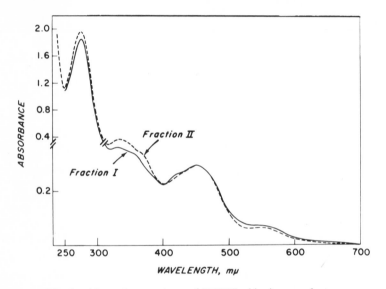

Fig. 1.   Absorption spectrum of NADH–ubiquinone reductase.

flavoprotein is known to be labile, the differences may be related to variable losses in its activity during isolation. Comparison of Q_6 reduction activity is particularly difficult, since this activity is even more labile (*7, 11*).

The absorption spectrum of ubiquinone reductase, shown in Fig. 1, has maxima at 550, 450, 333, and 276 mμ and a shoulder at 370 mμ (*11*). Reduction by excess NADH under anaerobic conditions produces a 36% decrease in absorbance at 450 mμ, the bleaching corresponding to reduction of virtually all of the bound flavin (Fig. 2). It is of interest to note that dithionite produces no further bleaching even in the 550 mμ region. The spectrum of the reduced enzyme shows maxima at 420, 452, and 550 mμ in the visible region and is reminiscent of the spectrum of an iron protein from chloroplasts (PPNR) (*20*) and the flavin-free iron protein component of aldehyde and xanthine oxidases (*21*). Rajagopalan and Handler (*21*) have estimated the molar absorption of protein-bound iron at 550 mμ in several iron flavo-proteins as approximately 2800. This value may not apply to the iron in the

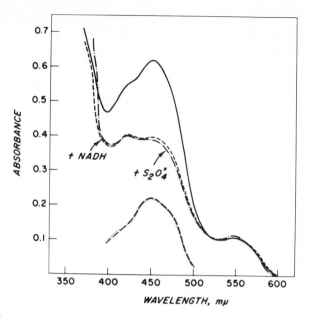

Fig. 2. Changes in spectrum of NADH–ubiquinone reductase on reduction.

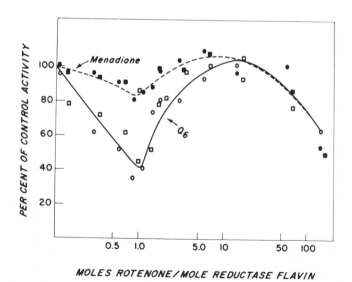

Fig. 3. Effect of rotenone on NADH–ubiquinone reductase activity of purified enzyme.

ubiquinone reductase, which, on the basis of the above value, should have approximately 2 Fe/FMN; chemical analysis has shown 4 Fe/FMN.

The role of the iron in the NADH–ubiquinone reductase is not known. It can be reduced by NADH as demonstrated by subsequent chelation with o-phenanthroline, but the kinetics of its reduction have not been studied. The reduced enzyme, after chelation of over 95% of iron, is still nearly as active in

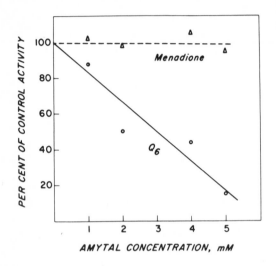

Fig. 4. Effect of Amytal on NADH–ubiquinone reductase activity of purified enzyme.

catalyzing the oxidation of NADH by menadione as the untreated enzyme (11, 22). Prolonged incubation with o-phenanthroline apparently leads to inactivation of the reductase (22a), possibly by secondary changes in the flavoprotein. It must be emphasized that the iron in the reductase flavoprotein is distinct from the iron protein (free of flavin) isolated by Hatefi and Stempel (18a). The two appear to be closely associated in the NADH dehydrogenase of Singer (23), in complex I (1), and in ETP-type preparations from mitochondria.

The ubiquinone reductase activity with Q_6 as acceptor is inhibited appreciably by rotenone when the rotenone to flavin ratio is approximately 1:1 (Fig. 3). Increasing the rotenone concentration leads to restoration of the activity, but further increase again produces inhibition. The biphasic response is apparent also with menadione as acceptor, but the extent of inhibition is less. Amytal also inhibits reduction of Q_6 (50% at 3 mM and 90% at 5.4 mM Amytal), but no inhibition is apparent if menadione is used as the oxidant in place of Q_6 (Fig. 4).

V. Relationship of Ubiquinone Reductase to NADH Dehydrogenases

Two classes of NADH dehydrogenases have been obtained from heart mitochondrial preparations. One class is a low-molecular-weight (approximately 1×10^5 g/mole) flavoprotein, represented initially by Mahler's cytochrome c reductase (*13*). These dehydrogenases seem to be closely related to the NADH–ubiquinone reductase in physical properties and many of the catalytic activities. The major significant difference is related to activity with long-chain ubiquinones as electron acceptors. The absent or minimal activity of the NADH dehydrogenases of Mahler, of Mackler, and of King with Q_6 (*22*) is believed to be due to a difference in the conformational state of the protein. Since the ubiquinones are soluble to any appreciable degree only in lipid solvents, it is possible that a nonpolar environment imposed by the tertiary structure of the polypeptide is necessary at the reactive site for Q_6 reduction activity. The site may have been altered in these dehydrogenases during the isolation procedures. In a recent communication, Biggs *et al.* (*10*) have reported ubiquinone reductase activity in cytochrome c reductase prepared from pH 5.4 particles and claimed to be Mahler's reductase. However, a number of modifications have been made in the original procedure of Mahler *et al.* (*13*). The modifications include omission of lyophilization, which "is destructive to CoQ reductase activity" [as quoted from Ref. (*10*)], and use of freshly prepared particles. It has not been specified whether the preparation was made from hog heart or beef heart. The stability of the Q reductase activity made from the two sources could well be different although the more stable cytochrome c reduction activity may be equally stable. Finally, Biggs *et al.* have also noted a preferential loss of Q_6 reduction activity with respect to cytochrome c reductase activity. Their results, in fact, support the hypothesis that maintenance of a specialized conformational state is necessary for activity with the lipid-soluble ubiquinones.

The second type of dehydrogenase is a high-molecular-weight (approximately 5 to 9×10^5 g/mole) preparation isolated by Singer and co-workers (*16, 23*) through digestion of submitochondrial particles with phospholipase A at $30°$. The preparation has a high Fe/FMN ratio (approximately $18:1$) (*23*) and exhibits the $g = 1.94$ electron spin resonance signal on reduction (*23a*). Singer's preparation has no detectable Q_6 or Q_{10} reduction activity, but an active ubiquinone reductase may be extracted from it by treatment with acidic aqueous–ethanol (*17*). The earlier failure (*22*) to obtain active extracts has been attributed to the presence of an inhibitor of ubiquinone reductase activity in the extract (*19*). From the ether-pretreated submitochondrial particles, by means of Triton X-100, Kaniuga *et al.* (*23b*) have extracted an enzyme with acceptor specificity similar to Singer's enzyme. Ubiquinone-1 and cytochrome c activities are generated when this enzyme is incubated at 34° in the presence of 9% ethanol at pH 6.0. Based on these observations, it has been proposed that the NADH–ubiquinone reductase (17) is a degradation product of the high-molecular-weight NADH dehydrogenase, much like the Mahler NADH–cytochrome c reductase is claimed to be. The pros and cons of whether the low-molecular-weight dehydrogenases are degradation products or distinct entities integrated within a complex have been discussed extensively in several articles, including Chapter 3 of this volume, and will not be considered here.

VI. Relationship of Isolated NADH–Ubiquinone Reductase to the Respiratory Chain

One of the reasons for the controversy on NADH dehydrogenase is the difficulty in assaying for only the flavoprotein activity in a reliable manner when it is integrated with the mitochondrial membrane and other components as part of a larger enzyme complex. It is well recognized, for example, that antimycin-insensitive cytochrome c reductase activity, which may be an inherent property of the flavoprotein, becomes apparent only after it is disengaged from other components. Some of the possible factors responsible for the seeming alterations in properties may be enumerated as follows.

(a) Modification of the activity of the enzyme as a consequence of being bound to a membranous structure. An example of a similar change is the ability of isolated cytochrome c to bind cyanide, while endogenous cytochrome c or cytochrome c reconstituted in deficient particles does not bind cyanide (24). A second example is the rise in redox potential of isolated cytochrome b from -340 mV to a more positive value as in the respiratory chain on binding to structural protein (25). Another is the masking of ATPase activity when isolated coupling factor 1 (F_1) binds to mitochondrial particles (26). A fourth is the marked alteration in the kinetic parameters K_m and V_{max} of purified malate dehydrogenase on binding to the mitochondrial membrane or structural protein (27).

(b) Modification of activity due to the lipid environment and consequent differences in concentration of lipophilic compounds, substrates, and inhibitors within the hydrophobic region, as compared to an aqueous soluble system.

(c) Blocking of access to the active site by steric factors and lipid environment in the complex.

(d) Change in the flavoprotein itself, produced during the isolation procedures.

Before considering some specific differences in properties of the flavoprotein in the isolated versus the particle-bound state, it should be remembered that the sequence of electron carriers from NADH to cytochrome b has not been established. For purposes of discussion, the following pathways may be considered.

Pathway A:

Pathway B:

Pathway C:

In pathways A and C, iron protein and ubiquinone respectively have been shown on a side path. There have been suggestions that such side paths, if present, may be concerned with electron transfer between different respiratory chain assemblies or electron flow from NADH to fumarate (or succinate to NAD in energy-linked reversed electron transfer) (6).

Rotenone and Amytal inhibition are important phenomena in understanding the exact sequence of the respiratory carriers between NADH and cytochrome b. In intact mitochondria, nearly complete inhibition of respiration is generally considered to occur with one mole of rotenone per mole of the sensitive unidentified binding site. The sensitivity could be modified, however, in different types of particles or under different experimental conditions. For example, in the presence of Q_6, the rotenone titer changes by a factor of 2 in submitochondrial particles with oxygen as the acceptor (7). Cunningham et al. (27a) have concluded that the presence of solvent in the ether-extracted particle changed the affinity of the carrier for the lipophilic rotenone. As mentioned earlier (Fig. 3), the inhibition of ubiquinone reductase by rotenone shows a biphasic response, but only a titration-type inhibition curve is obtained with particulate preparations. The sensitivity to Amytal is somewhat greater in particles [70% inhibition at 1 mM and complete at 2 mM; Ref. (7)] than in the soluble reductase (70% at 3 mM). Pharo et al. (11) are inclined to place considerable emphasis on the inhibition at low levels of rotenone. The biphasic effects with excess rotenone are considered to be related to reason (a) or (b), discussed earlier in this section. On the other hand, Singer and co-workers (19, 28) contrast the biphasic nature of the inhibition obtained with the soluble reductase with the titration effect in mitochondrial particles and conclude that the flavoprotein has been altered during the isolation with resultant exposure of new sites for Q_6 and rotenone binding [item (d) discussed earlier]. In support of this hypothesis, Horgan and Singer (28) have found that ^{14}C-rotenone added to ETP appeared in the particulate residue rather than in the soluble fraction containing the ubiquinone reductase during the extraction of the enzyme. It was assumed that the binding of the rotenone to the sensitive site is not altered during the extraction at 43° and pH 5.3 in 10% ethanol. The possibility that the rotenone is split off from the original site during extraction and is transferred to the lipid or some other components of the insoluble residue needs investigation.

There is general agreement that rotenone binds to a site on the substrate side of cytochrome b. Hatefi (29) placed the rotenone site between the flavoprotein and the iron protein. His conclusion was based on the observation that the decrease in absorbance on adding NADH to complex I-III or complex I is partially sensitive to rotenone. Gutman et al. (30) failed to reproduce Hatefi's experiments with complex I. Albracht and Slater's work supported Hatefi's findings but the rotenone sensitive absorbance was attributed to Q_{10} present in complex I* (31). The more definite conclusion seems to have been deduced from the electron spin resonance measurements. It has been demonstrated that the $g = 1.94$ signal produced by the NADH dehydrogenase nonheme iron protein, in both submitochondrial particles (32, 33) and complex I (34) is not inhibited by rotenone. These results would indicate that the rotenone site is on the oxygen side of the iron protein. Bois and Estabrook (35) suggested nonheme iron protein as a possible site of rotenone inhibition. The oxidation of NADH by ubiquinone catalyzed by complex I is inhibited by Amytal, which presumably acts at the same site as rotenone. Thus, it would appear that the rotenone binding site is on the substrate side of ubiquinone. These results are consistent with the indicated localization of the rotenone sites in pathways A, B, and C.

Pathway C seems unlikely since two sites of rotenone binding would have

---

* However, ubiquinone-10 present in complex 1 can only account for about 10% of the rotenone sensitive absorbance. Slater's conclusion was drawn perhaps over miscalculation.

to be involved in order to explain available data. Moreover, submitochondrial particles were found to lose their NADH oxidase activity after removal of Q_{10} by extraction with pentane, and the oxidase activity was restored on re-incorporating Q_{10} into the extracted particles (*31, 36*).

Hatefi and Stempel (*18a*) have devised an elegant procedure for separating a flavoprotein similar to the NADH–ubiquinone reductase and a flavin-free iron protein from complex I. They have also observed that the iron protein is capable of undergoing reduction by NADH in the presence of the flavoprotein and that chemically reduced iron protein can be oxidized by Q_1 or Q_2. Presumably, no flavoprotein is required for the latter reaction, although this does not appear to have been specifically examined. The oxidized form of the iron protein could not be reduced by reduced Q_2. Thus, the redox potential of the iron protein should lie between the potentials of the flavoprotein and ubiquinone. The results are consistent with pathways A, B, and C, particularly since the chemically reactive, oxidized quinone can oxidize iron chelates nonenzymically. The iron protein in this particular series of experiments may be capable of acting like any other simpler iron chelate. On the other hand, if the kinetics and inhibitor sensitivity of the oxidation of reduced iron protein by Q_1 or Q_2 are indicative of an enzymatic reduction, pathway B would be greatly favored. These studies could thus prove to be of major importance in deducing the actual sequence of carriers.

VII. Conclusions

The various low-molecular-weight NADH dehydrogenases, Mahler's cytochrome *c* reductase, Crane's NADH–menadione reductase and the NADH–ubiquinone reductase, all appear to be derived from the same segment of the respiratory chain and related to each other. The differences in electron acceptor and inhibitor responses between them may be reflections of the differences in isolation procedures. The ubiquinone reductase reaction, as studied with the purified enzyme, may be regarded as a close model of the reaction in mitochondria. The two reactions cannot be identical in all respects because the environment of the enzyme in the structurally integrated mitochondria is different, and binding to membrane and associated lipid could modify its responses to different acceptors and inhibitors. The site of rotenone action remains to be established unequivocally. Present data do not exclude the possibility that it acts directly on the NADH–ubiquinone reductase flavoprotein. Considerably more and perhaps new types of experimental evidence will be necessary to settle the exact sequence of the respiratory carriers in this region of the mitochondrial respiratory chain.

174 D. R. Sanadi, P.-K. C. Huang, and R. L. Pharo

REFERENCES

1. Y. Hatefi, A. G. Haavik, and D. E. Griffiths, *J. Biol. Chem.*, **237**, 1676 (1962).
2. A. Kröger and M. Klingenberg, in *Current Topics in Bioenergetics* (D. R. Sanadi, ed.) Vol. 2, Academic Press, New York, (1967).
3. H. Beinert and G. Palmer, *Advan. Enzymol.*, **27**, 105 (1965).
4. H. Beinert, G. Palmer, T. Cremona, and T. P. Singer, *Biochem. Biophys. Res. Commun.*, **12**, 432 (1963).
5. B. Chance and E. R. Redfearn, *Biochem. J.*, **80**, 632 (1961).
6. D. R. Sanadi, *Ann. Rev. Biochem.*, **34**, 21 (1965).
7. R. L. Pharo, L. A. Sordahl, S. R. Vyas, and D. R. Sanadi, *J. Biol. Chem.*, **241**, 4771 (1966).
8. R. L. Pharo, *Abstr. X-54, Intern. Cong. Biochem. 6th, New York, 1964*, IUB Vol. 32, p. 786.
9. J. M. Machinist and T. P. Singer, *J. Biol. Chem.*, **240**, 3182 (1965).
10. D. R. Biggs, J. Hauber, and T. P. Singer, *Biochem. Biophys. Res. Commun.*, **27**, 632 (1967).
11. R. L. Pharo, L. A. Sordahl, H. Edelhoch, and D. R. Sanadi, *Arch. Biochem. Biophys.* **125**, 416 (1968).
12. Y. Hatefi and K. E. Stempel, *J. Biol. Chem.*, **244**, 2350 (1969).
13. H. R. Mahler, N. K. Sarkar, L. P. Vernon, and R. A. Alberty, *J. Biol. Chem.*, **199**, 585 (1952).
14. B. Mackler, *Biochim. Biophys. Acta*, **50**, 141 (1961).
15. F. L. Crane, J. L. Glenn, and D. E. Green, *Biochim. Biophys. Acta*, **22**, 475 (1956).
16. R. L. Ringler, S. Minakami, and T. P. Singer, *J. Biol. Chem.*, **238**, 801 (1963).
17. J. Salach, T. P. Singer, and P. Bader, *Biochem. J.*, **104**, 22c (1967).
17a. T. E. King and R. L. Howard, *J. Biol. Chem.*, **237**, 1686 (1962).
18. C. L. Hall and F. L. Crane, *Biochem. Biophys. Res. Commun.*, **26**, 138 (1967).
18a. Y. Hatefi and K. E. Stempel, *Biochem. Biophys. Res. Commun.*, **26**, 301 (1967).
19. J. Salach, H. D. Tisdale, T. P. Singer, and P. Bader, *Biochim. Biophys. Acta*, **143**, 281 (1967).
20. K. T. Fry and A. San Pietro, *Biochem. Biophys. Res. Commun.*, **9**, 218 (1962).
21. K. V. Rajagopalan and P. Handler, *J. Biol. Chem.*, **239**, 1509 (1964).
22. D. R. Sanadi, R. L. Pharo, and L. A. Sordahl, in *A Symposium on Non-Heme Iron Proteins: Role in Energy Conversion* (A. San Pietro, ed.) Antioch Press, Yellow Springs, Ohio, 1965.
22a. Y. Hatefi, K. E. Stempel, and W. G. Hanstein, *J. Biol. Chem.*, **244**, 2358 (1969).
23. T. Cremona and E. B. Kearney, *J. Biol Chem.*, **239**, 2328 (1964).
23a. H. Beinert, W. Heinen, and G. Palmer, in *Enzyme Models and Enzyme Structure*, Brookhaven Symposia in Biology, **15**, 229 (1962).
23b. Z. Kaniuga and A. Gardas, *Biochim. Biophys. Acta*, **143**, 647 (1967).
24. C. L. Tsou, *Biochem. J.*, **50**, 493 (1952).
25. R. Goldberger, A. Pumphrey, and A. Smith, *Biochim. Biophys. Acta*, **58**, 307 (1962).
26. E. Racker, *Proc. Natl. Acad. Sci. U.S.*, **48**, 1659 (1962).
27. K. D. Munkres and D. O. Woodward, *Proc. Natl. Acad. Sci. U.S.*, **55**, 1217 (1966).
27a. W. P. Cunningham, F. L. Crane, and G. L. Sottocasa, *Biochim. Biophys. Acta*, **110**, 265 (1965).
28. D. J. Horgan and T. P. Singer, *Biochem. Biophys. Res. Commun.*, **27**, 356 (1967).
29. Y. Hatefi, *Proc. Natl. Acad. Sci. U.S.*, **60**, 733 (1968).

30. M. Gutman, T. P. Singer, H. Beinert, and J. E. Casida, *Proc. Natl. Acad. Sci. U.S.*, **65**, 763 (1970).

31. S. P. J. Albracht and E. C. Slater, *Biochim. Biophys. Acta*, **189**, 308 (1969).

32. D. D. Tyler, R. A. Butow, J. Gonze, and R. W. Estabrook, *Biochem. Biophys. Res. Commun.*, **19**, 551 (1965).

33. G. Palmer, D. J. Horgan, H. Tisdale, T. P. Singer, and H. Beinert, *J. Biol. Chem.*, **243**, 844 (1968).

34. P.-K. C. Huang, unpublished results.

35. R. Bois and R. W. Estabrook, *Arch. Biochem. Biophys.*, **129**, 362 (1969).

36. L. Szarkowska, *Arch. Biochem. Biophys.*, **113**, 519 (1966).

CHAPTER **5**

THE CHOLINE AND α-GLYCEROPHOSPHATE DEHYDROGENASES AND ELECTRON-TRANSFER FLAVOPROTEIN

W. R. Frisell

DEPARTMENT OF BIOCHEMISTRY
COLLEGE OF MEDICINE AND DENTISTRY OF NEW JERSEY
NEWARK, NEW JERSEY

J. R. Cronin

DEPARTMENT OF CHEMISTRY
ARIZONA STATE UNIVERSITY
TEMPE, ARIZONA

I. Introduction

In addition to the NADH and succinate dehydrogenases, mitochondria contain a number of other "respiratory chain-linked" flavoproteins, which are not only substrate specific, but which can also be distinguished by differences in their intramitochondrial location, sensitivities to the morphological states of the mitochondrion, and responses to inhibitors of electron and energy transfer. Three of these flavoenzymes, α-glycerophosphate dehydrogenase, choline dehydrogenase, and electron-transfer flavoprotein (ETF), will be discussed in this chapter, and the current status of their chemical characterization and roles in mitochondrial electron and energy transfer will be assessed. Earlier reviews have considered the α-glycerophosphate dehydrogenase (1, 2), choline dehydrogenase (1), and ETF (3–5).

II. α-Glycerophosphate Dehydrogenase

At least three different kinds of α-glycerophosphate dehydrogenase are found in living systems. An NAD-linked enzyme, located in the cytoplasm of animal cells, catalyzes the oxidation of L-glycerol-3-phosphate to dihydroxyacetonephosphate (L-glycerol-3-phosphate: NAD oxidoreductase, EC 1.1.1.8) (6). This enzyme is frequently referred to as the cytoplasmic α-glycerophosphate dehydrogenase or the von Euler–Baranowski enzyme. An FAD-containing dehydrogenase, induced in S. faecalis by aeration, oxidizes the same substrate to dihydroxyacetonephosphate with the direct reduction of molecular oxygen to hydrogen peroxide (L-glycerol-3-phosphate:oxygen oxidoreductase) (7). The enzyme to be discussed in this section is the mitochondrial α-glycerophosphate dehydrogenase, a respiratory-chain-linked enzyme catalyzing the oxidation of glycerol-3-phosphate to dihydroxyacetonephosphate [L-glycerol-3-phosphate: (acceptor) oxidoreductase, EC 1.1.99.5]. This enzyme is commonly referred to as the mitochondrial α-glycerophosphate dehydrogenase or the Meyerhof–Green enzyme.

Although first found by Meyerhof in frog muscle (8), the earliest systematic investigation of the dehydrogenase was that of Green (9) who isolated it from rabbit skeletal muscle. The dehydrogenase was recognized as being attached to particles, since the activity was easily sedimented. Moreover, it was concluded that the enzyme was linked to a respiratory carrier or carriers, since the reduced product was water rather than hydrogen peroxide. Although the prosthetic group was not identified, it appeared that NAD was not involved (10). Later Tung, Anderson, and Lardy (11) achieved a "temporary solubilization" of the rabbit-muscle α-glycerophosphate dehydrogenase by treatment with desoxycholate. Using this preparation, the reaction product was identified unequivocally as dihydroxyacetonephosphate, and it was also confirmed that the pyridine nucleotides are not involved in the catalytic function of the enzyme. Despite the fact that this preparation did not consume oxygen in the presence of substrate and cytochrome c and was free of several enzymatic activities present in the Green preparation, it did not have the usual properties of a "soluble" enzyme. Thus, the solubilized dehydrogenase retained cytochrome oxidase activity and could be precipitated with freezing and thawing and at a low ammonium sulfate concentration. Ling et al. (12) subsequently solubilized the dehydrogenase by treatment of an acetone powder of pig-brain mitochondria with desoxy-cholate and trypsin. Using ammonium sulfate a fraction was obtained with a specific activity about twentyfold higher than that of the starting homogenate. The preparation reduced 2,6-dichlorophenol indophenol, but not the purified cytochromes b or c.

To date the more extensively studied soluble dehydrogenase is that described by Ringler and Singer (13, 14). These workers found that treatment of an acetone powder of the particulate fraction of pig-brain mito-chondria with phospholipase A (snake venom) leads to true solubilization of the α-glycerophosphate dehydrogenase and that the enzyme can be isolated by ammonium sulfate fractionation. However, the purified preparation was heterogeneous as judged by ultracentrifugation and hydroxylapa-tite chromatography. It is interesting to note that the action of phospholipase A in the solubilization of α-glycerophosphate dehydrogenase appears to be "specific." This lipase is known to catalyze the hydrolysis of the β-ester linkage in phospholipids (15).

A. Properties

The mitochondrial α-glycerophosphate dehydrogenase is generally thought to be a flavoprotein. The purified enzyme of Ringler (14) yields one mole of flavin per 2.1×10^6 g of protein, and the flavin has been reported to be FAD (1). The absorption spectrum of mitochondrial α-glycerophosphate dehydrogenase is similar to the spectra of the other respiratory-chain-linked flavoproteins. As can be seen in Fig. 1, the absorption in the visible range is not that of a simple flavoprotein with maxima at about 380 and 460 mμ, but instead has a single maximum at 415 mμ with a shoulder around 450 mμ. The 415-mμ maximum apparently is not Soret-band absorption, since heme could not be detected in this preparation. The purified enzyme does, however, contain 1 mole of nonheme iron per 3.5×10^5 g of protein, and only a

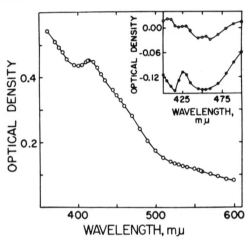

Fig. 1. Spectrum of purified α-glycerophosphate from pig-brain mitochondria. Reprinted from Ref. (*14*), p. 1196, by courtesy of *The Journal of Biological Chemistry*.

fraction of it is liberated by acid denaturation. Metals other than iron could not be detected by emission spectroscopy (*14*). The form of the absorption spectrum in the visible range is very likely related to the presence of this nonheme iron.

The effect of nonheme iron on the visible absorption spectrum of several metalloflavoenzymes has been studied by Rajagopalan and Handler (*16*). On the basis of their findings for aldehyde oxidase and xanthine oxidase, flavin would account for approximately 27% of the total absorbance at 450 mμ in an iron flavoprotein such as the α-glycerophosphate dehydrogenase in which the iron-to-flavin molar ratio is 6. Reduction of α-glycerophosphate dehydrogenase by its substrate gives only a 6% decrease in the absorbance measured at 450 mμ; however, one cannot be certain that this represents complete reduction of the dehydrogenase flavin. Dithionite reduction, on the other hand, yields a 44% decrease in absorbance at this wavelength.

The spectrum of the mitochondrial α-glycerophosphate dehydrogenase resembles that of the flavin-free, iron-containing xanthine and aldehyde oxidases. This similarity, as well as the low flavin-to-protein ratio, raises the question whether flavin is lost during solubilization and purification. The functional roles of the flavin and nonheme iron and the relative contributions of these chromophores to the absorption of the dehydrogenase in the visible range must be considered problematical at this time.

A variety of electron acceptors can be employed with the mitochondrial α-glycerophosphate dehydrogenase, and the propriety of a given choice depends largely on the nature of the enzyme preparation. Thus, with intact mitochondria, activity measurements can be based on rates of oxygen

TABLE 1

RELATIVE REACTION RATES OF α-GLYCEROPHOSPHATE DEHYDROGENASE WITH VARIOUS
ELECTRON ACCEPTORS[a,b]

	Electron Acceptor				
Preparation	Phenazine methosulfate	Methylene blue	Ferri-cyanide	Cytochrome c	O_2 (with excess cytochrome c)
Pig-brain mitochondria	100	54	103	65	63
Acetone powder brain mitochondria	100	58	—	0	0
Soluble preparation	100	66	71	0	0

[a] Reprinted from Ref. (*17*), p. 2212, by courtesy of *The Journal of Biological Chemistry.*
[b] All rates are V_{max} values based on initial reaction rates and were calculated as moles of substrate removed per unit time.

consumption. When submitochondrial particles are employed, maximal rates of oxygen uptake may require the addition of cytochrome c (*9*) or other redox dyes. Tung *et al.* (*11*) measured oxygen consumption in the presence of methylene blue, as did Green (*9*), who also employed benzyl viologen and nitrate. Ringler and Singer (*13, 17*) compared phenazine methosulfate, methylene blue, ferricyanide, cytochrome c, and oxygen as electron acceptors and found that the manometric assay with phenazine methosulfate gave the maximal rate of glycerol-3-phosphate oxidation regardless of the enzyme preparation. These data are shown in Table 1. Since phenazine metho-sulfate apparently intercepts the flow of electrons at the dehydrogenase, this acceptor has proved useful in comparing the α-glycerophosphate dehydro-genase activities of preparations differing in the extent to which the associated electron-transport chain has been disrupted.

The mitochondrial α-glycerophosphate dehydrogenase appears to be specific for the oxidation of *L*-glycerol-3-phosphate. K_m values have been determined in a variety of systems and are of the order of 10^{-2} M (*9, 13, 17,*

18). The D-isomer (*9, 11*), glycerol, and glycerol-2-phosphate (*9*) are not oxidized. The 3-phospho- and 2-phosphoglycerates (*9*) and the product, dihydroxyacetonephosphate (*14*), are competitive inhibitors of the enzyme derived from mammalian tissues. Dihydroxyacetonephosphate is without effect, however, on the α-glycerophosphate oxidase of insect flight-muscle sarcosomes (*18*). Ethylenediaminetetraacetate competitively inhibits the α-glycerophosphate oxidase of both rat-brain mitochondria (*19*) and flight-muscle sarcosomes (*18*).

pH optima have been determined, and the results vary considerably with different enzyme preparations and assay conditions. Maximal activity has been reported at pH values as low as 6.5 and as high as 10.1 (*9, 13, 17, 18*). The only estimate of the molecular weight of the α-glycerophosphate dehydrogenase is that based on flavin analysis of the enzyme solubilized with phospholipase A (*14*) and was calculated to be over 2 million. Such a large value suggests that this preparation might be fractionated further into smaller units. More likely, the preparation is far from pure or the flavin has been lost during the isolation. Measurements of the standard potential have not been made for the dehydrogenase. Attempts to demonstrate dihydroxy-acetonephosphate reduction have been unsuccessful using reduced FMN, leucobenzylviologen, and leucomethylviologen as electron donors (*17*).

B. Relation to the Respiratory Chain

The mitochondrial α-glycerophosphate dehydrogenase is tightly bound to the respiratory chain so as to form an efficient α-glycerophosphate oxidase system. The composition of the chain of electron carriers linking the flavo-enzyme to molecular oxygen has been investigated employing various respiratory inhibitors and differential spectrophotometric techniques. Although Amytal is without effect on oxygen consumption (*20*), the α-glycero-phosphate oxidase of pig-brain mitochondria shows the same sensitivity to antimycin A and 2-*n*-heptyl-4-hydroxyquinoline-*N*-oxide as succinate oxidase (*17*). The addition of glycerol-3-phosphate to intact liver mitochondria in the presence of Amytal leads to a slight reduction of cytochrome *b* upon establishment of a steady-state rate of oxidation. When the oxygen of the reaction mixture is exhausted, the components of the respiratory chain are completely reduced by glycerol-3-phosphate (*21*). As shown in Fig. 2, the spectral bands characteristic of reduced flavoprotein and the reduced forms of cytochromes *b*, *c*, *a*, and a_3 are all apparent.

It has been suggested that coenzyme Q is the link between flavodehydro-genases such as the α-glycerophosphate dehydrogenase and the cytochrome system. Glycerol-3-phosphate has been shown to reduce the endogenous

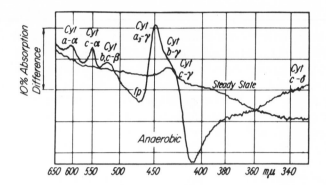

Fig. 2. Steady-state and anaerobic difference spectra for glycerol-3-phosphate oxidation by liver mitochondria in the presence of ADP, inorganic phosphate, and Amytal. Reprinted from Ref. (*21*), p. 570, by courtesy of *Angewandte Chemie*.

coenzyme Q of rat sarcosomes and to reduce exogenous coenzyme Q in either pig-brain mitochondria or with a soluble α-glycerophosphate dehydrogenase derived from pig-brain mitochondria (*22*).

From the work of Ringler and Singer (*17, 23*) it would appear that the α-glycerophosphate oxidase uses the terminal section of the respiratory chain in common with other mitochondrial oxidase systems or at least shares a rate-limiting section of the chain with them. That this is the case for the α-glycerophosphate and succinate oxidase systems is evident from the data of Table 2. Thus, a mutual interference can be demonstrated between the two substrates when either oxygen consumption or product formation is determined in the presence of pig-brain mitochondria. This competition is also evident when the reduction of added cytochrome *c* is measured, but is not seen when the autoxidizable acceptor phenazine methosulfate is employed.

TABLE 2

MUTUAL INTERFERENCE OF SUCCINATE AND GLYCEROL-3-PHOSPHATE OXIDASES IN PIG-BRAIN MITOCHONDRIA[a,b]

Substrate	Oxygen uptake (μatoms)	Products formed	
		DHAP (μmoles)	Fumarate (μmoles)
α-Glycero-P	4.0	4.8	—
Succinate	15.0	—	16.5
Succinate + α-glycero-P	13.8	3.3	8.4

[a] Reprinted from Ref. (*17*), p. 2215, by courtesy of *The Journal of Biological Chemistry*.
[b] Products determined after 21 minutes reaction in the presence of excess substrate.

Since phenazine methosulfate is believed to interact directly with the flavo-enzymes, the juncture of the pathways of electron flow presumably occurs between flavoprotein and cytochrome c. The data presented in Fig. 3 suggest that this convergence may occur quite close to the flavoenzymes, since a fumarate-dependent oxidation of glycerol-3-phosphate is found to occur in antimycin A or cyanide-blocked mitochondria of pig brain (17).

The rather surprising observation has been made that when intact liver mitochondria oxidize glycerol-3-phosphate, as much as 25% of the oxygen

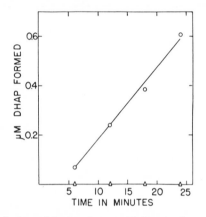

Fig. 3. Coupling of glycerol-3-phosphate oxidation to fumarate reduction in anti-mycin A inhibited pig-brain mitochondria. Circles, fumarate present; triangles, fumarate absent. Reprinted from Ref. (17), p. 2215, by courtesy of *The Journal of Biological Chemistry*.

consumption and a larger fraction of the phosphate uptake are sensitive to rotenone, a respiratory inhibitor considered specific for the NADH dehydro-genase. The rotenone-sensitive respiration is dependent on the presence of mitochondrial NAD and is accompanied by less than stoichiometric yields of dihydroxyacetonephosphate. This rotenone-sensitive oxidation has been interpreted as being due to the further oxidation of dihydroxyacetone-phosphate, although this conclusion is clouded by the finding that added dihydroxyacetonephosphate does not stimulate oxygen consumption (24). An alternative interpretation is that at least a part of the electron flux originating in the flavin-linked α-glycerophosphate dehydrogenase can go through site 1 of the electron-transfer chain, as in the case of choline oxidation (25).

The mitochondrial oxidation of glycerol-3-phosphate is accompanied by the synthesis of ATP from ADP and inorganic phosphate. P/O values between 1.0 and 1.7 have been observed for glycerol-3-phosphate and succinate with housefly sarcosomes under conditions that gave ratios greater than 2 for glutamate and α-ketoglutarate (26) and P/O values between 1.3 and 1.6 have been reported for glycerol-3-phosphate oxidation by flight-muscle mitochondria from the migratory locust (27). In the latter system the α-glycerophosphate oxidase has been shown to be subject to respiratory control, the sites of inhibition of electron flow or "crossovers" lying between cytochromes c and a and between cytochromes b and c. An Amytal-sensitive, energy-dependent reduction of NAD by glycerol-3-phosphate has been reported which presumably operates by reversal of oxidative phosphorylation at site 1 (27, 28).

C. Metabolic Role

The importance of glycerol-3-phosphate in lipid metabolism is well known, this compound being required for the biosynthesis of both neutral fats and

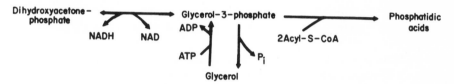

Fig. 4.   Major metabolic routes to and from glycerol-3-phosphate.

phospholipids. The major metabolic routes to and from glycerol-3-phosphate are shown in Fig. 4. The existence of the NAD-linked α-glycerophosphate dehydrogenase allows glycerol-3-phosphate to enter and be withdrawn from the Embden–Meyerhof pathway through the triose phosphates.

Since cytoplasmic NADH does not appear to pass the mitochondrial membrane readily, whereas glycerol-3-phosphate does, the suggestion has been made that the respiratory-chain-linked α-glycerophosphate dehydrogenase functions in conjunction with the cytoplasmic, NAD-dependent enzyme to constitute a "glycerol-3-phosphate cycle." It can be seen in Fig. 5 that the net effect of the operation of such a cycle is the oxidation of extramitochondrial NADH by way of the respiratory chain. It has been proposed that this mechanism is functional in insect flight muscle (18) and in vertebrate tissues such as brain and skeletal muscle, which are relatively rich in the mitochondrial α-glycerophosphate dehydrogenase (21).

The activity of the respiratory-chain-linked α-glycerophosphate dehydrogenase varies considerably from one tissue to another and also has been found by Lardy and his colleagues (29–31) to be affected significantly by

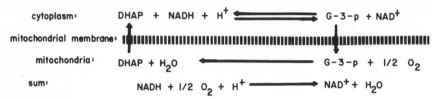

Fig. 5.   The α-glycerophosphate cycle.

thyroid hormone levels. As can be seen in Table 3, α-glycerophosphate oxidase activity is increased twentyfold in mitochondrial preparations obtained from the liver of rats fed thyroid extract. Similar effects are observed on injection of thyroxine, triiodothyronine, and triiodothyroacetic acid, whereas thyroidectomy depresses the α-glycerophosphate oxidase activity. The greatly enhanced oxidase activity parallels an increased activity of the respiratory-chain-linked α-glycerophosphate dehydrogenase as measured with phenazine methosulfate. Presumably this increase in oxidation reflects synthesis of new enzyme because the hormone effect is strongly inhibited by ethionine (29), puromycin, and actinomycin D (32). The action of thyroid hormone appears to be specific for the mitochondrial α-glycerophosphate dehydrogenase, since a variety of other dehydrogenases tested, including the NAD-linked α-glycerophosphate dehydrogenase, lactate dehydrogenase, and succinate dehydrogenase, are not affected significantly. The thyroid-hormone effect is also organ specific, and a correlation exists between change in oxygen consumption and mitochondrial α-glycerophosphate dehydrogenase activity following thyroid administration. Organs such as liver, kidney, and heart, which respond to thyroid hormone with an increased rate of oxygen consumption, show greater mitochondrial α-glycerophosphate dehydrogenase activity. The enzyme is not induced in an organ such as the brain, in which oxygen consumption is unaffected by thyroid hormone. These observations imply that the respiratory-chain-linked α-glycerophosphate dehydrogenase and the "glycerol-3-phosphate

TABLE 3

RATES OF OXIDATION OF GLYCEROL-3-PHOSPHATE BY LIVER MITOCHONDRIA OF NORMAL AND THYROID-FED RATS[a]

Source of liver mitochondria	Number of experiments	Average $Q_{O_2}(N)$
Normal rats	12	8.0
Rats on diet containing 2% desiccated thyroid	6	180.0

[a] Reprinted from Ref. (30), p. 507, by courtesy of The Annals of the New York Academy of Sciences.

cycle'' may be of considerable importance in the regulation of metabolic activity.

III. Choline Dehydrogenase

In the oxidation of choline to betaine the dehydrogenation of the alcohol group is catalyzed by a flavoenzyme, choline dehydrogenase; a separate enzyme, the NAD-linked betaine aldehyde dehydrogenase, is responsible for the second oxidative reaction. The choline dehydrogenase [choline: (acceptor) oxidoreductase, EC 1.1.99] is found exclusively in mitochondria (33–35), and is localized in the membranous compartment (36). The existence of separate mitochondrial and extramitochondrial betaine aldehyde dehydrogenases is still equivocal (37–40). Nonetheless, in those cases where evidence has been presented for the existence of a mitochondrial aldehyde dehydrogenase, fractionation procedures indicate that the enzyme(s) can be readily separated from membranous components and can be isolated in soluble form (41–44). It has been stated that "choline dehydrogenase is limited in distribution to cells from which mitochondrial cristae have not been isolated" (1). With recent developments in techniques for separating the "inner" and "outer" membranes of mitochondria (45, 46), it should be possible to determine more precisely the localization of choline dehydrogenase in the membranes.

A. Properties

Because the choline dehydrogenase is membrane-bound, its isolation in soluble form has been a difficult problem, comparable with that encountered with the α-glycerophosphate and succinate dehydrogenases. To date, the most successful procedure has involved treatment of mitochondria with phospholipase A from either *Naja naja* or *Crotalus terrificus* venom. All evidence suggests that the solubilized enzyme has a high molecular weight, and the purest preparations contain 1 mole of acid-liberated FAD and 8 g atoms of nonheme iron per 850,000 g of protein.

Although it is now generally accepted that choline dehydrogenase itself is a flavoprotein and that the isolated dehydrogenase cannot reduce NAD directly (36), there is evidence that NAD may function indirectly in the dehydrogenation of choline in intact mitochondrial preparations. In one of the earliest studies of choline oxidation, Mann, Woodward, and Quastel (33) concluded that pyridine nucleotides were not involved. Later, however, Strength, Christensen, and Daniel (47) found that in a rat-liver fraction free of betaine aldehyde dehydrogenase, choline oxidation was significantly stimulated by the addition of NAD. Subsequently, Rothschild, Cori, and Barron (*37*) showed that FAD enhanced choline oxidation in rat-liver mitochondria and that NAD was not reduced under the same conditions. The latter investigators also found that choline, like succinate and α-glycerophosphate, gave a P/O value approaching 2. These apparently conflicting conclusions concerning the role of NAD in choline oxidation have been resolved at least partially by recent findings that the pathway(s) of electron transfer linked to the dehydrogenase can be controlled by the morphological state of the mitochondria.

B. *Relation to the Respiratory Chain*

Evidence for a structural–functional interdependence in choline oxidation in mitochondria was first obtained by Williams (*48*), who showed that choline is oxidized at an appreciable rate only if the mitochondrial integrity is altered. More recent studies (*49, 50*) have demonstrated that choline oxidation in intact mitochondria can occur in a biphasic manner. The overall rate of oxidation is controlled by the intramitochondrial concentrations of adenosine di- or triphosphates and Mg^{2+}, which apparently affect membrane permeability to choline. The slow phase is coupled to phosphate uptake (*49*) and is largely sensitive to the respiratory inhibitor rotenone (*25*). During the subsequent rapid phase of oxidation, initiated when the mitochondria are swollen, the dehydrogenation of choline is uncoupled from phosphorylation and is also insensitive to rotenone (*25*). Similarly, choline oxidation is insensitive to rotenone in submitochondrial particles of rat liver that have been uncoupled with carbonyl cyanide *m*-chlorophenylhydrazone (*51*).

Several lines of evidence from studies with phosphorylating mitochondria (*25*) suggest that at least a part of the electron flux originating in the flavin-linked choline dehydrogenase can go through site 1 of the electron-transfer chain: (a) Exogenous NAD stimulates the oxidation of choline to betaine aldehyde, and the addition of choline to mitochondria under anaerobic conditions results in an energy-independent reduction of oxalacetate to malate. (b) A significant part of the initial rate of respiration and its accompanying phosphorylation can be abolished by rotenone in concentrations sufficient to block dehydrogenation of NAD-linked substrates, and this inhibition can be bypassed by vitamin K_3. These results indicate that the potentials of the choline dehydrogenase and the flavin-linked site 1 of the transfer chain may not be too disparate.

Whereas the inhibition of choline oxidation by rotenone is only partial and seems to be limited to tightly coupled mitochondria, Amytal blocks the oxidation of choline completely (but not the dehydrogenase itself) in both phosphorylating and nonphosphorylating mitochondria (25, 52) and also in uncoupled submitochondrial preparations. This difference in their activities suggests that the sites of action of the two inhibitors are not the same and that Amytal may be less specific (53). It has been shown by Ernster, Dallner, and Azzone (54) that rotenone duplicates all of the effects of Amytal on electron transport, including the energy-dependent reduction of acetoacetate by succinate via NAD. In contrast to Amytal, however, rotenone affects neither the phosphate–ATP exchange reaction nor the 2,4-dinitrophenol-induced ATPase reaction. Titration experiments with rotenone in intact mitochondria indicate that the rotenone-sensitive component occurs at the lowest molar ratio among the known catalysts of the electron-transport system of liver mitochondria and consequently may be the rate-limiting component during maximal respiratory activity (54).

With regard to the choline oxidase system, a further distinction between the effects of Amytal and rotenone was revealed by the demonstration that the rotenone-sensitive site can be bypassed by the addition of vitamin K_3. Oxygen uptake was restored completely and phosphorylation, partially. Amytal inhibition, on the other hand, was not relieved by addition of vitamin K_3 (25). This is in contrast to Amytal inhibition of NAD-linked oxidations in which vitamin K_3 can restore respiration completely and also phosphorylation, partially, to a P/O value of 2 (55).

It has been of interest to compare the pathway of electron transfer in choline oxidation beyond site 1 with that for succinate and NADH. Aside from the apparent implication of site 1 in oxidative phosphorylation with choline and some observations of a partially nonphosphorylating respiration (25, 56), all other current evidence suggests the respiratory pathways available for the oxidation of choline, NADH, and succinate are common between cytochrome b and oxygen. Thus, in submitochondrial particles or with aged or frozen–thawed mitochondria that have been freed of endogenous substrates (51), inhibition of either choline or succinate oxidation by antimycin is essentially complete. In the earlier studies of Kimura, Singer, and Lusty (57), comparisons of degrees of inhibition of various respiratory chain activities in intact mitochondria were made on a percentage basis only, and it was concluded that cytochrome b components of the choline and succinate oxidation systems of liver mitochondria are not the same. Subsequent work has indicated that the apparent inhibitor-resistant oxidation of choline in such preparations also includes an inhibitor-resistant oxidation of endogenous substrates. In submitochondrial particles free of endogenous substrates and with total anaerobiosis (51), the cytochrome b of rat-liver mitochondria is fully reduced with either choline or succinate. This finding implies that the respiratory chains responsible for the oxidation of choline and succinate are interlinked.

On the basis of all of the evidence discussed above, major pathways of electron transport operative in choline oxidation in intact liver mitochondria can be summarized schematically as shown in Fig. 6. By this sequence, reactions 1 → 2 → 3 may occur in both phosphorylating and nonphosphorylating (e.g., swollen or aged) mitochondria. The pathways involving reactions

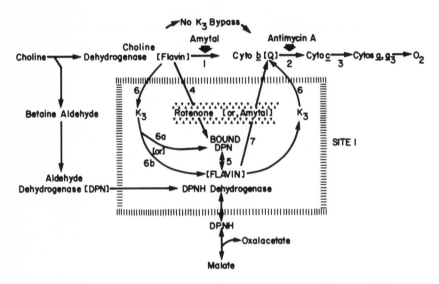

Fig. 6. Relations of choline dehydrogenase to the respiratory chain.

4 → 5 → 7 or 6 → 7, on the other hand, presumably are functional only in coupled mitochondria and would also require that site 1 be intact.

There is no evidence that a "mediator flavoprotein" such as ETF is required in the mitochondrial oxidation of choline to betaine. The oxidative metabolism of the N-methyl groups, on the other hand, is linked ultimately to the function of ETF. Thus, after the oxidation of choline to betaine, one of the methyl groups can be removed nonoxidatively by transmethylation to enter the "biologically labile methyl pool" (58). The remaining two methyl groups are oxidized successively, via dimethylglycine and sarcosine, to active formaldehyde and are metabolized through the "one-carbon cycle" (59). As will be discussed in Section IV, ETF is necessary for these two oxidative–demethylation reactions. Hence, insofar as choline is a component of the "one-carbon cycle," the oxidation of two of its three methyl groups can be considered to be *indirectly* dependent on the function of ETF.

IV. Electron-Transfer Flavoprotein

A. Functional Role in Oxidative Metabolism

In addition to the succinate and NADH dehydrogenases, and the α-glycerophosphate and choline dehydrogenases discussed in Sections II and III of this chapter, there is a fifth major route of entry for electrons into the respiratory chain. This pathway utilizes the electron-transfer flavoprotein (ETF), an enzyme required for the oxidation of the reduced form of the fatty acyl-Coenzyme A dehydrogenases (60–66) and of the reduced dimethylglycine and sarcosine dehydrogenases (5, 67–73) in the "one-carbon cycle" (58). A component of the respiratory chain, as yet unidentified, constitutes the only naturally occurring oxidant for the reduced form of ETF. The enzymatic role of ETF in fatty-acid and N-methyl oxidation can be summarized as follows:

Fatty-acyl-CoA → Dehydrogenase (flavin)
⟍
Dimethylglycine → Dehydrogenase (flavin) → ETF (FAD)
↗ ↓
Sarcosine → Dehydrogenase (flavin) ↓
Respiratory chain
↓
O_2

Purified ETF preparations active in fatty-acid metabolism have been obtained from pig liver (74), monkey liver (72), and mycobacteria (66); and the best characterized ETF of the N-demethylation system has been isolated from rat- (5, 69–75) and monkey-liver (72) mitochondria.

B. Properties

1. INTRAMITOCHONDRIAL LOCALIZATION, SOLUBILIZATION, AND PURIFICATION OF ETF

In contrast to the membrane-bound choline and α-glycerophosphate dehydrogenases described earlier in this chapter, the ETF of rat and monkey liver is localized exclusively in the soluble compartment of the mitochondrion (72, 76) and can be liberated by sonic irradiation, freeze–thawing, or osmotic lysis. The ETF of pig-liver and -heart mitochondria is likewise easily extracted. The details of the purification of ETF by ammonium sulfate fractionation, column electrophoresis, and chromatography, from both liver and heart mitochondria, are described elsewhere (72, 74, 75).

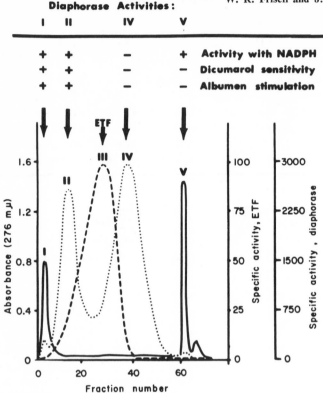

Fig. 7. Chromatographic separation of ETF and diaphorase activities. Solid curve, absorbance (276 mμ); dashed curve, specific activity of ETF; dotted curve, specific activity of diaphorases (5).

No specific, low-molecular-weight reductants for ETF are presently known. Consequently, the assay of the enzyme necessitates measuring reduction of an acceptor, such as 2,6-dichlorophenolindophenol, in the presence of one of the primary dehydrogenases and its substrate. The dehydrogenase, in turn, must be purified sufficiently to be free of ETF activity. The sarcosine-sarcosine dehydrogenase system of rat-liver mitochondria constitutes a convenient assay, since the dehydrogenase can be obtained free of ETF. Moreover, in contrast to the fatty acyl–coenzyme A derivative, sarcosine as the primary substrate is very stable.

As noted first by Beinert for the pig-liver ETF (62, 77) and later by Hoskins with the monkey-liver preparation (72), the most active ETF isolated from rat-liver mitochondria also exhibits a diphosphopyridine–triphosphopyridine nucleotide ("D–T") diaphorase activity (78). However, as shown in Fig. 7, ETF can be distinguished from the "D–T" diaphorase as well as from other diaphorases of the soluble compartment of the mitochondrion. It is of interest that the "D–T" diaphorase, like ETF, appears to be localized exclusively in the soluble compartment of the mitochondrion. Thus, in the soluble compartment both the NADH and NADPH diaphorase activities are inhibited by dicumarol. On the other hand, the NADH diaphorase activity of the membranes is not affected by dicumarol, and no NADPH diaphorase activity is detected in this fraction (5). The "D–T" diaphorase cannot substitute for ETF in the sarcosine dehydrogenase system, and there is as yet no evidence that the diaphorase has any functional role in either fatty-acid or N-methyl oxidation.

2. Flavoprotein Character of ETF

Generally when ETF is isolated by ammonium sulfate fractionation, flavin is detached from the protein and the activity that is lost can be restored by the addition of FAD, but not with FMN or riboflavin (3, 67, 68). Chromatographic isolation as well as reactivation of the apoenzyme of D-amino acid oxidase have supported the conclusion that FAD is the prosthetic group of ETF. To date it has not been possible to obtain the apoenzyme of ETF totally freed of the FAD, presumably because of the instability of such preparations (74).

3. Spectral Characteristics of ETF

As first observed by Beinert (3) and subsequently by Cronin et al. (5) and Hoskins (72), the spectrum of ETF has some distinctive characteristics. The pig-liver ETF (Fig. 8) shows maxima at 270, 375, 437.5, and 460 mμ, and

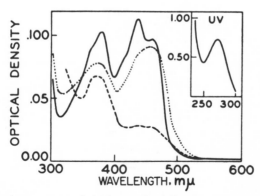

Fig. 8.   Absorption spectrum of pig-liver ETF. Solid line, oxidized form; dotted line reduced with palmityl CoA and acyl dehydrogenase. Reprinted from Ref. (62), p. 723, by courtesy of *The Journal of Biological Chemistry*.

minima at 310 and 398 mμ. The position of the major peak in the visible region at the relatively short wavelength of 437.5 mμ, the additional minor peak at 460 mμ, the deep trough at 310 mμ, and the shoulders at 360 and 410 mμ are not usually characteristic of flavoprotein spectra. The spectrum of ETF of rat liver (5, 69) (Fig. 9) resembles that of the pig-liver preparation in the 270- and 450–460-mμ regions, but the absorption at 410 mμ is more pronounced for the rat-liver ETF. Moreover, the 375- and 437.5-mμ peaks

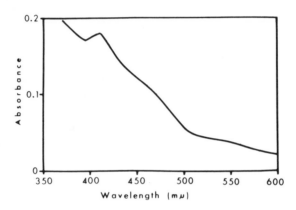

Fig. 9. Absorption spectrum of rat-liver ETF. Preparation eluted from calcium phosphate gel-cellulose (5, 75).

of the pig-liver ETF are missing or masked in the rat-liver enzyme. The ETF of monkey liver shows maxima at 375 and 410 mμ and a shoulder in the 440–480-mμ region (Fig. 10).

The shoulder or peak exhibited by all of the ETF preparations in the 410-mμ region is not without precedent and is not necessarily due to a hemoprotein impurity (3). Overall, the spectrum of the rat-liver ETF resembles the spectrum of succinate dehydrogenase (79) and α-glycerophosphate dehydrogenase (see Section II, Fig. 1), both of which contain nonheme iron. The spectral differences between the pig- and rat-liver ETF may be relevant to some of the difficulty encountered in reconstituting a fatty acyl–coenzyme A oxidase system with the pig-liver ETF (3). Crane and Beinert (62) reported that the purified ETF of pig liver contains less than one atom of iron per 5 molecules of flavin. It would be of interest to determine whether loss of iron during the fractionation procedures might account for both the decreased reactivity of the pig ETF toward the terminal electron-transport chain and also for the spectral differences relative to the ETF of rat liver (5).

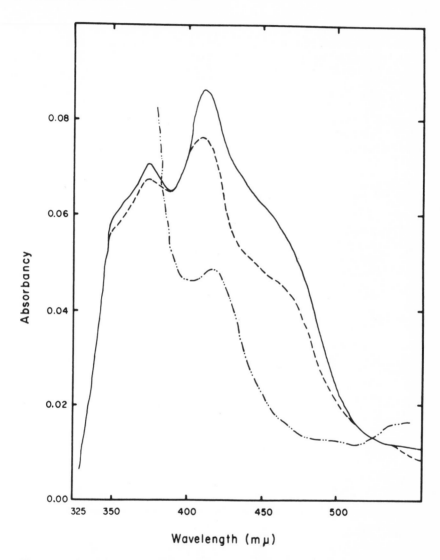

Fig. 10. Absorption spectrum of monkey-liver ETF. Solid curve, oxidized enzyme; dashed curve, reduced with monkey-liver sarcosine dehydrogenase; dot-dash curve, reduced with excess sodium hydrosulfite. Reprinted from Ref. (*72*), p. 2205, by courtesy of *The Journal of Biological Chemistry.*

4. KINETIC PROPERTIES OF ETF AND MECHANISM OF REACTION WITH
DEHYDROGENASE SUBSTRATES

As is the case with the reduced dehydrogenases that serve as its substrates, ETF_{red} is reoxidized by oxygen very slowly $(3, 5)$. Reduced ETF does not reduce free FAD, nor does ETF_{ox} oxidize free FAD2H (3). All evidence thus far supports the conclusion, therefore, that the coenzymic function of FAD with ETF is only elicited when it is bound to the apoenzyme, and that the reaction between the ETF and its reduced dehydrogenase substrates results in electron transfer between two flavin moieties bound to their respective proteins.

The multicomponent system acyl–coenzyme A substrate → dehydrogenase → ETF → terminal acceptors provides an interesting example of complex oxidation–reduction kinetics. Hauge (80) has considered two alternative mechanisms for the fatty-acyl oxidation system [see also Ref. (5)]:

Mechanism I:

(1)   acyl–CoA + dehydrogenase $\xrightleftharpoons[k_2]{k_1}$

acyl CoA–dehydrogenase $\xrightleftharpoons[k_4]{k_3}$ dehydrogenase·2H + product

(2)   dehydrogenase·2H + ETF $\xrightleftharpoons[k_2']{k_1'}$

dehydrogenase·2H·ETF $\xrightleftharpoons[k_4']{k_3'}$ ETF·2H + dehydrogenase

(3)   ETF·2H + acceptor $\xrightleftharpoons[k_2'']{k_1''}$ ETF·2H·acceptor $\xrightleftharpoons[k_4'']{k_3''}$ ETF + acceptor·2H

Mechanism II:

(1)   dehydrogenase + ETF ⇌ dehydrogenase·ETF

(2)   acyl–CoA + dehydrogenase·ETF ⇌
              acyl–CoA·dehydrogenase·ETF ⇌ product + dehydrogenase·2H·ETF

(3) dehydrogenase·2H·ETF + acceptor ⇌
              acceptor·2H·dehydrogenase·ETF ⇌ acceptor·2H + dehydrogenase·ETF

Spectral studies would favor mechanism I: first, the dehydrogenase can be reduced by the acyl–coenzyme A substrate in the absence of ETF; and second, substrate quantities of ETF can be reduced by the acyl–coenzyme A in the presence of catalytic amounts of dehydrogenase $(62, 63)$.

As would be expected for the system described by mechanism I, the observed rate of reaction increases toward a maximum when, one at a time, the concentration of the substrate, acceptor, dehydrogenase, or ETF is increased, while the concentrations of the others are held constant. It has been shown with the sarcosine-dehydrogenase-ETF-2,6-dichlorophenol-indophenol system (5) that the concentrations of the primary dehydrogenase and ETF can be widely varied with respect to each other without affecting the observed rate of dye reduction (Fig. 11).

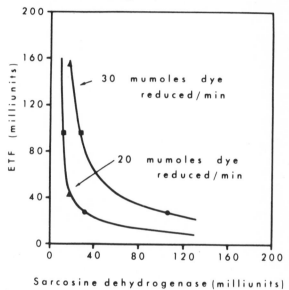

Fig. 11.   Constancy of rate of reduction of 2,6-dichlorophenol indophenol with varying amounts of sarcosine dehydrogenase and ETF (5).

As observed with the reduced acyl dehydrogenase–ETF interaction (80), when the concentration of either sarcosine dehydrogenase or ETF is held constant and the concentration of the other is varied, the overall rate of reaction as measured by dye reduction or oxygen consumption is consistent with the usual Michaelis–Menten kinetics. The estimated K_m of $1–2 \times 10^{-6}$ M for the reduced sarcosine dehydrogenase as a substrate for the rat-liver ETF compares with the value of $1.6–4.5 \times 10^{-7}$ M obtained by Hauge for three fatty acyl–CoA dehydrogenases and the pig-liver ETF (80). K_m values of this magnitude suggest a high degree of affinity between ETF and the reduced dehydrogenases.

Accounting for the kinetic characteristics of the total substrate(s)–enzyme(s)–acceptor system given in mechanism 1, the turnover number for the oxidation of the reduced primary dehydrogenase by ETF and the

TABLE 4

PROPERTIES OF α-GLYCEROPHOSPHATE DEHYDROGENASE, CHOLINE DEHYDROGENASE, AND ETF [a]

	Choline dehydrogenase [Ref. (1)]	α-Glycerophosphate dehydrogenase [Ref. (1)]	ETF			
Source	Rat liver	Pig brain	Pig Liver [Ref. (74)]	Beef heart [Ref. (74)]	Rat liver [Ref. (5)]	Monkey liver [Ref. (72)]
Method of extraction	Phospholipase A digestion	Phospholipase A digestion	Tris-acetate	Tris-acetate	Sonic irradiation or osmotic lysis	Sonic irradiation or osmotic lysis
Purification procedures	Iso-octane extraction of lipid; calcium phosphate and hydroxy-apatite absorption	$(NH_4)_2SO_4$ and hydroxy-apatite	Zn^{2+}+ethanol and $(NH_4)_2SO_4$ fractionation	$(NH_4)_2SO_4$ and Zn^{2+}-alcohol fractionation; column electrophoresis; $(NH_4)_2SO_4$ fractionation	$(NH_4)_2SO_4$, hydroxy-apatite or DEAE cellulose fractionation	$(NH_4)_2SO_4$, calcium phosphate gel, and DEAE cellulose fractionation
Mol. wt.	—	$>2 \times 10^6$	80,000	80,000	—	—
Flavin	1 mole of acid-extractable FAD/850,000 g protein	1 mole of flavin per 2×10^6 g protein	1 mole FAD/80,000 g protein	1 mole FAD/80,000 g protein	—	1 mole FAD/490,000 g protein
Metal ions	1 g atom nonheme $Fe/2–2.5 \times 10^5$ g protein	1 g atom $Fe/3.5 \times 10^5$ g protein	—	—	—	—
Acceptors	$PMS \rightarrow O_2$ or $PMS \rightarrow DCIP$ or cyto c	PMS or MB or $Fe(CN)_6^{3-} \rightarrow O_2$	DCIP, $Fe(CN)_6^{3-}$, cyto c, mito-chondrial particles	DCIP, $Fe(CN)_6^{3-}$, cyto c, mito-chondrial particles	DCIP, cyto c, mitochondrial particles	DCIP, mitochondrial particles
Optimum pH	7.6–8.2	7.5–7.7 (PMS) 7.8–8.0 (MB)	6.8–6.9	6.8–6.9	7.5 (DCIP)	7.5
Turnover No.	—	—	100–500/min (DCIP, 30°)	100–500/min (DCIP), 30°	—	—
K_m	$6.5–7.0 \times 10^{-3}$ M $(PMS \rightarrow O_2)$	9.5×10^{-3} M $(PMS \rightarrow O_2)$	$1.6–4.5 \times 10^{-7}$ M	—	$1–2 \times 10^{-6}$ M	—
Competitive inhibitors	Betaine aldehyde ($K_i = 2.0 \times 10^{-3}$ M)	Dihydroxyacetone phosphate ($K_i = 1.8 \times 10^{-4}$ M)	—	—	—	—

[a] Abbreviations: phenazine methosulfate, PMS; 2,6-dichlorophenolindophenol, DCIP; methylene blue, MB.

apparent K_m for the reduced dehydrogenase, considered as a substrate of ETF, are given by the following expressions:

$$\text{turnover no.} = \frac{k_3 k_3'}{k_3 + k_3'}$$

$$K_m = \frac{(k_2 + k_3)k_3'}{(k_3 + k_3')k_1}$$

The relationship between the ETF active in fatty-acid oxidation and that functioning in N-methyl dehydrogenation is a significant question. Beinert and Frisell (70) showed that the ETF of pig liver, as a component of the fatty-acid dehydrogenase system, is functionally interchangeable with the rat-liver ETF that is required in sarcosine oxidation. More recently, an interesting integrated study of butyryl coenzyme A dehydrogenase, sarcosine dehydrogenase, and ETF, all isolated from monkey-liver mitochondria, has been reported by Hoskins (71–73). Accepting the evidence presented that the purified ETF of monkey liver is a single functional species of protein, these studies give further support to the concept that ETF can serve as a common intermediate in fatty-acyl and N-methyl oxidation. However, the monkey-liver ETF may possess separate binding sites that are at least partially specific for each dehydrogenase. This possibility was suggested by the following observations: (a) The temperature lability curves for the ETF–sarcosine dehydrogenase and the ETF–butyryl dehydrogenase systems are different. (b) When compared with the individual rates, the rate of reduction of indophenol in the presence of *both* dehydrogenases, plus a limiting amount of ETF, is greater than either of the individual rates; i.e., even higher than the more rapid one. That the two dehydrogenase-binding sites on the ETF may be proximal, or capable of interaction, is suggested by the finding that reduced butyryl dehydrogenase effects a depression in the formation of $H^{14}CHO$ from sarcosine-$^{14}CH_3$, while the oxidized acyl dehydrogenase fails to do so. This result would not be expected if more than one ETF were involved in the two dehydrogenation systems.

For purposes of comparison, Table 4 summarizes some of the properties of α-glycerophosphate dehydrogenase, choline dehydrogenase, and ETF that have been discussed in Sections II.A, III.A, and IV.B.

C. The Relation of ETF to the Respiratory Chain

1. ACCEPTOR SPECIFICITY

In contrast to the oxidative specificity toward its substrates, ETF in its reduced form can react with a variety of acceptors, including 2,6-dichlorophenolindophenol, ferricyanide, quinones, and cytochrome c, and can also mediate the reduction of cytochromes b, c_1, and coenzyme Q of mitochondrial particles (*3, 5, 72*). As was shown by Beinert (*62, 77*), the reactivity of reduced ETF with cytochrome *c* is variable and is lost upon storage. Activity can be regained by treatment with ammonium sulfate under acidic conditions, suggesting that there is reexposure of an active site.

2. RECONSTITUTION OF THE FATTY-ACYL AND SARCOSINE OXIDASES

The ability of ETF to reduce a wide variety of acceptors *in vitro* has underscored the central question as to which compound serves as the natural

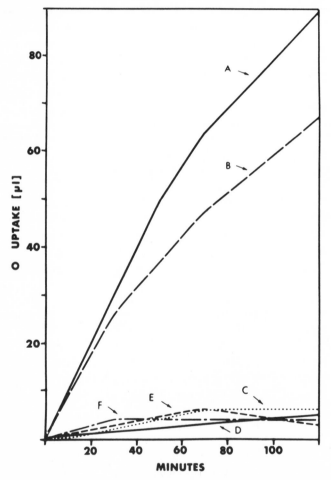

Fig. 12.   Reconstitution of sarcosine oxidase. Curve A, sarcosine dehydrogenase, ETF, and electron-transport particles; curve B, unfractionated soluble compartment of mitochondria plus electron-transport particles; curve C, sarcosine dehydrogenase and ETF; curve D, ETF and electron-transport particles; curve E, sarcosine dehydrogenase and electron-transport particles; curve F, electron-transport particles (*5*).

acceptor in the mitochondrial transport chain. In their early studies, Beinert and Crane (*62*) and Beinert and Lee (*74*) showed that ETF can reduce cytochromes b, c_1, and $a–a_3$ in mitochondrial particles under anaerobic conditions in the presence of butyryl dehydrogenase and butyryl–coenzyme A. Presumably owing to the slow rate of reduction, steady-state levels of cytochrome reduction were not observed. Nonetheless, the low oxidative activity of the reconstituted system could be measured by microrespirometry, and the acyl product crotonyl–coenzyme A was also detected. The rate of cytochrome c_1 reduction observed in this system was approximately the same as that obtained with indophenol. Coenzyme Q was also reduced by ETF in the mitochondrial particles.

TABLE 5

Rates of Electron Flow from Sarcosine Dehydrogenase to Terminal Acceptors via ETF or Phenazine Methosulfate[a]

System	Relative rates
ETF → electron-transport particles → O_2	100
ETF → 2,6-dichlorophenolindophenol	63
Phenazine methosulfate → 2,6-dichlorophenolindophenol	34

[a] See also Ref. (*5*).

It has been possible to reconstitute a sarcosine oxidase composed only of the purified rat-liver ETF and sarcosine dehydrogenase, plus a terminal electron-transport system devoid of these two activities (Fig. 12). When the two enzymes were recombined in the same proportions in which they occurred in the original unfractionated supernatant of the mitochondria, the recovered oxidase activity of the reconstituted system was found to be equal to that of the unfractionated system. From a comparison of the velocities of the reactions initiated by sarcosine dehydrogenase, summarized in Table 5, it is apparent that the rate of reaction between ETF and the mitochondrial particles is significantly greater than either the rate of reduction of indophenol with ETF or the rate of reduction of the dye with dehydrogenase plus phenazine methosulfate. These results indicate that the site of reaction of ETF with the terminal electron-transport chain remains intact during the isolation procedure and also that the linkage of ETF to the terminal-transport particle is a direct one and requires no other mitochondrial constituents, at least under the conditions employed for the purification of the components.

The sarcosine oxidase activity of monkey-liver mitochondria has also been reconstituted (*72*). With sarcosine-[14]CH_3 as the substrate, good agreement was obtained between the oxygen utilization and the H [14]CHO isolated.

3. Functional Relations of ETF with Other Flavoenzymes
   and with the Terminal Electron-Transport Chain

With regard to its function of mediating electron transfer between the reduced fatty-acyl and N-methyl dehydrogenases and the respiratory chain, ETF appears to interact with the terminal chain at about the same level as the NADH, succinate, choline, and α-glycerophosphate dehydrogenases. That is to say, electrons originating in all of these flavoenzymes share a common pathway, $(Q, b) \rightarrow c_1 \rightarrow a, a_3$. Three lines of evidence support this conclusion: (a) The fatty-acyl and sarcosine oxidases of both intact mitochondria and the reconstituted systems exhibit the same sensitivities to antimycin A as found for the oxidation of succinate and NADH. (b) Comparable to the cases of succinate and NADH (5, 81), and succinate and α-glycerophosphate (17), competition is also exhibited between the sarcosine, succinate, and NADH oxidase systems (5). (c) The oxidation of sarcosine results in a P/O ratio of 2 (82).

Reductive reactions, initiated by ETF·2H, within the fumarate/succinate and NAD/NADH couples have provided further evidence for cross communication between these electron transport systems. Beinert and Lee, employing submitochondrial particles of heart mitochondria plus acyl dehydrogenase and ETF, showed that fumarate can be reduced by butyryl–coenzyme A (74). Cronin was able also to demonstrate an ATP dependent reduction of NAD with a cyanide treated system consisting of a preparation of phosphorylating submitochondrial particles, sarcosine dehydrogenase, and ETF (5). Subsequently, an energy-dependent reduction of NAD with sarcosine was observed in intact mitochondria under aerobic conditions. These results support the conclusion that NAD is reduced by ETF·2H via the first site of energy conservation and that the energy for this endergonic process can be transferred from the two terminal sites of energy conservation without the participation of inorganic phosphate (83).

The reaction sequences shown in Fig. 13 summarize current knowledge of the role of ETF in mitochondrial electron transport. Since sarcosine oxidation is Amytal-insensitive (5, 72) but is inhibited by antimycin A, it is proposed that the point of interaction of ETF with the respiratory chain is located between the Amytal- and antimycin-sensitive sites.

Evidence thus far (5) suggests that those NADH dehydrogenases occurring with ETF in the soluble compartment of the mitochondrion are not derived from the DPNH dehydrogenase(s) of the respiratory chain in the membranes. Two of the soluble diaphorases reported by Cronin (5) appear to be similar to those described by Stein and Kaplan (84). One of the dehydrogenases obtained in their procedures had the characteristics of a "D–T" diaphorase, while the other exhibited transhydrogenase activity as well as a NADH-specific diaphorase activity. The question remains whether such NADH dehydrogenases of the soluble compartment of the mitochondrion may function in the binding of pyridine nucleotides or in reactions involving some form of NAD as a high energy intermediate in oxidative phosphorylation in site 1. The avid association of ETF with these NADH dehydrogenases, as evident from the work of Beinert and Cronin (3, 5), raises the question whether ETF may also be a reaction component of such an energy-transfer system. With regard to a more general role in mitochondrial electron transport, it has been suggested that ETF might serve as a "mobile" link between electron carriers (5). It will be of particular interest to assess the possibility of such a function in intermembrane reactions of mitochondria (85).

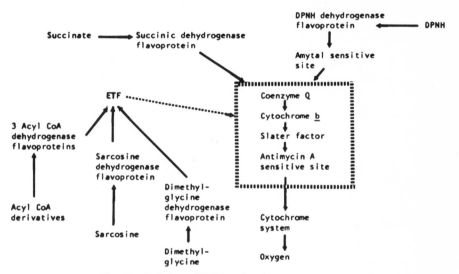

Fig. 13. Relation of ETF to the electron-transport system.

REFERENCES

1. T. P. Singer, in *The Enzymes* (P. D. Boyer, H. Lardy, and K. Myrbäck, eds.), Vol. 7, Academic Press, New York, 1963, p. 354.
2. T. Bücher, *Sitzber. Ges. Befoerder. Ges. Naturw. Marburg*, **83/84**, 383 (1961/62).
3. H. Beinert, in *The Enzymes* (P. D. Boyer, H. Lardy, and K. Myrbäck, eds.), Vol. 7, Academic Press, New York, 1963, p. 467.
4. H. Beinert and F. C. Crane, in *Inorganic Nitrogen Metabolism* (W. D. McElroy and B. Glass, eds.), Johns Hopkins Press, Baltimore, 1956, p. 601.
5. W. R. Frisell, J. R. Cronin, and C. G. Mackenzie, in *Flavins and Flavoproteins* (E. C. Slater, ed.), American Elsevier, New York, 1966, p. 367.
6. T. Baranowski, in *The Enzymes* (P. D. Boyer, H. Lardy, and K. Myrbäck, eds.), Vol. 7, Academic Press, New York, 1963, p. 85.
7. N. J. Jacobs and P. J. VanDemark, *Arch. Biochem. Biophys.*, **88**, 250 (1960).
8. O. Meyerhof, *Arch. ges. Physiol.*, **175**, 20 (1919).

9. D. E. Green, *Biochem. J.*, **30**, 629 (1936).

10. J. G. Dewan and D. E. Green, *Biochem. J.*, **31**, 1074 (1937).

11. T. Tung, L. Anderson, and H. A. Lardy, *Arch. Biochem. Biophys.*, **40**, 194 (1952).

12. K.-H. Ling, S.-H. Wu, S.-M. Ting, and T.-C. Tung, in *Proc. Intern. Symp. Enzyme Chemistry, Tokyo-Kyoto, 1957* (K. Ichihara, ed.), Academic Press, New York, 1958, p. 260.

13. R. L. Ringler and T. P. Singer, *Biochim. Biophys. Acta*, **29**, 661 (1958).

14. R. L. Ringler, *J. Biol. Chem.*, **236**, 1192 (1961).

15. N. H. Tattrie, *J. Lipid Res.*, **1**, 60 (1959).

16. K. V. Rajagopolan and P. Handler, *J. Biol. Chem.*, **239**, 1509 (1964).

17. R. L. Ringler and T. P. Singer, *J. Biol. Chem.*, **234**, 2211 (1959).

18. R. W. Estabrook and B. Sacktor, *J. Biol. Chem.*, **233**, 1014 (1958).

19. B. Sacktor, L. Packer, and R. W. Estabrook, *Arch. Biochem. Biophys.*, **80**, 68 (1959).

20. B. Chance and B. Sacktor, *Arch. Biochem. Biophys.*, **76**, 509 (1958).

21. T. Bücher and M. Klingenberg, *Angew. Chem.*, **70**, 552 (1958).

22. L. Szarkowska and A. K. Drabikowska, *Life Sciences*, **7**, 519 (1963).

23. R. L. Ringler and T. P. Singer, *Arch. Biochem. Biophys.*, **77**, 229 (1958).

24. G. Bianchi, G. Penso, and G. F. Azzone, *Biochim. Biophys. Acta*, **92**, 154 (1964).

25. G. Bianchi and G. F. Azzone, *J. Biol. Chem.*, **239**, 3947 (1964).

26. B. Sacktor and D. G. Cochran, *Arch. Biochem., Biophys.*, **74**, 266 (1958).

27. M. Klingenberg and T. Bücher, *Biochem. Z.*, **334**, 1 (1961).

28. M. Klingenberg and P. Schollmeyer, *Biochem. Z.*, **333**, 335 (1960).

29. Y.-P. Lee, A. E. Takemori, and H. Lardy, *J. Biol. Chem.*, **234**, 3051 (1959).

30. H. A. Lardy, Y.-P. Lee, and A. Takemori, *Ann. N.Y. Acad. Sci.*, **86**, 506 (1960).

31. Y.-P. Lee and H. A. Lardy, *J. Biol. Chem.*, **240**, 1427 (1965).

32. O. Z. Sellinger and K.-L. Lee, *Biochim. Biophys. Acta*, **91**, 183 (1964).

33. P. J. G. Mann, H. E. Woodward, and J. H. Quastel, *Biochem. J.*, **32**, 1025 (1938).

34. C. J Kensler and H. Langeman, *J. Biol. Chem.*, **192**, 551 (1951).

35. J. N. Williams, Jr., *J. Biol. Chem.*, **194**, 139 (1952).

36. G. Rendina and T. P. Singer, *J. Biol. Chem.*, **234**, 1605 (1959).

37. H. A. Rothschild, O. Cori, and E. S. G. Barron, *J. Biol. Chem.*, **208**, 41 (1954).

38. J. L. Glenn and M. Vanko, *Arch. Biochem. Biophys.*, **82**, 145 (1959).

39. K. T. N. Yue, P. J. Russell, and D. J. Mulford, *Biochim. Biophys. Acta*, **118**, 191 (1966).

40. V. G. Erwin and R. A. Deitrich, *J. Biol. Chem.*, **241**, 3533 (1966).

41. G. R. Rendina and T. P. Singer, *Biochim. Biophys. Acta*, **30**, 441 (1958).

42. M. Korzenovsky and B. V. Auda, *Biochim. Biophys. Acta*, **29**, 463 (1958).

43. G. R. Rendina and T. P. Singer, *J. Biol. Chem.*, **234**, 1605 (1959).

44. T. Kimura and T. P. Singer, in *Methods of Enzymology* (S. P. Colowick and N. O. Kaplan, eds.), Vol. V, Academic Press, New York, 1962, p. 562.

45. A. I. Caplan and J. W. Greenwalt, *J. Cell. Biol.*, **31**, 455 (1966).

46. D. Parsons, G. Williams, and B. Chance, *Ann. N.Y. Acad. Sci.*, **137**, 643 (1966).

47. D. R. Strength, J. R. Christensen, and L. J. Daniel, *J. Biol. Chem.*, **203**, 63 (1953).

48. G. R. Williams, *J. Biol. Chem.*, **235**, 1192 (1960).

49. T. Kagawa, D. R. Wilken, and H. A. Lardy, *J. Biol. Chem.*, **240**, 1836 (1965).

50. D. R. Wilken, T. Kagawa, and H. A. Lardy, *J. Biol. Chem.*, **240**, 1843 (1965).

51. D. D. Tyler, J. Gonze, and R. W. Estabrook, *Arch. Biochem. Biophys.*, **115**, 373 (1966).

52. L. Packer, R. W. Estabrook, T. P. Singer, and T. Kimura, *J. Biol. Chem.*, **235**, 535 (1960).

53. A. Giuditta, and G. Prisco, *Biochim. Biophys. Acta*, **77**, 394 (1963).

54. L. Ernster, G. Dallner, and G. F. Azzone, *J. Biol. Chem.*, **238**, 1124 (1963).

55. T. Conover, and L. Ernster, *Biochim. Biophys. Acta*, **58**, 189 (1962).
56. L. Ernster, O. Jalling, H. Löw, and O. Lindberg, *Exptl. Cell Research*, Suppl. 3, 124 (1955).
57. T. Kimura, T. P. Singer, and C. J. Lusty, *Biochim. Biophys. Acta*, **44**, 284 (1960).
58. W. R. Frisell and C. G. Mackenzie, in *Radioactive Isotopes in Physiology, Diagnostics, and Therapy* (H. Schwiegk and F. Turba, eds.), Springer, Berlin, 1961, p. 920.
59. C. G. Mackenzie, in *Amino Acid Metabolism* (W. D. McElroy and B. Glass, eds.), Johns Hopkins Press, Baltimore, 1955, p. 718.
60. D. E. Green, S. Mii, H. R. Mahler, and R. M. Bock, *J. Biol. Chem.*, **206**, 1 (1954).
61. F. L. Crane and H. Beinert *J. Am. Chem. Soc.*, **76**, 4491 (1954).
62. F. L. Crane and H. Beinert, *J. Biol. Chem.*, **218**, 717 (1956).
63. F. L. Crane, S. Mii, J. G. Hauge, D. E. Green, and H. Beinert, *J. Biol. Chem.*, **218**, 701 (1956).
64. J. G. Hauge, F. L. Crane and H. Beinert, *J. Biol. Chem.*, **219**, 727 (1956).
65. H. Beinert and E. Page, *J. Biol. Chem.*, **225**, 479 (1957).
66. A. Gelbard and D. S. Goldman, *Arch. Biochem. Biophys.*, **94**, 228 (1961).
67. G. P. Mell and F. M. Huennekens, *Federation Proc.*, **19**, 411 (1960).
68. D. D. Hoskins and C. G. Mackenzie, *J. Biol. Chem.*, **236**, 177 (1961).
69. W. R. Frisell, J. R. Cronin, and C. G. Mackenzie, *J. Biol. Chem.*, **237**, 2975 (1962).
70. H. Beinert and W. R. Frisell, *J. Biol. Chem.*, **237**, 2988 (1962).
71. D. D. Hoskins and R. A. Bjur, *J. Biol. Chem.*, **239**, 1856 (1964).
72. D. D. Hoskins and R. A. Bjur, *J. Biol. Chem.*, **240**, 2201 (1965).
73. D. D. Hoskins, *J. Biol. Chem.*, **241**, 4472 (1966).
74. H. Beinert and W. Lee, in *Methods in Enzymology* (S. P. Colowick and N. O. Kaplan, eds.), Vol. VI, Academic Press, New York, 1963, p. 424.
75. J. R. Cronin, W. R. Frisell, and C. G. Mackenzie, in *Methods in Enzymology* (R. W. Estabrook and M. E. Pullman, eds.), Vol. 10, Academic Press, New York, 1967, p. 302.
76. W. R. Frisell, M. V. Patwardhan, and C. G. Mackenzie, *J. Biol. Chem.*, **240**, 1829 (1965).
77. F. L. Crane, J. G. Hauge, and H. Beinert, *Federation Proc.*, **14**, 199 (1955).
78. L. Ernster, L. Danielson, and M. Ljunggren, *Biochim. Biophys. Acta*, **58**, 171 (1962).
79. T. P. Singer and E. B. Kearney, in *The Enzymes* (P. D. Boyer, H. Lardy, and K. Myrbäck, eds.), Vol. 7, Academic Press, New York, 1963, p. 383.
80. J. G. Hauge, *J. Am. Chem. Soc.*, **78**, 5266 (1956).
81. C. Y. Wu and C. L. Tsou, *Sci. Sinica* (*Peking*), **4**, 137 (1955).
82. W. R. Frisell and N. C. Sorrell, *Biochim. Biophys. Acta*, **131**, 207 (1967).
83. W. R. Frisell and J. J. Van Buskirk, *J. Biol. Chem.*, **242**, 312 (1967).
84. A. M. Stein and N. O. Kaplan, *Biochim. Biophys. Acta*, **29**, 452 (1958).
85. D. W. Allman, L. Galzigna, R. E. McCamann, and D. E. Green, *Arch. Biochem. Biophys.*, **117**, 413 (1966).

CHAPTER 6

THE *b*-GROUP CYTOCHROMES part A

ISOLATION AND PROPERTIES OF THE *b*-GROUP CYTOCHROMES

Ichiro Sekuzu

DEPARTMENT OF BIOLOGY, FACULTY OF SCIENCE
OSAKA UNIVERSITY, TOYONAKA, OSAKA, JAPAN

I. The *b*-Group Cytochromes

The *b* group cytochromes contain protohaem as the prosthetic group and show their reduced α-band around 554–564 mμ. Generally speaking, the

TABLE 1

PROPERTIES OF THE b-GROUP CYTOCHROMES

Component	Sources	Reduced			Oxidized	E_0' (volts)	Molecular weight per haem	Autoxidizability	References
		α	β	γ	γ				
b	Beef heart	561.5	531	428.5	414.5	−0.021	21,300	+	(9, 14)
b	Beef heart	562.5	532.5	429	418	−0.340[c]	28,800	+	(8, 21)
b5	Calf liver	557[b]	526	421	412	+0.02	14,400	+	(39)
b5	Pig liver mitochondria	556[b]	528	423	412	+0.014	15,000	+	(76)
b7	Aroid mitochondria	560	529	—	—	−0.03	—	+	(47)
P-420	Liver microsome	559	530	426	414	−0.020	—	+	(43, 44)
b-561	Mung bean seedlings	561	531	427	418	+0.002	—	+	(43–45)
b-555	Mung bean seedlings	555[b]	526	423	413	−0.030	13,500	+	(73)
b3	Mung bean seedlings	559	529	425	415	—	28,000	+	(71, 72)
b3	Plant microsome	559	529	425	—	—	—	:	(75)
b6	Chloroplast	563	—	—	—	+0.04 / −0.06	—	+	(75)
b2[a]	Baker's yeast	557	528	422	411	+0.12	20,000	:	(63)
b2	Baker's yeast	557	528	423.5	413		80,000	:	
b-560	Rhodospirillum	560	525	430	—		—	:	(70)
b-559	Rhodopseudomonas spheroides	559	529	426	419		—	+	(78)
b-561–554	Sclerotina libertiana	561:554	528	424	413		—	:	(77)
b-562	Bact. anitratum	562	532	428	419		—	:	(78)
b-557	Thiobacillus sp.	557	525	419	408	+0.155	—	+	(80)
b-558[c]	Streptomyces griseus	558	527	424	410		—	:	(81)
b-560	Micrococcus sp.	560	530	428	—		—	+	(82)
Haemoprotein 558	Acetobacter suboxydans	558	—	437	402		—	−	(85)
b-562[a]	E. coli	562	531	427	418	+0.113	12,000	+	(15)
b1[a]	E. coli	557.5	527.5	425	418	−0.34	62,000–66,000	−	(18)
b	E. coli	559	530	427	415	−0.02	160,000	+	(20)
b	M. denitrificance / P. denitrificance	559	528	426	415	−0.05	—	+	(69)
b-556	Garden snail	556	526	424	412	+0.20	—	:	(84)
b	Helix pomatia	556	526.5	422	408	+0.20	18,500	:	(83)
Helicorubin	Helix pomatia	561.5[b]	530	427	408		—	:	(83)
b-555	Larvae of housefly	555[b]	526	424	414	+0.100	18,000	+	(12, 74)
b-562[a]	Larvae of housefly	563	530	428.5	418.5		23,400	−	(13, 14)

Band position (mμ)

[a] The preparation was reported to be obtained in a crystalline form.

[b] The α-band has a shoulder around 560 mμ.

[c] The value of the cytochrome b-structural protein was reported to be higher than the present one.

b-group components have low redox potentials of nearly 0 V, and the components are more or less autoxidizable, although far more slowly than is the cytochrome oxidase system. Many kinds of the *b*-type components have been found in various organisms, and some of them have been obtained in the purified states, as shown in Table 1. As can be seen from the table, the properties of these *b*-group cytochromes are similar in various respects. However, their functions in various electron-transfer systems seem to be not identical. For instance, it has been established that there exist two kinds of *b*-type cytochrome in the liver microsome, namely, cytochrome b_5 and P-450. Both components can catalyze the oxidation of either NADH or NADPH, but the role of components in the oxidase system is greatly different, as will be discussed later.

This chapter will deal mainly with the properties and functions of mitochondrial cytochrome *b*, microsomal cytochrome b_5 and P-450, and baker's-yeast cytochrome b_2 in order to outline the general properties of the *b*-group cytochromes.

II. Mitochondrial Cytochrome *b*

A. Purification

As shown in Table 1, cytochrome *b* shows its reduced α-band around 562–564 mμ. The pigment was first discovered by Keilin (*1*) in 1925 as a component of the electron-transfer system in the particulate preparation derived from beef-heart muscle. Attempts to isolate and purify the component have been made by many workers (*2–6*), and recently two procedures were reported to yield a highly purified preparation from beef-heart muscle (*7–9*). Since cytochrome *b* is very tightly bound to other proteins in mitochondria as well as to cytochromes *a* and c_1, cholate or deoxycholate has been used to solubilize the component from the membranous structure. Checking carefully the concentration of bile acid to be added to the particulate fraction, the cytochrome *b*–c_1 complex could be isolated without much contamination of cytochrome *a*. Cytochrome *b* could be separated from the c_1 component in the presence of high concentration of cholate, forming an insoluble cholate–cytochrome-*b* complex. In this case, heat treatment was reported to accelerate the separation effectively (*8, 10*). From the precipitated fraction, cytochrome *b* has been recovered by two procedures. Goldberger *et al.* (*7, 8*) used sodium dodecyl sulfate as the solubilizing detergent. The component was solubilized in the presence of sodium dodecyl sulfate and repeatedly fractionated with ammonium sulfate. The final preparation was dissolved in the detergent as a polymerized form or in a

TABLE 2

COMPARISON OF PROPERTIES OF CYTOCHROME b

Cytochrome b preparation	Beef-heart muscle [Goldberger et al. (8, 21)]	Beef-heart muscle [Ohnishi (9, 14)]	Larvae of housefly [Ohnishi (13, 14)]	E. coli [Itagaki and Hager (15)]
Absorption peaks (mμ)				
Reduced α	562.5	561.5	563.0	562.0
β	532.5	531.0	530.0	531
γ	429.0	428.5	428.5	427
Oxidized γ	418.0	414.5	418.5	418
Shift of reduced γ-peak by CO	429-426	428.5-424	No shift	No shift
$A_{\gamma\text{(reduced)}}/A_{\alpha\text{(reduced)}}$	—	6.0	6.5	5.7
mM Extinction coefficient (mM^{-1} cm^{-1})				
α (reduced)	—	24.7	28.2	31.6
α (reduced–oxidized)	13.2	14.4	17.9	21.9
Iron content (mμmoles per mg protein)	36.1	47.0	46.2	35.1 (87.8[a])
Minimum molecular weight	28,000	21,300	23,400	12,000
Molecular weight estimated by hydrodynamic determination	4,000,000	Several million	23,000	12,000
Autoxidizability	Yes	Yes	No	No

[a] Values based on the dry weight.
[b] Values obtained in the presence of structural protein are higher than that indicated.

cationic detergent such as cetyldimethylethylammonium bromide as a monomeric form. On the other hand, Ohnishi (9) solubilized the insoluble fraction by treatment with a bacterial proteinase, Nargarse, according to the method of Sekuzu and Okunuki (5). After the treatment, the solubilized fraction was further purified by means of ammonium sulfate fractionation, gel-filtration on Sephadex G-75, and chromatography on DEAE-cellulose. The final preparation was dispersed in a synthetic nonionic detergent, Emasol 1130.

These purified preparations of cytochrome b from beef-heart muscle were found to be water-insoluble in nature and obtained in a highly polymerized form, as shown in Table 2. It has not yet been possible to obtain a water-soluble cytochrome b preparation from mitochondria. However, crystalline preparations of soluble cytochrome b have been reported recently from two laboratories. One was obtained from larvae of a housefly, *Musca domestica L.*, by Ohnishi and Okunuki (11). Yamanaka *et al.* (12) first observed that the component as well as cytochromes b-555 and c can be extracted readily with 40% saturated ammonium sulfate at neutral pH from the larvae, but not from the adult flies. Ohnishi (13, 14) studied various properties of the crystalline larvae cytochrome b, but its function still remains obscure. The other preparation was obtained from *Escherichia coli* by Itagaki and Hager (15) and called cytochrome b-562. The component has been shown by a number of workers (16, 17) to be quite distinct from membrane-bound cytochrome b_1, which was also obtained in a crystalline form by Deeb and Hager (18). Cytochrome b_1 of *E. coli* has been known to be a major haemochromogen constituent of the electron-transfer system associated with succinate, NADH, formate, and pyruvate dehydrogenases as well as nitrate respiration (19, 20), while the function of the soluble cytochrome b-562 is not yet established.

B. General Properties

General properties of the mitochondrial cytochrome-b preparations obtained by Goldberger *et al.* (8, 21) and Ohnishi (9, 13, 14) are summarized in Table 2 in comparison with those of soluble preparations obtained from housefly larvae (14) and *E. coli* (15). The table illustrates large differences between properties of the mitochondrial preparations and the soluble crystalline preparations, though these preparations are all highly purified, as can be judged from their minimum molecular weights. The mitochondrial preparations exist in a highly polymerized form and are autoxidizable, showing an ability to combine with carbon monoxide. Moreover, the mM extinction coefficient of the reduced α-band is less than those for the soluble preparations.

Goldberger *et al.* (*21*) showed that the oxidation–reduction potential of their purified preparation ($E_0' = -0.340$ V at pH 7.0) is raised significantly by formation of a complex between cytochrome b and mitochondrial structural protein. The value given for the component in mitochondria is +0.077 V, as reported by Holton and Colpa-Boonstra (*22*). The complex

TABLE 3

COMPARISON OF AMINO ACID COMPOSITION OF CYTOCHROME b

Residues	Beef heart (*14*)	Larvae of housefly (*14*)	E. coli (*15*)
Lys	4	18	11
His	3	4	2
Arg	5	3	4
Asp	10	20	16
Thr	(11)	13	6
Ser	(9)	13	3
Glu	9	18	14
Pro	7	9	5
Gly	15	14	4
Ala	(16)	16	15
Cys/2	(3)	(3)	0
Val	9	13	5
Met	(6)	(2)	2
Ile	12	9	3
Leu	20	13	9
Tyr	7	(3)	2
Phe	11	8	2
Try	—	—	0
NH₃	—	—	11
Total residues	157	179	103
Molecular weight	19,797 [a]	22,426 [a]	11,954
Hydrophilic/hydrophobic[b]	36.9/63.1	51.3/48.7	56.3/43.7

[a] Lipid, haem amide, and Try residue are not included.

[b] The ratio of numbers of content of hydrophilic residues (Lys, Arg, His, Glu, Ser, Thr, Tyr, Asp) to hydrophobic residues (Gly, Ala, Pro, Val, Ile, Leu, Phe, Try, Met, Cys).

formation is also accompanied by a change in sedimentation pattern in ultracentrifugation and in reducibility by reduced quinones. Their results seem to suggest the possibility that properties of mitochondrial cytochrome b are modified after isolation from the membrane structure owing to breakdown of a strong hydrophobic interaction. As a matter of fact, Ohnishi (*14*) showed that the protein itself of mitochondrial cytochrome b is quite hydrophobic, as indicated by comparison of its amino-acid residues composition with those of soluble cytochrome b. The results are given in Table 3.

From the above reasons, the function of cytochrome *b* in the electron-transfer system has not been finally established by using the purified preparations. For instance, dithionite has been observed to be only one of the effective powerful chemical reductants for the purified preparation. Yeast L(+)-lactate dehydrogenase has been shown to reduce the preparation anaerobically at a slow rate, but other reductase systems derived from mitochondria cannot. Subsequent attempts to discover the true electron donor and acceptor of cytochrome *b* have been unsuccessful.

C. Function

For the aforementioned reasons, it has been hard to elucidate the function of cytochrome *b* in the electron-transfer system. Therefore information concerning the function has been obtained mainly from studies on the behavior of the absorption spectra in mitochondria at the various stages. Chance and Schoener (*23, 24*) have performed many significant observations on the properties of mitochondrial cytochrome *b* in the absence or in the presence of various inhibitors as well as in various stages of respiration. Their results suggest that cytochrome *b* can be modified to shift the position of the reduced α-peak in the presence of antimycin A and so forth. According to their observations, the conformation of the cytochrome-*b* segment in mitochondria seems to be highly complicated. In any case, the low redox potential of cytochrome *b* indicates that it is the cytochrome located nearest the site of the succinate and NADH dehydrogenases.

Green and his colleagues (*25–27*) have attempted to isolate various enzyme complexes of the electron-transfer system from beef-heart mitochondria. According to their scheme, as shown in Fig. 1, cytochrome *b* is contained in both complexes II and III. Cytochrome *b* in complex II (succinate–coenzyme-Q reductase) appeared to be inert to enzymic reduction, and thus the function in the complex is obscure. Cytochrome *b* in complex III (reduced coenzyme-Q–cytochrome-*c* reductase) was observed to be native, since the component was reduced enzymically and electron transfer from the component to cytochrome c_1 was completely blocked in the presence of antimycin A. As mentioned previously, however, the results cannot provide direct evidence concerning the true electron donor and acceptor of cytochrome *b*, though it was concluded that reduced coenzyme Q is the electron donor of the component. Thus we should await further developments of studies on the purification and function of cytochrome *b* for the final elucidation of the problems.

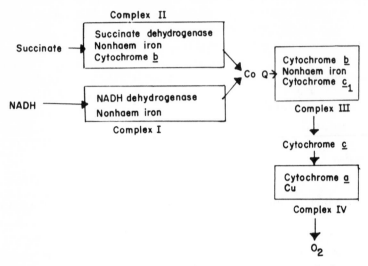

Fig. 1. Schematic representation of mitochondrial electron-transferring system of beef-heart muscle by Green and Wharton (*27*).

III. Microsomal Cytochrome b_5 and P-450

In liver microsomes, the existence of two b-type cytochromes, cytochrome b_5 and P-450, has been shown. The former was first called "cytochrome b'" by Yoshikawa (*28*) and later "cytochrome m" by Strittmatter and Ball (*29*). The name "cytochrome b_5" was first given to the component found in the midgut of larvae of the *Cecropia* silkworm by Pappenheimer and Williams (*30*). Now the name "cytochrome b_5" is generally used for the liver component having its reduced α-peak around 557 mμ with a shoulder at longer wavelength. The P-450 component has been shown to be a CO binding pigment in liver microsomes (*31, 32*); when CO is bubbled through a microsome suspension in the reduced state, a new peak is observed at 450 mμ in the difference of [CO reduced] − [reduced]. Omura and Sato (*33*) later found that ethylisonitrile, one of the reagents reacting like oxygen, can combine with the pigment in a competitive way with CO. Based on these results, they concluded that a protohaem containing component reactive with both oxygen and carbon monoxide exists in liver microsome and named it P-450. In this section, properties of both cytochrome components will briefly be discussed.

A. Cytochrome b_5

Cytochrome b_5 has been obtained in a highly purified state by many workers from rabbit liver (34), calf liver (35, 36), and pig liver (37–39). A brief report (40) claimed that a crystalline preparation was obtained. Kajihara (41) recently designed a procedure for preparation of crystalline cytochrome b_5 from rabbit liver.

Strittmatter and Velick (34), purified the pigment until it was homogeneous when tested by ultracentrifugation. The molecular weight was calculated to be 14,400. The results are reviewed in detail by Strittmatter (42). The most distinctive feature of cytochrome b_5 in its absorption spectrum is the asymmetric shape of its reduced α-band. Components having such asymmetric α-bands at 557 mμ are found in various tissues, and some of them are extracted with aqueous solution without treatment with proteinase or phospholipase. Examples of such components are cytochrome b-556 from pig liver mitochondria, obtained by Mahler et al. (39); cytochrome b-556 from larvae of a housefly, Musca domestica L., (12); and cytochrome b-555 from Mung bean seedlings (43–45), as shown in Table 1.

Calf liver cytochrome b_5 can be split into the apoprotein and protohaem by a usual acid–acetone treatment, and the original pigment can readily be reconstituted from the components, as reported by Strittmatter (36). This finding enabled him to examine the nature of haem binding in cytochrome b_5. First he prepared an iodinated or acetylated cytochrome b_5, but these preparations showed no change in the absorption spectrum, indicating essentially no alteration of the haem binding. From the chemically modified pigment, protohaem could also be isolated and reintroduced, though reconstitution with the native apoprotein was completed in less than 5 seconds while that with the chemically modified apoprotein required several minutes. The chemical modification of the apoprotein caused strong inhibition of the protohaem recombination. Based on these results, he concluded that neither sulfhydryl groups nor tyrosine residues are involved in the interaction, but one specific and highly reactive imidazole may be directly involved in the binding. He also suggested that the second site will be a histidyl group too.

A partial resolution of the primary structure of calf-liver cytochrome b_5 was reported by Strittmatter and Ozols (46).

B. P-450

P-450 has been found as a component whose Soret band is at 450 mμ in the presence of carbon monoxide in the reduced state. Omura and Sato (33, 47, 48) solubilized the component from rabbit-liver microsome by means of heated snake venom and deoxycholate treatments and obtained a purified preparation after subsequent purification. The preparation showed absorption maxima at 414 mμ in the oxidized state and at 559 mμ, 530 mμ, and 427 mμ in the reduced state. The reduced pigment–CO complex showed its Soret peak at 420 mμ, indicating that some modification of the pigment occurred during the purification procedure. This purified preparation was designated as P-420 in order to differentiate it from P-450. The redox potential of the purified P-420 was estimated to be approximately −0.020 V at pH 7.0 and 20°.

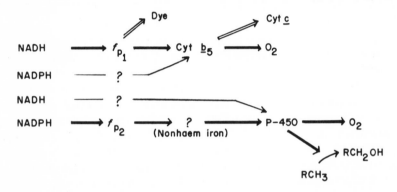

Fig. 2. Schematic representation of microsomal electron-transferring system by Sato (*51*).

On the basis of measurements of action spectrum of carbon monoxide inhibition of the oxidation of steroid C-21 hydroxylation in adrenal-cortex microsomes, Estabrook *et al.* (*49, 50*) calculated the partition coefficient between CO and oxygen of P-450 to be 0.5–2.0. This value was less than that of cytochrome oxidase (the value given was 10–20) and greater than that of hemoglobin (0.005).

The above results all indicate that P-450 is a unique component in its spectral properties as well as in other properties. The function of both cytochrome b_5 and P-450 has been studied by many workers in relation to the function of microsome. As an example, a scheme proposed by Sato and his colleagues (*51*) is shown in Fig. 2.

IV. Cytochrome b_2

The pigment called "cytochrome b_2" has been found in baker's yeast as an entity of *L*-acetate dehydrogenase (*L*-lactate:ferricytochrome *c* oxidoreductase, EC 1.1.2.3). The enzyme was purified by Bach *et al.* (*52*) and later

obtained as a crystalline preparation by Appleby and Morton (*53*). The
latter investigators treated dried yeast with *n*-butanol to remove lipids,
extracted the enzyme with sodium lactate solution, fractionated the extract
with acetone at low temperatures, and crystallized the enzyme by dialysis
against the sodium lactate solution of low ionic strength.

The crystalline preparation was homogeneous by electrophoretic and
ultracentrifugal criteria (*53*), and the molecular weight was calculated to be
80,000 (*52*) on the basis of haem and flavin contents and to be 183,400 ±
3,900 by physical measurements (*54*). The enzyme preparation contained
FMN and protohaem in the ratio of 1:1 (*56, 57*). Some of the crystalline
preparations were reported to contain DNA (6% of the dry weight) (*58*), but
the DNA was shown to have no relation to the activity. The enzyme can
oxidize $L(+)$-lactate at the highest rate (*55*) and α-hydroxybutyrate and
glycolate at a far slower rate. Cytochrome *c* is the best electron acceptor
of the enzyme, while ferricyanide, 2,6-dichlorophenolindophenol, and
1,2-naphthoquinone-4-sulfate are also found to serve as acceptors. Participation of FMN and protohaem on the enzymic activity appeared to be highly
complicated. Since the absorption spectrum of FMN is largely masked by
that of protohaem, it was hard to detect oxidation and reduction of FMN
spectrophotometrically during the enzymic reaction. By the oxidation of
the reduced enzyme with methylene blue, Hasegawa and Ogura (*59, 60*) found
absorption maxima at 385 mμ and 455 mμ, indicating that FMN was oxidized
but the haem group remained in the reduced state under the conditions.
Chance et al. (*61*) also found that the back-titration of the reduced enzyme
with ferricyanide caused no reaction between the haem group and the dye.
These results strongly indicated that sites at which electrons are received by
various acceptors are not identical, but vary according to the acceptors used.

On the other hand, Yamashita et al. (*62*) obtained a FMN-deficient
cytochrome b_2 from baker's yeast in a crystalline form. The properties of this
preparation were extensively studied by Yamashita and Okunuki (*63*).
The component was extracted at pH 8.0–8.5 from baker's yeast liquefied with
ethyl acetate, fractionated with ammonium sulfate, chromatographed on a
cation exchanger, Amberlite CG-50, and obtained as a crystalline preparation
from the ammonium sulfate solution. The cytochrome b_2 thus obtained
(molecular weight 20,000) was autoxidizable and was slowly reduced by
$L(+)$-lactate dehydrogenase and lactate. The reduced pigment could
combine with CO, producing a pronounced shift of the absorption peaks.
Their preparation was considered by Morton and Shepley (*64*) to be the
denatured product of the native enzyme containing FMN. As demonstrated
by Boeri and Rippa (*65*) and Baudras (*66*) as well as by Morton and Shepley
(*64*), however, the dehydrogenase could be split into two subunits, FMN-
protein and haemoprotein. Therefore, it may be concluded that the

dehydrogenase consists of two subunits and shows a complex mechanism of the enzyme action.

The kinetic investigation of lactate–ferricyanide reaction was reported by Morton and Sturtevant (67). The results supported the hypothesis that electrons are transferred from substrate to the FMN and from the FMN to the haem groups by an intramolecular mechanism.

Morton and Shepley (68) also showed the modification of the protohaem to haem that of a c-type by treatment of the enzyme with pCMB and acetone.

ACKNOWLEDGMENT

The author is greatly indebted to Professor K. Okunuki for his valuable advice and discussion during the preparation of the manuscript.

REFERENCES

1. D. Keilin, *Proc. Roy. Soc.* (*London*), *Ser. B*, **98**, 312 (1925).
2. B. Eichel, W. W. Wainio, T. Person, and S. J. Cooperstein, *J. Biol. Chem.*, **183**, 89 (1950).
3. G. Hübscher, M. Kiese, and R. Nicholas, *Biochem. Z.*, **325**, 223 (1954).
4. C. Widmer, H. Clark, H. A. Neufeld, and E. Stotz, *J. Biol. Chem.*, **210**, 861 (1954).
5. I. Sekuzu and K. Okunuki, *J. Biochem.*, **43**, 107 (1956).
6. D. Feldman and W. W. Wainio, *J. Biol Chem.*, **235**, 3635 (1960).
7. R. Bomstein, R. Goldberger, and H. D. Tisdale, *Biochem. Biophys. Res. Commun.*, **2**, 234 (1960).
8. R. Goldberger, A. L. Smith, H. D. Tisdale, and R. Bomstein, *J. Biol. Chem.*, **236**, 2788 (1961).
9. K. Ohnishi, *J. Biochem.*, **59**, 1 (1966).
10. I. Sekuzu, Y. Orii, and K. Okunuki, *J. Biochem.*, **48**, 214 (1960).
11. K. Ohnishi and K. Okunuki, *Biochim. Biophys. Acta*, **99**, 574 (1965).
12. T. Yamanaka, S. Tokuyama, and K. Okunuki, *Biochim. Biophys. Acta*, **77**, 592 (1963).
13. K. Ohnishi, *J. Biochem.*, **59**, 9 (1966).
14. K. Ohnishi, *J. Biochem.*, **59**, 17 (1966).
15. E. Itagaki and L. P. Hager, *J. Biol. Chem.*, **241**, 3687 (1966).
16. C. T. Gray, J. W. T. Wimpenny, D. E. Hughes, and J. Ranlett, *Biochim. Biophys. Acta*, **67**, 157 (1963).
17. T. Fujita and R. Sato, *Biochim. Biophys. Acta*, **77**, 690 (1963).
18. S. S. Deeb and L. P. Hager, *J. Biol. Chem.*, **239**, 1024 (1964).
19. A. W. Linnane and C. W. Wrigley, *Biochim. Biophys. Acta*, **77**, 408 (1963).
20. T. Fujita, E. Itagaki, and R. Sato, *J. Biochem.*, **53**, 282 (1963).
21. R. Goldberger, A. Pumphrey, and A. L. Smith, *Biochim. Biophys. Acta*, **58**, 307 (1962).
22. F. A. Holton and J. Colpa-Boonstra, *Biochem. J.*, **76**, 179 (1960).
23. B. Chance, *J. Biol. Chem.*, **233**, 1223 (1958).
24. B. Chance and B. Schoener, *J. Biol. Chem.*, **241**, 4567, 4574, 4577 (1966).
25. K. A. Doeg, S. Krueger, and D. M. Ziegler, *Biochim. Biophys. Acta*, **41**, 491 (1960).

26. Y. Hatafi, A. G. Haavik, and P. Jurtshuk, *Biochim. Biophys. Acta*, **52**, 119 (1961).
27. D. E. Green and D. C. Wharton, *Biochem. Z.*, **338**, 325 (1963).
28. H. Yoshikawa, *J. Biochem.*, **38**, 1 (1951).
29. C. F. Strittmatter and E. G. Ball, *Proc. Natl. Acad. Sci. U.S.*, **38**, 19 (1952).
30. A. M. Pappenheimer and C. M. Williams, *J. Biol. Chem.*, **209**, 915 (1954).
31. M. Klingenberg, *Arch. Biochem. Biophys.*, **75**, 376 (1958).
32. D. Garfinkel, *Arch. Biochem. Biophys.*, **77**, 493 (1958).
33. T. Omura and R. Sato, *J. Biol. Chem.*, **237**, PC1375 (1962).
34. P. Strittmatter and S. F. Velick, *J. Biol. Chem.*, **221**, 253 (1956).
35. P. Strittmatter and S. F. Velick, *J. Biol. Chem.*, **228**, 785 (1957).
36. P. Strittmatter, *J. Biol. Chem.*, **235**, 2492 (1960).
37. D. Garfinkel, *Arch. Biochem. Biophys.*, **71**, 111 (1957).
38. K. Kirsch and H. Staudinger, *Biochem. Z.*, **331**, 37 (1958).
39. I. Raw, R. Molinari, D. F. deAmoral, and H. R. Mahler, *J. Biol. Chem.*, **233**, 225 (1958).
40. I. Raw and W. Colli, *Nature*, **184**, 1798 (1959).
41. T. Kajihara, personal communication.
42. P. Strittmatter, in *The Enzymes*, Vol. 8, 2nd ed. (P. D. Boyer, H. Lardy, K. Myrbäck, eds.) Academic Press, New York and London, 1963, p. 113.
43. H. Shichi and D. P. Hackett, *J. Biol. Chem.*, **237**, 2955 (1962).
44. H. Shichi and D. P. Hackett, *J. Biol. Chem.*, **237**, 2959 (1962).
45. H. Shichi, D. P. Hackett, and G. Funatsu, *J. Biol. Chem.*, **238**, 1156 (1963).
46. P. Strittmatter and J. Ozols, *J. Biol. Chem.*, **241**, 4787 (1965).
47. T. Omura and R. Sato, *Biochim. Biophys. Acta*, **71**, 224 (1963).
48. T. Omura and R. Sato, *J. Biol. Chem.*, **239**, 2379 (1964).
49. R. W. Estabrook, D. Y. Cooper, and O. Rosenthal, *Biochem. Z.*, **338**, 741 (1963).
50. T. Omura, R. Sato, D. Y. Cooper, O. Rosenthal, and R. W. Estabrook, *Federation Proc.*, **24**, 1181 (1965).
51. R. Sato, *Protein, Nucleic Acid and Enzyme* (in Japanese), **10**, 475 (1965).
52. S. T. Bach, M. Dixon, and L. G. Zerfas, *Biochem. J.*, **40**, 229 (1946).
53. C. A. Appleby and R. K. Morton, *Nature*, **173**, 749 (1954); *Biochem. J.*, **71**, 492 (1954).
54. J. McD. Armstrong, J. H. Coates, and R. K. Morton, *Nature*, **186**, 1032 (1960); *Biochem. J.*, **86**, 136 (1963).
55. E. Boeri, E. Cutolo, M. Luzzati, and L. Tosi, *Arch. Biochem. Biophys.*, **56**, 487 (1955).
56. A. P. Nygaard, *Biochim. Biophys. Acta*, **33**, 517 (1959).
57. C. A. Appleby and R. K. Morton, *Biochem. J.*, **73**, 539 (1959).
58. C. A. Appleby and R. K. Morton, *Biochem. J.*, **75**, 258 (1960).
59. H. Hasegawa and Y. Ogura, in *Haematin Enzymes* (J. E. Falk, R. Lemberg, and R. K. Morton, eds.) Pergamon, Oxford, 1961, p. 534.
60. H. Hasegawa, *J. Biochem.*, **52**, 5, 12, 207 (1962).
61. B. Chance, M. Klingenberg, and E. Boeri, *Federation Proc.*, **15**, 231 (1956).
62. J. Yamashita, T. Higashi, T. Yamanaka, M. Nozaki, H. Mizushima, H. Matsubara, T. Horio, and K. Okunuki, *Nature*, **179**, 959 (1957).
63. J. Yamashita and K. Okunuki, *J. Biochem.*, **52**, 117 (1962).
64. R. K. Morton and K. Shepley, *Biochem. Z.*, **338**, 122 (1963).
65. E. Boeri and M. Rippa, *Arch. Biochem. Biophys.*, **94**, 336 (1961).
66. A. Baudras, *Biochem. Biophys. Res. Commun.*, **7**, 310 (1962).
67. R. K. Morton and J. M. Sturtevant, *J. Biol. Chem.*, **239**, 1614 (1964).
68. R. K. Morton and K. Shepley, *Biochim. Biophys. Acta*, **96**, 349 (1965).

69. L. P. Vernon, *J. Biol. Chem.*, **222**, 1035 (1956).
70. S. R. Elsden, M. D. Kamen, and L. P. Vernon, *J. Am. Chem. Soc.*, **238**, 1162 (1963).
71. R. Hill and R. Scarisbrick, *New Phytologist*, **50**, 98 (1951).
72. E. M. Martin and R. K. Morton, *Nature*, **176**, 113 (1953).
73. H. Shichi, H. E. Kasinsky, and D. P. Hackett, *J. Biol. Chem.*, **238**, 1162 (1963).
74. Y. Okada, personal communication.
75. R. Hill, *Nature*, **174**, 501 (1954).
76. D. S. Bendall and R. Hill, *New Phytologist*, **55**, 206 (1956).
77. T. Yamanaka, T. Horio, and K. Okunuki, *Biochim. Biophys. Acta*, **40**, 349 (1960).
78. J. A. Orlando and T. Horio, *Biochim. Biophys. Acta*, **50**, 367 (1961).
79. J. G. Hauge, *Biochim. Biophys. Acta*, **45**, 250 (1960).
80. P. A. Trudinger, *Biochem. J.*, **78**, 673, 680 (1961).
81. Y. Inoue and H. Kubo, *Biochim. Biophys. Acta*, **110**, 57 (1965).
82. K. Hori, *J. Biochem.*, **50**, 440, 481 (1961).
83. J. Keilin, *Biochem. J.*, **64**, 663 (1956); *Nature*, **180**, 427 (1957).
84. K. Kawai, *Biochim. Biophys. Acta.*, **43**, 349 (1960); **52**, 241 (1961).
85. H. Iwasaki, *Plant Cell Physiol.*, **7**, 199 (1966).

THE *b*-GROUP CYTOCHROMES part B

OXIDATION-REDUCTION POTENTIALS AND "ENERGY STATES" OF *b* CYTOCHROMES

David F. Wilson and P. Leslie Dutton

DEPARTMENT OF BIOPHYSICS AND PHYSICAL BIOCHEMISTRY
JOHNSON RESEARCH FOUNDATION, UNIVERSITY OF PENNSYLVANIA
PHILADELPHIA, PENNSYLVANIA

I. The Oxidation-Reduction Midpoint Potentials of b Cytochromes in Uncoupled Mitochondria

A. Mammalian Mitochondria

The b cytochromes in mitochondria are a part of a complex respiratory chain which has a central role in cellular oxygen consumption and energy metabolism [see for example Refs. (*1–3*)]. The majority of studies of the oxidation-reduction properties of mammalian b cytochromes have been done with a variety of submitochondrial preparations from beef heart. The reported values of the midpoint potential range from -60 mV to $+90$ mV at pH values in the region of 7.0: such as, -40 mV as reported by Ball (*4*); -60 mV by Sekuzu and Okunuki (*5*); $+73$ mV by Chance (*6*); $+68$ mV by Feldman and Wainio (*7*); $+60$ to $+90$ mV by Holton and Colpa-Bonnstra (*8*); $+70$ mV by Hill and King (*9*); $+60$ mV by Straub and Colpa-Bonnstra (*10*); -340 mV for the isolated, and 0 mV for the recombined with membrane components by Goldberger *et al.* (*11*); $+30$ mV by Reiske (*12*); and $+72$ mV by Urban and Klingenberg (*13*). Table 1 shows the reported midpoint potentials of the b cytochromes of intact beef heart mitochondria and submitochondrial particles (*14*) obtained by simultaneous and continuous readout of absorbance change as a function of oxidation-reduction potential from $+300$ to -200 mV under strict anaerobic conditions. This technique [for further experimental details see Refs. (*15–17*)] has revealed three species of b cytochromes which contribute to the absorbance changes at 562–575 nm and which have differing midpoint potentials. The values shown encompass the earlier reported numbers. This would be expected if these components were not spectrally resolved and if titrations were conducted over limited oxidation-reduction potential ranges (as in the case of the earlier values). For example, determinations done by equilibration of the b cytochrome, via succinate dehydrogenase, with ratios of succinate and fumarate, would take into account only those cytochromes which are succinate-reducible (a succinate:fumarate ratio of 99:1 would exert a potential of about -30 mV on the electron transport chain). The midpoint potential determined by this approach with beef heart submitochondrial preparations (about $+70$ mV) may, therefore,

TABLE 1

OXIDATION-REDUCTION POTENTIALS OF b-TYPE CYTOCHROMES
IN MAMMALIAN MEMBRANES

| | | (a) Mitochondrial Respiratory Chain | | | | |
Source	Component	Approximate contribution to absorbance (%)	Wavelengths meas. minus ref. (nm)	E_m (mV)[a]	pH	Ref
Beef heart	b	20–30	562–575	$+125 \pm 20$		
	b	45–55		$+38 \pm 10$	7.2	(14)
	b	15–35		-103 ± 15		
Pigeon heart	b	60	561.5–575	$+30 \pm 10$		
	b_T	40		-30 ± 10	7.2	(25)
+ATP	b	60	561.5–575	$+30 \pm 10$		
	b_T	40		$+250$		
Rat liver	b	50	430–412	$+35 \pm 15$		
	b_T	50		-40 ± 20	7.2	(17)
+ATP	b	50	430–412	$+35 \pm 10$		
	b_T	50		$+245$		

| | (b) Nonrespiratory Chain Cytochromes | | | |
Source	Component	E_m (mV)[a]	pH	Ref
Rat liver endoplasmic reticulum	P_{460}	-410	7.0	(70)
Beef adrenal mitochondrion	P_{450}	-400	7.0	(71, 72)
Rat liver microsome	P_{450}	-335	7.0	(73)
Dog and rat liver preparations	b_5	-130	7.0	(82)
Rat liver preparation	b_5	$+7$	7.0	(79, 83)
Rat liver microsome	b_5	$+7$	7.0	(73)

[a] E_m represents the oxidation-reduction midpoint potential of the cytochrome at the given pH.

be considered to be only the half-reduction point of the component b cytochromes with midpoint potentials $+125$ mV and $+38$ mV shown in Table 1. The three b cytochromes of beef heart mitochondria are present to similar extents and with similar midpoint potentials in both submitochondrial particles and their parent intact mitochondria. There is some variation in the content of the three cytochromes from one mitochondrial preparation to another (14).

B. Pigeon Heart Mitochondria

These mitochondria do not possess detectable quantities of the $+125$ mV component and the midpoint of what would be regarded as the "classical" cytochrome b is clearly $+30$ mV at pH 7.2. There is, however, a second and lower potential b cytochrome evident at this pH with a midpoint potential of about -30 mV. Differential pH dependencies (Wilson and Dutton, unpublished results) of these two cytochromes have allowed clear potentiometric separation at pH 8.1 (*14*); at this pH the midpoint potentials are -15 mV and -100 mV, respectively, for the $+30$ mV and -30 mV (pH 7.2) cytochromes. The -100 mV component of beef heart mitochondria is also not present in significant amounts in pigeon heart mitochondria.

C. Mung Bean Mitochondria

The three mitochondrial cytochromes b_{553}, b_{557}, and b_{562} which are reported to be membrane-bound in mung bean and several other plant mitochondria

TABLE 2

OXIDATION-REDUCTION POTENTIALS OF b-TYPE CYTOCHROMES IN
MEMBRANES OF PLANTS AND MICROORGANISMS

Source	Component	E_m (mV)[c]	pH	Ref
Mung bean mitochondrion	$b_{553}{}^a$	$+75 \pm 5$	7.2	(*19*)
Mung bean mitochondrion	$b_{557}{}^a$	$+42 \pm 5$	7.2	(*19*)
Mung bean mitochondrion	$b_{562}{}^a$	-77 ± 5	7.2	(*19*)
Barley chloroplast	b_{563} (b_6)	0	7.0	(*84*)
Spinach chloroplast	b_{563} (b_6)	-180 ± 20	7.0	(*85*)
Spinach chloroplast	b_{559}	$+80 \pm 20$	7.0	(*85*)
Spinach chloroplast	b_{559}	$+55$	7.0	(*92*)
Spinach chloroplast	b_{559}	$+330$	8.2	(*89*)
Pea chloroplasts	b_{559}	$+370$	6.5–7.5	(*93*)
Euglena chloroplast	b_{558}	$+320$	6–9	(*94*)
Rhodopseudomonos spheroides	b?	$+160 \pm 20$	7.2	—[b]
	b_{562}	$+65 \pm 10$	7.2	—[b]
Rhodospirillum rubrum	b?	$+175 \pm 20$	7.2	—[b]
	c_{428} (RHP, cc')	$+10 \pm 10$	7.2	—[b]
	b_{562}	-100 ± 15	7.2	—[b]

[a] Wavelength maxima of the reduced minus oxidized α band measured at 77°K. The other subscripts listed are from room temperature determinations.

[b] P. L. Dutton and J. B. Jackson, to be published.

[c] E_m represents the oxidation-reduction midpoint potential of the cytochrome at the given pH.

(*18*) were identified from their metabolic behavior and spectral characteristics to be of the b-group cytochromes. The subscripts indicating the wavelength maxima of the ferrocytochrome are, after the suggestion of Chance *et al.* (*3*), from difference spectra at 77°K. At room temperature the reduced minus oxidized difference spectra of these cytochromes have respective maxima at approximately 556–557 nm, 560 nm, and 565 nm (*18*). The oxidation-reduction midpoint potentials at pH 7.2 of the cytochromes b_{553}, b_{557}, and b_{562} are +75 mV, +42 mV, and −77 mV, respectively (*19*), as shown in Table 2.

The readily soluble b cytochromes isolated from mung bean seedlings, b_3 (*20*), b_{555}, and b_{561}, have been studied in some depth (*21–23*). Some of the physical properties are listed in Table 1 of Part A, this chapter. However, Moore (*24*) has concluded that cytochrome b_{555} is located in the microsomal membrane. Chance *et al.* (*3*) consider that cytochrome b_{561} is also a microsomal b cytochrome, but the location of the b_3 cytochrome remains unknown.

For further detailed information on plant b cytochromes, the reader is referred to a review by Chance *et al.* (*3*).

II. ATP-Induced Oxidation-Reduction Midpoint Potential Changes of Cytochrome *b*

The addition of ATP to strictly anaerobic coupled mitochondria from rat liver and beef heart (*17*) and pigeon heart (*25*) has been reported to generate a form of cytochrome b which has a measured oxidation-reduction midpoint potential of about +240 mV at pH 7.2. In each case, as shown in Table 1, whether assayed in the γ- or the α-band region, about half the complement of b cytochrome assumes the high midpoint potential; the remainder appears to undergo no change in midpoint potential, which remains at approximately +30 mV at pH 7.2. Evidence so far (*17, 25*) suggests the "energizable" cytochrome b has a resting midpoint potential of −30 to −50 mV. In comparison to rat liver and beef heart, the absence of high rates of electron flow from endogenous substrates and the presence of a less complex cytochrome complement has made the pigeon heart mitochondria an attractive experimental material. Kinetic studies (see following section) with coupled pigeon heart mitochondria (*25, 26*) have led to a spectral separation of the two b cytochromes, suggesting the reduced minus oxidized spectrum of the higher and lower potential b cytochromes (in the absence of ATP) to be maximal at 560 nm and 564 nm, respectively.

It may be summarized therefore that, of the two b cytochromes, the lower potential and longer wavelength form whose midpoint potential is energy dependent (becoming more positive with increasing "phosphate potential") may be involved in energy transduction at site II of the mitochondrial respiratory chain. This transducing cytochrome has been designated cytochrome b_T.

With respect to the plant mitochondria, Dutton and Storey (19) were not able to detect any ATP-induced change in the measured oxidation-reduction potentials of three membrance-bound b cytochromes in coupled mung bean mitochondria.

The above interpretation of the involvement of a form of b cytochrome with a high oxidation-reduction midpoint potential (as high as cytochromes $c + c_1$) in energy conservation in mammalian mitochondria relies on the cytochromes in question being at or near equilibrium with the measuring platinum electrode which is in contact with the mitochondrial suspension. Redox dyes have been employed to act as electron mediators between the electrode and the membrane bound cytochrome in order to facilitate an accurate electrometric assay of the oxidation-reduction potential of the cytochrome. This parameter together with the simultaneously monitored spectrophotometric assay of the degree of reduction of the cytochrome has been used (16, 17, 25, 27) to reveal any alterations in the electronic structure of the heme which result in changes in the oxidation-reduction midpoint potential of the cytochrome.

As a guard against the observed change being the result of ATP-induced reversed electron flow, combinations of redox mediating dyes were chosen such that there is interaction at both ends (cytochrome b and $c + c_1$) of the potential span which forms the energy conserving site. A number of criteria suggest that the observed ATP induced midpoint potential change was not due to reversed electron transfer: first, the measured midpoint potentials of one of the b cytochromes, cytochromes $c + c_1$, and cytochrome a were not measurably changed by ATP addition; second, a fivefold change in the concentration of added mediators had no effect on the measured midpoint potentials of the cytochrome in the presence or absence of ATP; and finally, preliminary studies on submitochondrial particles give qualitatively the same ATP induced midpoint changes as do studies on intact mitochondria (submitochondrial particles are thought to be "inside out" and have greatly altered rates of interaction with the individual redox mediators as compared to intact mitochondria).

Mitchell (28) has postulated that in intact mitochondria an ATP induced membrane potential will cause apparent changes in the oxidation-reduction midpoint potentials of any cytochrome which is not in direct electrical contact with the outer aqueous phase of the mitochondrial suspension. This postulate requires that inside the mitochondrial vesicle there is an increase in the number of negative charges relative to the positive charges and that the resulting negative electrical field will favor oxidation of any cytochrome on the inside of the mitochondrial inner membrance (thus the cytochrome would assume a more negative apparent midpoint potential). Hinkle and Mitchell (27) have reported data consistent with this hypothesis obtained using rat liver mitochondria. They reported that ATP induced apparent changes of the midpoint potentials of cytochrome a_3 (and to a lesser extent cytochrome a) but not in cytochromes $c + c_1$.

The results of Hinkle and Mitchell (27) differ from those of Wilson and Dutton (16, 17, 25) in that the latter workers found no effect of ATP on the cytochrome a midpoint potential. In addition, the ATP induced positive movement of the midpoint potential of cytochrome b_T is opposite in sign to that predicted by Mitchell (28). The preliminary observation that the ATP induced changes in the midpoint potentials of cytochromes b_T and a_3 in submitochondrial particles are qualitatively the same as those observed in intact mitochondria (Dutton, Wilson, and Lee; Wilson, Lindsay, and Dutton, unpublished results) suggest that they are not the result of imposed membrane potentials.

III. The Effect of the Mitochondrial "Energy State" on the *b* Cytochromes in Mitochondria

A. The Kinetics of the b Cytochromes

The kinetic behavior of cytochrome b in tightly coupled mitochondria has recently been reported to be dependent on the "energy state" of the mitochondria (25, 26). When pigeon heart mitochondria were made anaerobic and the oxidation-reduction components reduced by substrate, activation of respiration by rapid addition of oxygen yielded kinetics of oxidation of cytochrome b which were different for ATP treated mitochondria than for uncoupled mitochondria. In ATP treated mitochondria, the oxidation was biphasic. The fast phase (20–200 msec) consisted of the oxidation of a b cytochrome with an alpha absorption maximum at 560 nm, while the slow phase (200–2000 msec) consisted of the oxidation of a b cytochrome with an alpha maximum at approximately 564 nm. When uncoupler was added, both b cytochromes were oxidized in a fast monophasic transition which was complete in approximately 20 msec. The experiments of Chance et al. (25) also suggest that ATP does not induce a spectral shift in cytochrome b_T from a shorter wavelength to 564 nm since the sum of the absorbancies due to b and b_T in the presence of ATP forms a curve essentially identical to that of b plus b_T in the uncoupled state. The detection of a long wavelength b cytochrome in the presence of ATP has been reported by Slater et al. (29), but this was considered to be the result of a red shift in the spectrum of the *b* cytochrome (see Section IV, C).

B. The Spectral Properties of the b Cytochromes

Chance and co-workers (*30–32*) reported a novel form of cytochrome *b*, the properties of which were dependent on the energy state of the mitochondria. This *b* cytochrome had absorption maxima at 430 nm and 560 nm at room temperature and 425 nm and 555 nm at 77°K. It was postulated to be an unstable compound which was present in the "high energy" state of the mitochondria and submitochondrial particles but was not present in the "unenergized" state. Slater and co-workers (*29, 33, 34*) have reported an ATP induced red shift in the spectrum of reduced cytochrome *b* in mitochondrial and in submitochondrial particles. The absorption maximum of the alpha band of the cytochrome *b* reduced in the presence of ATP is 565 nm, a similar wavelength to that reported for the long wavelength, low potential cytochrome b_T (*17, 25, 26*). The difference between cytochrome b_{555} (*30–32*) and the cytochrome absorbing at 565 nm may be resolved by the report of Sato and Chance (personal communication) that at liquid nitrogen temperature spectra of the low potential cytochrome *b* reveal two maxima, one at 555 nm and the other at 563 nm. A disagreement exists in the literature over the possible existence of an actual ATP induced shift in the spectrum of reduced cytochrome *b*. The ATP induced changes reported by Wilson and Dutton (*17*), Dutton *et al.* (*14*), Chance and co-workers (*25, 26*), and Sato and Chance (personal communication) are consistent with an ATP induced reduction of a *b* cytochrome rather than a shift in the absorption maximum to a new wavelength (*29, 35*).

IV. The Properties of the Antimycin A Binding Site and Its Effects on the b Cytochromes of the Respiratory Chain

The study of the mitochondrial respiratory chain has been greatly aided by the discovery of compounds which specifically inhibit partial reactions within the respiratory chain. Strong and co-workers (*36–39*) discovered the inhibitory effect of antimycin A on biological oxidation, then purified the antibiotic and elucidated its chemical structure. Since the site of inhibition is between cytochrome *b* and cytochrome c_1 (*40, 6*) it has been extensively used to study this region of the respiratory chain. A number of unique properties have contributed to the interest in antimycin A, including (1) its high activity as an inhibitor, (2) its pronounced effect on the kinetic properties

and degree of reduction of cytochrome *b*, (3) its ability to modify the properties of cytochrome *b* including the spectral properties of the reduced cytochrome, and (4) the sigmoidicity of many of the titration curves for these
effects of antimycin A.

A. The Stoichiometry and Reversibility of the Antimycin A Binding

Antimycin A is bound very strongly and specifically to its inhibitory site
between cytochromes *b* and c_1 and the antimycin A required for 100%
effect determined as a ratio of the cytochrome $a + a_3$ or cytochrome *b*
concentration to the concentration of added antimycin A. This value has
been remarkably reproducible and is between 2 and 3 (*40–45*). The binding
site is, therefore, present at about one for each two *b* cytochromes (one for
each cytochrome c_1 molecule). Mention may be made that, according to
Takemori and King (*45a, b*), the ratio of antimycin A required for complete
inhibition is much lower in a particulate succinate cytochrome *c* reductase
than in submitochondrial and mitochondrial preparations. The coenzyme
Q content of this reductase has been found less than one-tenth of that in
mitochondria (*45b*). The interesting point lies in the fact that the degree of
antimycin inhibition is dependent on the exogenous coenzyme Q present and
the inhibition is completely reversed by added coenzyme Q (*45a, b*).

Although the antimycin A is very tightly bound, it has been reported that
the inhibition is reversible (*43, 44, 46*), and that its exchange between binding
sites is stimulated by some detergents (*43*). The antibiotic can be extracted
from its binding site by organic solvents and the extracted material retains
its inhibitory properties (*43*).

Antimycin A has a pronounced effect on the structural properties of
cytochrome *b* in the membrane. It prevents the change in spectrum and
oxidation-reduction potential induced by high concentrations of cholate (*42*)
and the separating the cytochrome $b - c_1$ complex into free cytochrome *b*
and cytochrome c_1 (*43*).

Evidence has been obtained by Yamashita and Racker (*47*) and Berden
and Slater (*45*) that when cytochromes *b* and c_1 are separated by detergent
treatment the antimycin A binding site is present in the cytochrome *b*
fraction and not in the cytochrome c_1 fraction.

B. The Effect of Antimycin A on the Thermodynamic and Kinetic Properties of the b Cytochromes

The addition of antimycin A to submitochondrial particles causes an
increase in the substrate reducible cytochrome *b* (*6, 42, 48, 49*) and dramati-
cally increases the rate of reduction of cytochrome *b* by substrates such as

succinate and NADH (6, 49). These properties are dependent on the type of mitochondrial preparation used. In intact mitochondria antimycin A does not increase the amount of substrate reducible cytochrome b (6), while in submitochondrial particles (ETP) the reported increase in reduction ranges from 5% (49) to 30% (41). In cholate clarified ETP the reported increase was 25 to 30% (42) while in submitochondrial particles prepared differently, such as the Keilin-Hartree heart muscle preparation, still other values are obtained (6, 48). The effect of antimycin A on the kinetics of cytochrome b reduction by substrate has been observed only in submitochondrial particles. The cytochrome b in intact mitochondria has been reported to be kinetically rapid (25, 50) while in submitochondrial particles it is much slower (6). The effect of antimycin A is to cause the behavior of the cytochrome b in submitochondrial particles to approach that of the cytochrome b in intact mitochondria.

The addition of antimycin A has no effect on the measured oxidation-reduction midpoint potential of the b cytochromes of either ETP or intact mitochondria (Dutton, Erecinska, and Wilson, unpublished results) in agreement with the observation of Pumphrey (42) that it does not change the amount of cytochrome b reducible by menadiol in ETP. In the experiments by Pumphrey (42) the addition of antimycin A increased the amount of cytochrome b reduced by substrate (NADH, succinate, or coenzyme Q_2) by 20% to 30%, but had no effect on the amount of cytochrome b reduced by a nonenzymatic reductant (menadiol). The interpretation of Chance (6) that in submitochondrial particles the antimycin A causes a kinetic activation of an inactive form of cytochrome b thus appears to be correct.

C. The Effect of Antimycin A on the Spectral Properties of the b Cytochromes

The cytochrome b in submitochondrial particles which is reduced by substrate in the presence of antimycin A has been reported to have an absorption spectrum which differs from that of the cytochrome b reduced in the absence of antimycin A(6, 29, 40, 42, 44, 48, 49), but considerable differences in opinion exist over the origin of this spectral change. Chance (6) concluded that the spectral effect resulted from a heterogeneity of the cytochrome b. The cytochrome b which could be reduced by substrate had an alpha absorption maximum at 562 nm, while the cytochrome b which was reduced by substrate in the presence of antimycin A had an alpha maximum at 564 nm. Pumphrey (42), using ETP in the presence of 1–2 mg cholate/mg protein, observed an antimycin induced shift in the absorption spectra of both oxidized cytochrome b and of dithionite reduced cytochrome b. The presence of cholate in her assay mixture makes these observed spectral shifts less interpretable since she also demonstrated that at higher cholate concentrations (10–12 mg/mg protein) the spectral and enzymatic properties of the cytochrome *b* were strongly modified. This modification by high cholate was prevented by prior addition of antimycin A, and the cholated induced spectral shift is opposite in direction to the antimycin A induced spectral shift. The antimycin could therefore be reversing the effect of cholate on the cytochrome *b* spectrum.

More recently Slater and co-workers (*33–35, 44, 45*) have undertaken an extensive evaluation of the effects of antimycin A on the properties of cytochrome *b*. They report that the antibiotic causes a red shift in the spectrum of one-half of the cytochrome *b* both in submitochondrial particles (*35, 44*) and in intact mitochondria (*33, 34*). The reported red shift is not observed in dithionite reduced particles and is thus not similar to that reported by Pumphrey (*42*). The method of measuring the red shift was to use the spectral wavelengths 566 nm minus 560 nm in a dual wavelength spectrophotometer. Spectra of the red shift were obtained using 560 nm as a reference wavelength and show an absorption band with a maximum at 565 nm. A comparison of these spectra with those obtained for the fast and slow phases of cytochrome *b* oxidation is pigeon heart mitochondria (*25, 26*) indicates that an identical spectrum is obtained on oxidation and reduction of the cytochrome *b* which is kinetically slow in "energized" mitochondria (no antimycin A). This cytochrome *b* (b_T) has an alpha absorption maximum at 564 nm and an oxidation-reduction midpoint potential of +245 mV at pH 7.0 in the "energized" form (*17, 25*).

An attractive unification of the available data would be possible if the red shift reported by Slater and co-workers (*35, 44*) were in fact primarily a reduction of this long wavelength cytochrome b_T together with an antimycin induced change in the spectrum of one of the three *b* cytochromes present in beef heart submitochondrial particles. Support for such an interpretation arises from examination of the oxidation-reduction potentials of the *b* cytochromes using various wavelength pairs for measurement. The fraction of the total absorbance change contributed by the low and high potential components was wavelength dependent (Wilson and Dutton, unpublished observations). A detailed study was made of this wavelength dependence of the cytochrome *b* in preparation of complex I–III from beef heart prepared by Hatefi (Erecinska and Wyer, unpublished results). In titrations at pH 7.2 (wavelength pair 560 nm minus 575 nm) the absorbance change was 77% due to high potential cytochromes *b* (2 components, half-reduction potential +70 mV) and 23% due to low potential (midpoint potential −60 mV) cytochrome *b* while at 566 nm minus 575 nm the respective contributions were 58% and 42%. The half-reduction potentials of the two components could be measured directly at the wavelength pair 560 nm minus 566 nm. In this case the absorbance change due to the high potential component was positive and that of the low potential component was negative resulting in a very small net absorbance change on proceeding from the condition where both components were fully oxidized to the condition where both components were fully reduced.

Antimycin A did induce a spectral change in the high potential cytochromes b but had no effect on the midpoint potential of either component. When the wavelength pair 565.5 nm minus 558 nm was used, antimycin A decreased by one-half the absorbance change due to the high potential components but had no effect on the absorbance change due to the low potential component. The use of other wavelength pairs gave results consistent with the interpretation that antimycin A induced "red shift" of the absorption spectrum of at least one of the reduced high potential cytochromes b but had no effect on the spectrum of reduced low potential cytochrome b. The spectral shift observed by this experimental technique is directly analogous to that reported by Pumphrey (42) because it occurs in chemically reduced cytochrome b. This is in some disagreement with the red shift, reported by Slater and co-workers (33, 35, 44), which could not be induced in chemically reduced cytochrome b.

D. The Sigmoidicity of the Antimycin A Titration Curves

The reported titration curves for the inhibition of mitochondrial respiration by antimycin A are reported to be sigmoidal in shape (34, 35, 40–42, 44, 46, 51). In addition, sigmoid titration curves were obtained for the antimycin A induced reduction of cytochrome b by substrate (42) and for the red shift in the spectrum of cytochrome b as observed by Slater and co-workers (35, 44) but not in the spectral shift observed by Pumphrey (42) in dithionite reduced cytochrome b. It has been reported that the titration curves are sigmoid in "unenergized" mitochondria but not in "energized" mitochondria (34). In Pumphrey's systems (42), however, a more complex behavior is reported in the sense that in the presence of 1.1% cholate the titration of the reducibility of cytochrome b by substrate and the inhibition of DPNH oxidase activity were sigmoid while the titration curve for the red shift in the spectrum of dithionite reduced cytochrome b was hyperbolic.

Several attempts have been made to explain the sigmoid shape of the titration curves (41, 44, 46, 52). Bryla and co-workers (44) have proposed an allosteric model to explain the titration behavior of antimycin A. Such an allosteric model is interesting in the sense that the minimum unit which could show cooperative behavior (with sigmoid titration curves) is one containing at least two antimycin A binding sites. Such a unit would contain at least four molecules of cytochrome *b* and would thus represent an inter-chain cooperativity rather than a cooperativity between the two *b* cytochromes of a single respiratory chain (intrachain cooperativity). The long range forces involved in such allosteric interaction could also be responsible for major changes in membrane conformation and permeability.

V. Cytochrome P_{450}

A. General Properties and Function of Cytochrome P_{450}

The cytochrome P_{450} is included in this discussion of the *b* cytochromes although it does not have spectral properties normally associated with *b* cytochromes. Some justification for its inclusion can be derived from the observation that its prosthetic group is protoheme IX (53), but more properly it should be classified separately by function. This heme protein has been shown to be present in animal, plant and microbial cells [for recent reviews see, e.g., Refs. (54, 55)] and is an active component in cellular steroid hydroxyl-ations, oxidative demethylations, and drug metabolism (56–60). In general it appears to be a mixed function oxidase (61) and catalyzes reactions of the general type:

$$H^+ + AH_2 + TPNH + O_2 \rightarrow AHOH + TPN^+ + H_2O$$

in which one of the oxygen atoms of molecular oxygen is incorporated into the hydroxylated product. The activity is inhibited by CO and the CO inhibition is reversed by light (56, 62). The action spectrum for the reversal by light shows a maximum at 450 nm and clearly identifies the cytochrome P_{450} as the species inhibited by CO.

Various substrates and other compounds bind to the cytochrome P_{450} and cause changes in the optical (60, 63) and electron paramagnetic resonance (63–65) spectra of the cytochrome P_{450}. The substrate induced changes in the light absorption spectrum of oxidized cytochrome P_{450} have been classified as Type I (a shift in the Soret absorption maximum to shorter wavelengths) and Type II (a shift in the Soret absorption maximum to longer wavelengths). Estabrook and co-workers (63, 64) have reported a correlation between the effect of various compounds on the light absorption and EPR spectra of cytochrome P_{450}.

The rate and extent of reduction of cytochrome P_{450} by TPNH in the presence of CO have been found to be greatly increased by the presence of some substrates (66). A direct correlation was obtained among the increase in reduction rate, the increase in the extent of reduction, the increase in rate of substrate oxidation, and the increase in P_{450} absorbance at 390 nm as these parameters were measured as a function of substrate concentration (Narasimhulu, unpublished results).

Cytochrome P_{450} is identical to "reticulochrome" (67) and to Fe_x, a low spin ferric hemoprotein whose EPR spectrum was detected and characterized by Mason and co-workers (68, 69).

B. The Oxidation-Reduction Midpoint Potential of Cytochrome P_{450}

Waterman and Mason (70) first reported an oxidation-reduction midpoint potential for cytochrome P_{450}. Their value of -410 mV was obtained using preparations of endoplasmic reticulum from the livers of rats which had been treated with phenobarbitol to induce high concentrations of P_{450}. More recently, reports have been made of a value of -400 mV for the cytochrome P_{450} of beef adrenal mitochondria (71, 72) and a value of -335 mV for the cytochrome P_{450} of rat liver microsomes (73). Waterman and Mason (70) were unable to determine the number of electrons accepted per P_{450} molecule from their titrations, but both Schleyer and co-workers (71, 72) and Cohen and Wilson (73) reported that this cytochrome behaved as a one-electron acceptor in their titrations.

The very negative oxidation-reduction midpoint potential of cytochrome P_{450} is an important consideration in both the mechanism of the hydroxylation reaction and in the cellular control of drug metabolism. Mechanistically it raises the possibility of participation of the O_2^- radical in the hydroxylation (70), an interesting and useful idea. In terms of cellular control of drug metabolism, however, it is important to know the value of the midpoint potential relative to the oxidation-reduction potential of $NADP^+/NADPH + H^+$ couple which is the reductant for cytochrome P_{450}. In perfused livers the NADP/NADPH couple has an oxidation-reduction potential of between -345 mV and -390 mV (74, 75). In the perfused liver, therefore, NADPH could effectively reduce a cytochrome P_{450} having a midpoint potential of -335 mV (73) but not a cytochrome P_{450} with a midpoint potential of -410 mV (70). It is possible that a control mechanism exists in which the substrate interacts with cytochrome P_{450} to form a complex in which the oxidation-reduction midpoint potential is shifted to more positive values. There is no convincing evidence for such a control mechanism, and preliminary experiments show that if substrates affect the midpoint potential of cytochrome P_{450} in rat liver microsomes the effect is small (73).

C. The Heterogeneity of Cytochrome P_{450}

Several lines of evidence suggest that microsomal cytochrome P_{450} is heterogeneous. These include reports that the reduction of cytochrome P_{450}

in the presence of CO is biphasic (*66*); that rabbit liver microsomes contain two cytochromes, P_{450} and P_{446} (*76*); that bovine adrenal microsomes contain two cytochrome P_{450} entities (*77*); that unsaturated fatty acids inhibit Type I hydroxylations but not Type II hydroxylations (*78*); and that the curves for the oxidation-reduction potential dependence of the reduction of cytochrome P_{450} in the presence of CO deviates from the theoretical curve for a single component (*73*). This heterogeneity could explain the different oxidation-reduction midpoint potentials reported (see previous section) inasmuch as in each case a different measuring technique (or material) was used.

VI. Cytochrome b_5

Cytochrome b_5 is a pigment found in microsomes in high concentrations (see, e.g., *69, 79*). It is readily reduced by either NADP or NADPH but has no proven function in the cell. This cytochrome has been solubilized, purified, and its properties extensively studied [see Part A, this chapter; extensive reviews are provided by Strittmatter (*80, 81*)]. The oxidation-reduction midpoint potential of cytochrome b_5 in the microsomal membrane has been reported to be -130 mV at pH 7.0 (*79, 82, 83*) and $+7$ mV at pH 7.4 (*73*). These values differ by more than 100 mV, and the reason for the difference is not yet clear.

VII. The *b* Cytochromes of Chloroplasts

The oxidation-reduction midpoint potentials of the two important *b* cytochromes of the photosynthetic apparatus of plant chloroplasts are shown in Table 2. There are large variations in the reported values.

A. Cytochrome b_{563}

The difference (0 mV and -180 mV) in the reported midpoint potentials of cytochrome b_{563} (formerly designated cytochrome b_6) cited by Hill and Bendall (*84*) and Fan Cramer (*85*), respectively, may result, as Fan and Cramer (*85*) suggest, from the incomplete spectral separation of cytochromes b_{563} and b_{559} in the determination of Hill and Bendall (*84*). Their value lies in between those of cytochromes b_{563} and b_{559} reported by Fan and Cramer (*85*). The experimental approach of Fan and Cramer (*85*) (that of simultaneous readout of oxidation-reduction potential and absorbance change) is not sensitive, as was that of Hill and Bendall (*84*), to an incomplete spectral separation of the cytochromes measured. An analogous situation which supports the suggestion, and hence the value reported by Fan and Cramer (85), has been encountered with b cytochromes of beef heart mitochondria (14) (see Section I,A on mammalian mitochondria). The midpoint potential of cytochrome b_{563} has been reported as being independent of pH from values of 6 to 8 (85).

Cytochrome b_{563} has been considered, from its rapid reduction by "photosystem I" light and its light-induced responses in the presence of inhibitors, as not being involved in the transfer of electrons from "photosystem II" (geared to the oxidation of water) to photosystem I (geared to ferredoxin and NADP reduction) (86, 87), but rather to be concerned in the cyclic electron transfer system from the primary electron acceptor of photosystem I back to its reaction center chlorophyll P_{700} (see also 85, 88, 89). Its close association with photosystem I is supported by the fractionation studies of Boardman and Anderson (90), Vernon et al. (91), and Hind and Nakatani (92).

B. Cytochrome b_{559}

The oxidation-reduction midpoint potential of +80 mV reported by Fan and Cramer (85) for cytochrome b_{559} in the spinach chloroplast membrane agrees fairly well with that reported by Hind and Nakatani (92), who used a detergent solubilized preparation. Hind and Nakatani (92) noted that an acetone extraction of this detergent preparation causes the cytochrome to assume a midpoint potential about 100 mV more negative. However, in contrast to these values for cytochrome b_{559}, oxidation-reduction midpoint potentials of +320 mV and above have also been reported (89, 93, 94) (see Table 2). These differences in the measured properties of cytochrome b_{559} are, at present, hindering progress in understanding its position, from a thermodynamic standpoint, in the sequences of dark electron transport induced by the multiple light reactions which exist in the chloroplast. Of relevance to this, however, may be the observation of Bendall (93) who reported that besides the presence of a high potential cytochrome b_{559}, there also appeared to be a spectral component at 559 nm which was not reducible with ascorbate. This would suggest the possibility of there being two b cytochromes with similar spectral properties but differing oxidation-reduction potentials. The pH dependency of the midpoint potential of cytochrome b_{559} as measured by Fan and Cramer (85) was −40 mV per pH unit from pH 6 to 8.

The closer association of cytochrome b_{559} with photosystem II than with photosystem I seems clear from fractionation studies (90, 91, 95). From its reduction by red light (photosystem II) and reoxidation by far-red light (photosystem I) in uncoupled spinach chloroplasts (*87, 96*) and in the alga *C. reinhardi* (*97*) which was devoid of cytochrome b_{563}, cytochrome b_{559} would appear to be important in the transfer of electrons from photosystem II to photosystem I. Photosystem I light was reported to be effective for the oxidation of cytochrome b_{559} in the presence of 10 μM 3(3,4-dichlorophenyl)-1,1-dimethylurea (DCMU) (*87, 97*) and of antimycin A (*96*). In conflict with the results of Cramer and Butler (*87*) and Levine and Gorman (*97*) are the findings of Knaff and Arnon (*89*) that cytochrome b_{559} oxidation is impeded with 1 μM DCMU. Further, Knaff and Arnon (*98*) have shown preferential cytochrome b_{559} oxidation by photosystem II light in Tris-treated chloroplasts.

Fan and Cramer (*85*) consider cytochrome b_{559} (their value for the mid-point potential is +80 mV; see Table 2) to be a suitable candidate for the electron acceptor from the primary electron acceptor of photosystem II, midpoint potential about −50 mV. They suggest that it may form the low potential end of a noncyclic phosphorylation site between cytochromes b_{559} and f ($\Delta E_m \approx 270$ mV). Knaff and Arnon (*89*), however, working with subchloroplast preparations of spinach which are enriched in either photosystem I or II activity, have presented evidence which indicates that photosystem II may actually be two light reactions (IIb, IIa) linked in series which can generate reduced ferredoxin and NADPH from water independently of photosystem I (*89, 99*). They consider cytochrome b_{559} (their value for the midpoint potential is +330 mV; see Table 2) to act as an electron carrier between the two-part light reactions (IIb for oxidation of water and IIa for ferredoxin reduction) of photosystem II, and to function as the higher potential end of a phosphorylation site, with plastoquinone (midpoint potential about 0 mV) functioning as the lower potential end. Plastocyanin has been shown necessary in the photosystem II-enriched subchloroplast preparations of Arnon and co-workers (*99, 100*) for both the photooxidation of cytochrome b_{559} and for NADP reduction from water.

Clearly the major experimental differences concerning cytochrome b_{559} and other components involved in electron transport and energy conservation can only be resolved by further careful studies on both intact chloroplasts and the subchloroplast preparations derived therefrom.

C. Reactions at Liquid Nitrogen Temperatures

Chance and Bonner (*101*) described in spinach chloroplasts the non-reversible photooxidation at liquid nitrogen temperatures (77°K) of what they considered to be cytochrome *f*. Floyd *et al.* (*102*) have pointed out that the action spectra of the reaction at 77°K, obtained by Chance and Bonner (*101*), however, is typical of photosystem II, a result which appears anomalous since at room temperature cytochrome f appears to be oxidized by photosystem I. (Furthermore, fractionation studies suggest that cytochrome f is associated with the components of this light reaction.) Knaff and Arnon (*98*) have confirmed that cytochrome f in chloroplasts from spinach and romaine lettuce, and leaves of *Amaranthus*, undergoes photooxidation at 77°K by photosystem II light (664 nm) but not by photosystem I light (714 nm). They have reported further that in these chloroplasts the cytochrome b_{559} is also nonreversibly photooxidized at these low temperatures. Their reported light-induced spectrum is approximately that expected from the known difference spectra obtained at 77°K for the two cytochromes by chemical oxidation and reduction (*90, 103*); the absorbance minimum (oxidized minus reduced) of cytochrome b_{559} is at about 556 nm at liquid nitrogen temperatures. Floyd *et al.* (*102*) have presented further evidence for cytochrome b_{559} photooxidation at 77°K in Swiss chard and spinach leaves, and spinach chloroplasts. The photooxidation of cytochrome f was not clearly apparent from their spectra of the light-induced changes at 77°K. However, Floyd *et al.* (*102*) verified that the light-induced absorbance minimum at 556 nm in the 77°K spectra was due to cytochrome b_{559} by demonstrating that at a temperature of 194°K the minimum was approximately 558 nm. Further evidence that the primary oxidant of cytochrome b_{559} was the oxidized chlorophyll (P_{680}^+) of photosystem II was obtained using 20-nsec pulses of laser light. In these experiments, the kinetics of cytochrome b_{559} oxidation (half-time 4.6 msec) were found to closely match one phase of the biphasic re-reduction of P_{680}^+ (photosystem II) (half-time 4.5 sec). The nature of the physiological donor for the oxidized chlorophyll of photosystem II is of major importance in our understanding of photosynthesis, and for this reason the relevance of the photosystem II light-induced oxidation of cytochromes b_{559} and f at 77°K has been the subject of considerable discussion. Since the interpretations are necessarily dependent on working hypotheses of the discussant the readers are referred to the recent literature (*98, 102*) for these interpretations.

Employing a method (*15*) for determination of the oxidation-reduction midpoint potential of photooxidizable components at 77°K, Dutton, Larkum, and Bonner (unpublished results) determined the cytochrome species (552–540 nm) which undergoes photooxidation at 77°K in intact spinach chloroplasts to have a midpoint potential of +360 mV ($n = 1$) at pH 7.2. The measured midpoint could be a mixture of both cytochromes f and b_{559} if indeed cytochrome b_{559} has a high midpoint potential (see Table 2) similar to that of cytochrome f [midpoint potential of about +350 mV at pH 7 (*93, 104*)]. However, the photooxidation of a cytochrome b_{559} with a midpoint potential of between $+50$ and $+100$ mV was not detected at the wavelengths used, since there was no further increase of light-induced absorbance changes as the oxidation-reduction potential of the chloroplasts was lowered from $+250$ to -150 mV.

VIII. The Oxidation-Reduction Midpoint Potentials of the *b* Cytochromes of *Rhodopseudomonas spheroides* and *Rhodospirillum rubrum*

The oxidation-reduction midpoint potentials of the *b* cytochromes, measured in chromatophore particles using the method of simultaneous oxidation-reduction potential measurement and optical assay at 560–540 nm (Dutton and Jackson, unpublished results), are shown in Table 1(b). In *Rps. spheroides* and *R. rubrum* the b_{562} cytochromes have midpoint potentials of $+65$ mV and -100 mV, respectively. These values are appropriately more positive than the value generally accepted for the respective primary electron acceptors from bacteriochlorophyll. The midpoint potential of the primary acceptor has been reported in *Rps. spheroides* to be approximately -50 mV (*105, 106*) and in *R. rubrum* is -145 mV (*106*), although in *R. rubrum* other values, -22 mV and -60 mV, obtained by other techniques have been reported (*107*). In *R. rubrum* the cytochrome with the midpoint potential $+10$ mV (Table 2) is probably *Rhodospirillum* heme protein (RHP) which has a midpoint potential in the isolated form of -8 mV at pH 7.0 (*108*). HRP has also been called cytochromoid *c*, cytochrome *cc'*, or cytochrome c_{428}. Fowler and Sybesma (*109*) determined the *in situ* midpoint potential of cytochrome c_{428} to be -10 mV at pH 7.0, which is in close agreement with the quoted values. However, they reported its oxidation and reduction to involve the transfer of four electrons ($n = 4$), which contrasts with the n value of one found for the course of oxidation and reduction for all of the other cytochromes of *Rps. spheroides* and *R. rubrum* listed in Table 1(b). Both *Rps. spheroides* and *R. rubrum* contain cytochromes, which also could be of the *b* group, with midpoint potentials of approximately $+160$ and $+175$ mV.

REFERENCES

1. M. Klingenberg, in *Biological Oxidations* (T. P. Singer, ed.), Wiley (Interscience), New York, 1968, p. 3.
2. E. C. Slater, *Comprehensive Biochem.*, **14**, 327 (1966).
3. B. Chance, W. D. Bonner, Jr., and B. T. Storey, *Ann. Rev. Plant Physiol.*, **19**, 295 (1968).
4. E. G. Ball, *Biochem. J.*, **295**, 262 (1938).

5. L. Sekuzu and K. Okunuki, *J. Biochem.*, **43**, 107 (1956).
6. B. Chance, *J. Biol. Chem.*, **233**, 735 (1958).
7. D. Feldman and W. W. Wainio, *J. Biol. Chem.*, **235**, 3635 (1960).
8. F. A. Holton and J. P. Colpa-Boonstra, *Biochem. J.*, **76**, 179 (1960).
9. R. Hill and T. E. King, unpublished experiments for the Keilin-Hartree preparation; the potentials were determined at Oregon State University, Corvallis (1961).
10. J. P. Straub and J. P. Colpa-Boonstra, *Biochim. Biophys. Acta*, **60**, 650 (1962).
11. R. Goldberger, A. Pumphrey, and A. Smith, *Biochim. Biophys. Acta*, **58**, 367 (1962).
12. J. S. Rieske, *Federation Proc.* **28**, 471 (1969).
13. P. F. Urban and M. Klingenberg, *Eur. J. Biochem.*, **9**, 519 (1969).
14. P. L. Dutton, D. F. Wilson, and C. P. Lee, *Biochemistry*, **9**, 5077 (1970).
15. P. L. Dutton, *Biochim. Biophys. Acta*, **226**, 63 (1971).
16. D. F. Wilson and P. L. Dutton, *Arch. Biochem. Biophys.*, **136**, 583 (1970).
17. D. F. Wilson and P. L. Dutton, *Biochem. Biophys. Res. Communs.*, **39**, 59 (1970).
18. C. Lance and W. D. Bonner, *Plant Physiol.*, **43**, 756 (1968).
19. P. L. Dutton and B. T. Storey, *Plant Physiol.*, **47**, 228 (1971).
20. R. Hill and R. Scarisbrick, *New Phytol.*, **51**, 98 (1951).
21. H. Shichi and D. P. Hackett, *J. Biol. Chem.*, **237**, 2955 (1962).
22. H. Shichi and D. P. Hackett, *J. Biol. Chem.*, *Japan*, **59**, 84 (1966).
23. H. Shichi, D. P. Hackett, and G. Funatsu, *J. Biol. Chem.*, **238**, 1156 (1963).
24. C. W. D. Moore, Ph.D. Dissertation, Univ. of Cambridge, Cambridge, England (1967).
25. B. Chance, D. F. Wilson, P. L. Dutton, and M. Erecinska, *Proc. Natl. Acad. Sci., U.S.*, **66**, 1175 (1970).
26. M. Erecinska and B. Chance, in *Colloquium on Bioenergetics*, Pugnochioso, Italy, 1970, 1971.
27. P. Hinkle and P. Mitchell, *J. Bioenergetics*, **1**, 45 (1970).
28. P. Mitchell, in *Chemiosmotic Coupling and Energy Transduction*, Glynn Research Ltd., Bodmin, 1968.
29. E. C. Slater, C. P. Lee, J. A. Berden, and H. J. Wegdam, *Nature*, **226**, 1246 (1970).
30. B. Chance and B. Schoener, *J. Biol. Chem.*, **241**, 4567 (1966).
31. B. Chance and B. Schoener, *J. Biol. Chem.*, **241**, 4577 (1966).
32. B. Chance, C. P. Lee, and B. Schoener, *J. Biol. Chem.*, **241**, 4574 (1966).
33. H. J. Wegdam, J. A. Berden, and E. C. Slater, *Biochim. Biophys. Acta*, **223**, 365 (1970).
34. W. D. Bonner and E. C. Slater, *Biochim. Biophys. Acta*, **223**, 349 (1970).
35. E. C. Slater, C. P. Lee, J. A. Berden, and H. J. Wegdam, *Biochim. Biophys. Acta*, **223**, 354 (1970).
36. B. R. Dunshee, C. Leben, G. W. Keitt, and F. M. Strong, *J. Am. Chem. Soc.*, **71**, 2436 (1949).
37. K. Ahmad, H. G. Schneider, and F. M. Strong, *Arch. Biochem.*, **28**, 281 (1950).
38. F. M. Strong, J. P. Dickie, M. E. Loomas, E. E. van Tamelen, and R. S. Dewey, *J. Am. Chem. Soc.*, **82**, 1513 (1960).
39. E. E. van Tamelen, J. P. Dickie, M. E. Loomas, R. S. Dewey, and F. M. Strong, *J. Am. Chem. Soc.*, **83**, 1639 (1961).
40. B. Chance, *Nature*, **169**, 215 (1952).
41. R. W. Estabrook, *Biochim. Biophys. Acta*, **61**, 236 (1962).
42. A. M. Pumphrey, *J. Biol. Chem.*, **237**, 238 (1962).
43. J. S. Rieske, S. H. Lipton, H. Baum and, H. I. Silman, *J. Biol. Chem.*, **242**, 4888 (1967).

44. J. Bryla, Z. Kaniuga, and E. C. Slater, *Biochim. Biophys. Acta*, **189,** 317 (1969).

45. J. A. Berden and E. C. Slater, *Biochim. Biophys. Acta*, **216,** 237 (1970).

45a. S. Takemori and T. E. King, *Science*, **144,** 852 (1964).

45b. S. Takemori and T. E. King, *J. Biol. Chem.*, **239,** 3546 (1964).

46. M. B. Thorn, *Biochem. J.*, **63,** 420 (1956).

47. S. Yamashita and E. Racker, *J. Biol. Chem.*, **244,** 1220 (1969).

48. E. C. Slater and J. P. Colpa-Boonstra, in *Haematin Enzymes* (J. E. Falk, R. Lemberg, and R. K. Morton, eds.), Vol. 2, Pergamon, London, 1961, p. 575.

49. B. T. Storey, *Arch. Biochem. Biophys.*, **121,** 271 (1967).

50. B. Chance and G. R. Williams, *J. Biol. Chem.*, **217,** 429 (1955).

51. V. R. Potter and A. E. Reif, *J. Biol. Chem.*, **194,** 287 (1952).

52. R. W. Estabrook and A. Kravitz, *Biochim. Biophys. Acta*, **60,** 249 (1962).

53. T. Omura and R. Sato, *J. Biol. Chem.*, **239,** 2370 (1964).

54. O. Hayaishi, *Ann. Rev. Biochem.*, **38,** 21 (1969).

55. L. Ernster and S. Orrenius, *Federation Proc.*, **24,** 1190 (1965).

56. R. W. Estabrook, D. Y. Cooper, and O. Rosenthal, *Biochem. Z.*, **338,** 741 (1963).

57. A. H. Conney, *Pharmacol. Rev.*, **19,** 317 (1967).

58. J. R. Gillette, *Adv. Pharmacol.*, **4,** 219 (1967).

59. J. L. Holtzman, T. E. Gram, P. L. Gigon, and J. R. Gillette, *Biochem. J.*, **110,** 407 (1968).

60. S. Narasimhulu, D. Y. Cooper, and O. Rosenthal, *Life Sciences*, **4,** 2101 (1965).

61. H. S. Mason, *Adv. Enzymol.*, **19,** 79 (1957).

62. D. Y. Cooper, S. Narasimhulu, O. Rosenthal, and R. W. Estabrook, *Science*, **147,** 400 (1965).

63. R. W. Estabrook, J. B. Schenkman, W. Cammer, D. Y. Cooper, S. Narasimhulu, and O. Rosenthal, in *Hemes and Hemoproteins* (B. Chance, R. W. Estabrook, and T. Yonetani, eds.), Academic, New York, 1966, p. 511.

64. W. Cammer, J. B. Schenkman, and R. W. Estabrook, *Biochem. Biophys. Res. Communs.*, **23,** 264 (1966).

65. R. Tsai, C. A. Yu, I. C. Gunsalus, J. Peisach, W. Blumberg, W. H. Orme-Johnson, and H. Beinert, *Proc. Natl. Acad. Sci., U.S.*, **66,** 1157 (1970).

66. P. L. Gigon, T. E. Gram, and J. R. Gillette, *Mol. Pharmacol.*, **5,** 109 (1969).

67. R. Sato, T. Omura, and H. Nishiboyashi, in *Oxidases and Related Redox Systems* (T. E. King, H. S. Mason, and M. Morrison, eds.), Vol. II, Wiley, New York, 1965, p. 861.

68. Y. Hashimoto, T. Yamano, and H. S. Mason, *J. Biol. Chem.*, **237,** PC3843 (1962).

69. H. S. Mason, T. Yamano, J. C. North, Y. Hashimoto, and P. Sakagishi, in *Oxidases and Related Redox Systems* (T. E. King, H. S. Mason, and M. Morrison, eds.), Vol. II, Wiley, New York, 1965, p. 879.

70. M. R. Waterman and H. S. Mason, *Biochem. Biophys. Res. Communs.*, **39,** 450 (1970).

71. H. Schleyer, D. Y. Cooper, and O. Rosenthal, *Federation Proc. Abstr.*, 3871 (1970).

72. D. Y. Cooper, H. Schleyer, and O. Rosenthal, *Ann. N.Y. Acad. Sci.*, **174,** 205 (1970).

73. B. S. Cohen and D. F. Wilson, *Arch. Biochem. Biophys.* (1971), in press.

74. J. R. Williamson, in *The Energy Level and Metabolic Control in Mitochondria* (S. Papa, J. M. Tager, E. Quagliariello, and E. C. Slater, eds.), Adriatica Editrice, Bari, Italy, 1969.

75. R. L. Veech, L. V. Eggleston, and H. A. Krebs, *Biochem. J.*, **115,** 609 (1969).

76. A. Hildebrandt, H. Remmer, and R. W. Estabrook, *Biochem. Biophys. Res. Communs.* **30,** 607 (1968).

77. M. L. Sweat, R. B. Young, and M. J. Bryson, *Biochim. Biophys. Acta*, **223,** 105 (1970).

78. R. P. DiAugustine and J. R. Fouts, *Biochem. J.*, **115**, 547 (1969).
79. C. F. Strittmatter and E. G. Ball, *Proc. Natl. Acad. Sci., U.S.*, **38**, 19 (1952).
80. P. Strittmatter, in *The Enzymes* (P. D. Boyer, H. Lardy, and K. Myrback, eds.), Vol. 8, Academic, New York, 1963, p. 113.
81. P. Strittmatter, in *Biological Oxidations* (T. P. Singer, ed.), Wiley (Interscience), New York, 1968, p. 171.
82. H. Yoshikawa, *J. Biochem., Tokyo*, **38**, 1 (1951).
83. Y. Kawai, Y. Yoneyama, and H. Yashikawa, *Biochim. Biophys. Acta*, **67**, 522 (1963).
84. R. Hill and D. S. Bendall, in *Biochemistry of Chloroplasts* (T. W. Goodwin, ed.), Vol. 2, Academic, London, 1967, p. 559.
85. H. N. Fan and W. A. Cramer, *Biochim. Biophys. Acta*, **216**, 200 (1970).
86. G. Hind and J. R. Olsen, *Brookhaven Symp. Biol.*, **19**, 188 (1966).
87. W. A. Cramer and W. L. Butler, *Biochim. Biophys. Acta*, **143**, 332 (1967).
88. K. Tagawa, H. Y. Tsujimoto, and D. I. Arnon, *Proc. Natl. Acad. Sci., U.S.*, **49**, 567 (1963).
89. D. B. Knaff and D. I. Arnon, *Proc. Natl. Acad. Sci., U.S.*, **64**, 715 (1969).
90. N. K. Boardman and J. M. Anderson, *Biochim. Biophys. Acta*, **143**, 187 (1967).
91. P. Vernon, B. Ke, H. H. Mollenhauer, and E. R. Shaw, in *Progress in Photosynthesis Research* (H. Metzner, ed.), Vol. 1, Intern. Union Biol. Sci. Tubingen, 1969, p. 137.
92. G. Hind and H. Y. Nakatani, *Biochim. Biophys. Acta*, **216**, 223 (1970).
93. D. S. Bendall, *Biochem. J.*, **109**, 46P (1968).
94. I. Ikegami, S. Katoh, and A. Takamiya, *Biochim. Biophys. Acta*, **162**, 604 (1968).
95. D. I. Arnon, H. Y. Tsujimoto, B. D. McSwain, and R. K. Chain, in *Comparative Biochemistry and Biophysics of Photosynthesis* (K. Shibata *et al.*, eds.), Univ. of Tokyo Press, Tokyo and University Part Press, State College, Pa., 1968, p. 113.
96. G. Hind, *Photochem. Photobiol.*, **7**, 369 (1968).
97. R. P. Levine and D. S. Gorman, *Plant Physiol.*, **41**, 1293 (1966).
98. D. B. Knaff and D. I. Arnon, *Proc. Natl. Acad. Sci., U.S.*, **63**, 956 (1969).
99. D. I. Arnon, R. K. Chain, B. D. McSwain, H. Y. Tsujimoto, and D. B. Knaff, *Proc. Natl. Acad. Sci., U.S.*, **67**, 1404 (1970).
100. D. B. Knaff and D. I. Arnon, *Biochim. Biophys. Acta*, **223**, 201 (1970).
101. B. Chance and W. D. Bonner, in *Photosynthetic Mechanisms in Green Plants*, NAS-NCR Publication, **1145**, 1963, p. 66.
102. R. A. Floyd, B. Chance, and D. DeVault, *Biochim. Biophys. Acta*, **226**, 103 (1971).
103. R. Hill and W. D. Bonner, in *Light and Life* (W. D. McElroy and B. Glass, eds.), Johns Hopkins Press, Baltimore, Maryland, 1961, p. 424.
104. H. E. Davenport and R. Hill, *Proc. Roy. Soc., London*, **B139**, 327 (1952).
105. D. W. Reed, K. L. Zankel, and R. K. Clayton, *Proc. Natl. Acad. Sci., U.S.*, **63**, 42 (1969).
106. W. A. Cramer, *Biochim. Biophys. Acta*, **189**, 54 (1969).
107. P. A. Loach, *Biochemistry*, **5**, 592 (1966).
108. R. G. Bartsch and M. D. Kamen, *J. Biol. Chem.*, **230**, 41 (1958).
109. C. F. Fowler and C. Sybesma, *Biochim. Biophys. Acta*, **197**, 276 (1970).

CHAPTER 7

CYTOCHROMES c^*

Martin D. Kamen, Karl M. Dus,† Torgeir Flatmark,‡
and Henk de Klerk§

UNIVERSITY OF CALIFORNIA AT SAN DIEGO
LA JOLLA, CALIFORNIA
AND
LABORATOIRE DE PHOTOSYNTHÈSE
CENTRE NATIONAL DE LA RECHERCHE SCIENTIFIQUE
GIF-SUR-YVETTE, FRANCE

* Authors of various sections are indicated in the Table of Contents. Some of the material presented is based on work supported by grants-in-aid from the National Institutes of Health (C-5592 and HD-01262) and the National Science Foundation (GB-2892), and by a Research Award from the C. F. Kettering Research Foundation to the senior author. See also cross references in articles by M. D. Kamen and T. Horio in *Annual Reviews of Biochemistry* (Vol. 39, 1970) and *Annual Reviews of Microbiology* (Vol. 24, 1970).

† Present address: Department of Chemistry, University of Illinois, Urbana, Illinois.

‡ Public Health Service International Fellow; present address: Department of Biochemistry, University of Bergen, Bergen, Norway.

§ Present address: Shell Oil Company, Amsterdam, The Netherlands.

I. Introduction

Studies of cytochrome c and systems associated with it have provided, and continue to provide, insights into the nature of many basic problems of biochemistry, such as clarification at the molecular level of relations between structure and function in proteins, and conversion and storage of chemical free energy in biological oxidations. Recently, two uniquely relevant publications have appeared that can be recommended as aids in deepening the reader's comprehension of the manner in which researches on cytochromes, and on cytochrome c in particular, have been intertwined with the development of modern biochemistry. The first of these is David Keilin's history of his researches on respiratory pigments (*1*). This volume, by the discoverer of the cytochrome system and prime mover in the early researches on its nature and function, has already taken an honored place among the classics of biological literature. The second (*2*) is an extensive and learned review, entitled "Cytochrome c" by E. Margoliash and A. Schejter. In addition to these, we include references to the classic text by R. Lemberg and J. W. Legge (*3*), a very useful and informative review by the late R. K. Morton (*4*), and two volumes that describe contributions to two symposia concerned especially with haem proteins (*5, 6*). Also, by the time this monograph appears there will be available the proceedings of an international symposium on cytochromes, held as part of the program for the International Congress of Biochemistry (*7*).

We propose in this chapter to develop our discussion on the basis that "cytochromes c" is not simply the plural form of "cytochrome c." We may begin by noting some definitions, recommended in 1964 by a subcommission on cytochrome nomenclature, acting under the aegis of the International Union of Biochemistry, and incorporated in the official volume on enzyme nomenclature (8). Thus, "haem" (or "heme") is taken to be any tetra-pyrrolic chelate of iron, not specifically the iron chelate of protoporphyrin IX. "Cytochome c" is defined as a haem protein present in mitochondria, which possesses an isoelectric point above pH 7 and a ferro form oxidized by O_2 in the presence of a specific catalyst, cytochrome oxidase (E.C. 1.9.3.1). Further, it is relatively heat stable at pH 7; its oxidation–reduction potential over most of the physiological pH range is close to +0.25 V; and the characteristic absorption bands of the ferro form are at 550 mμ (α), 520 mμ (β), and 415 mμ (γ). However, one notes a list of cytochromes in the "c" category that do not answer to this description of "cytochromes c," defined simply as "cytochromes with covalent linkage between the side chains and protein." This group includes all cytochromes with prosthetic groups linked in this way.

A thoroughgoing revision of concepts about function—and correlatively, structure—of c-type cytochromes underlies these definitions. This is best understood by a brief consideration of the history of researches on cytochrome c structure and function.

The original direction of cytochrome research was determined by Keilin's demonstration in the mid-twenties that the cytochrome-c–cytochrome oxidase system was universally present in all aerobic (respiratory) tissues. In essence, the question asked was: Is cytochrome c and its oxidase the unique catalyst for the biologically controlled reduction of oxygen to water? The answer given by results of a great mass of researches over the years intervening since Keilin's classic discoveries, and including efforts by O. Warburg, H. Theorell, and many other eminent biochemists, is that cyto-chrome c, as it occurs in strictly aerobic tissues, *is* the unique substrate in the terminal steps of aerobic respiration. There is no need to labor this point on which, as we have remarked, the literature references cited above give adequate testimony.

There is another question that could well have been asked: Is the function of cytochrome c *solely* that of a respiratory catalyst? Here, history testi-fies that the emphasis on the study of respiratory systems has led to the tacit belief that the answer to this question is in the affirmative. However, to ascertain that cytochrome c might exist in variant forms *unrelated* to function-ality in the oxidase system of aerobes, it would have been necessary to examine tissues in which anaerobiosis was the mode of life, i.e., bacteria. It should be remembered that, in the late twenties and early thirties, bacteriology was not dominated by the concepts of comparative biochemistry, just then being developed largely by the Dutch school of microbiologists, including A. J. Kluyver and C. B. van Niel (and, apparently independently, by K. Shibata in Japan). Clinical bacteriology provided the background for most researches on bacteria. It may be this circumstance that dictated the choices of bacteria made in the various surveys of cytochrome distribution that followed in the wake of Keilin's original demonstrations. At any rate, one finds that all examinations of microorganisms for cytochrome c content made in the decade 1925–1935 included almost exclusively those species of bacteria that exhibited, on the one hand, a strictly aerobic pattern, or, on the other, a strictly fermentative metabolism. The few exceptions were some facultative organisms among the coliform bacteria, which—by hindsight—we can only now appreciate as unlikely to yield definitive information on functionality of c-type cytochromes other than that ascribed strictly to cytochrome c in aerobic metabolism.

Other limitations were set by the small quantities of pure culture material available, nutritional variations, heterogeneities in cell population, and low resolution and sensitivity of the assay methods employed. Thus, the necessary sophistication both in approach and methodology, which would have demonstrated "cytochromes c" in contrast to "cytochrome c," were absent at the time the future of cytochrome research was being determined. The natural preoccupation of biochemists with the problem of human and animal respiration completed the list of factors responsible for the delay in recognizing the existence of a general category of c-type cytochromes, of which cytochrome c is a highly specialized member.

Some examples of early surveys—notably by Tamiya and his co-workers (9) and by Fujita and Kodama (10)—may be noted. These, and others of the time, were remarkably complete, despite inherent limitations noted above, in establishing the general pattern of cytochrome distribution among bacteria. In particular, they reinforced the findings of Keilin and others, that, in aerobic systems, cytochrome c was invariably present associated with an oxidase whose spectroscopic properties were closely similar to those of the mammalian system. Moreover, in strictly nonaerobic systems, particularly in the pathogenic clostridia and streptococci, no cytochromes, c or otherwise, could be detected.

The tendency to depend on spectroscopic evidence as the sole basis for cytochrome identifications has persisted until relatively recent times. Thus, until 1953, it appears that no effort was made—or at least no successful attempt was reported—to actually isolate cytochromes from bacteria and so establish the identity of cytochrome c from bacterial sources with specimens from the classical systems of beef heart, horse heart, or yeast. It was assumed tacitly that moieties in bacterial tissue that, upon deaeration of cell suspensions, exhibited the usual absorption maxima of the characteristic alpha band at 549–551 mμ, would be revealed as essentially identical to the same moieties identified as cytochrome *c* by isolation and spectrochemical characterization from muscle mitochondria. Thus, by 1950 most microbiologists, as well as biochemists, had come to believe that cytochrome *c*, as such, exhibited only its assigned aerobic function in bacteria, as well as in plants and in animals.

However, in 1953, simultaneously and independently in three laboratories, as far apart as possible (in Japan, England, and the United States), there were found haem proteins with the characteristic spectrochemical properties of cytochrome *c*, present and readily extractable in relatively large yields from certain strictly anaerobic bacteria. Thus, Postgate in England (*11*, *12*) and Ishimoto and his co-workers in Japan (*13*), demonstrated a *c*-type cytochrome, "cytochrome c_3" (*8*), in extracts from various strains of the sulfate-reducing bacterium, *Desulfovibrio desulfuricans*—probably the most anaerobic tissue known. At the same time, Kamen and Vernon (*14*), showed that extracts prepared from a variety of photosynthetic bacteria, and in particular from a strictly anaerobic species of the genus *Chlorobium* (*15*), contained large amounts of another *c*-type cytochrome. These findings can be said to mark the beginning of a comparative biochemistry of cytochrome *c* (*16*).

It may be noted that the particular families of bacteria in which these discoveries were made—the chemosynthetic anaerobic sulfate-reducers and the anaerobic photosynthetic bacteria—were not included among the anaerobes investigated in the early surveys. Thus, the question about a possible multiplicity of function for cytochromes *c*, posed earlier and answered in the negative, required reformulation. Subsequent work, continuing to the present, has established the following functions for cytochromes *c*:

(1) Substrate for terminal oxidation in the anaerobic reduction of sulfate to sulfide (*17*).

(2) Substrate for terminal oxidations in the anaerobic reduction of nitrites and nitrates to lower oxidation states of nitrogen (*18–20*).

(3) Participation in the photo-activated electron transport systems of plants and photosynthetic bacteria (*21–25*).

If, to these, we add the classic function of cytochrome *c* as the terminal substrate for the mitochondrial oxidation system that catalyzes the controlled reduction of oxygen to water, we arrive at the conclusion that *c*-type cytochromes can exist in a variety of forms, each appropriate to any one of at least four general types of function. We may expect, then, that considerable variance in structure can exist in the general class of cytochromes *c*. For this reason, it has appeared expedient to adopt the very general definition of structural features, given above for cytochromes c, specifying merely covalent binding of the haem group, though, to the present, no covalent binding other than the thio–ether linkage to cysteine residues, characteristic of mammalian and yeast cytochrome c, has been found. Detailed chemical examination of bacterial cytochromes c has been accomplished for just a few of the total of such proteins available (see Sections II.E and V.D); hence, it is possible that in certain variant forms of cytochrome c, covalent linkages through bonds other than thio–ether bonds may exist. Suggestions as to a more chemically based classification scheme for cytochromes can be expected as knowledge of structure increases, and at least one has already been proposed (26). The need to reappraise the definition of cytochrome c, and of cytochromes in general, will undoubtedly become increasingly apparent in the next few years (8).

The material presented in this chapter is, in a sense, complementary to that given by Margoliash and Schejter (2). These authors have considered at length, and most completely, much of the knowledge about cytochrome c that should have been included otherwise in some detail in this chapter. We have attempted, therefore, to place before the reader a review that emphasizes more the general comparative approach to cytochrome c biochemistry with less emphasis on cytochrome c itself. It is recommended that the reader consult the review of Margoliash and Schejter concomitantly with his perusal of this chapter.

II. Cytochrome c and Some Cytochromes c

A. General Properties

Practically all that we know about cytochromes c at the molecular level is derived from studies of the protein as extracted and purified in soluble form from muscle or yeast cells, wherein it is attached invariably to mitochondria as part of the oxidative phosphorylation system. In Tables 1–9 we have attempted a compilation of the various specimens of cytochrome c, as described in reports available to the present. Included are data on some cytochromes c of bacterial origin. These data include the gross character- istics specified as required for definitions of cytochrome c—absorption maxima of the ferro form, molecular weight, oxidation potential, isoelectric point, and source material. As discussed elsewhere (see Sections IV, V), uncertainties as to content of contaminants, variant forms (denatured components, "isocytochromes"), adventitious ions, adsorbed peptides, and other plagues of protein chemists discourage attempts to evaluate the quantitative significance of such data. However, certain general properties can be said to be shared by all the mammalian forms of cytochrome *c*, which we will refer to as "cytochrome *c*, viz.:

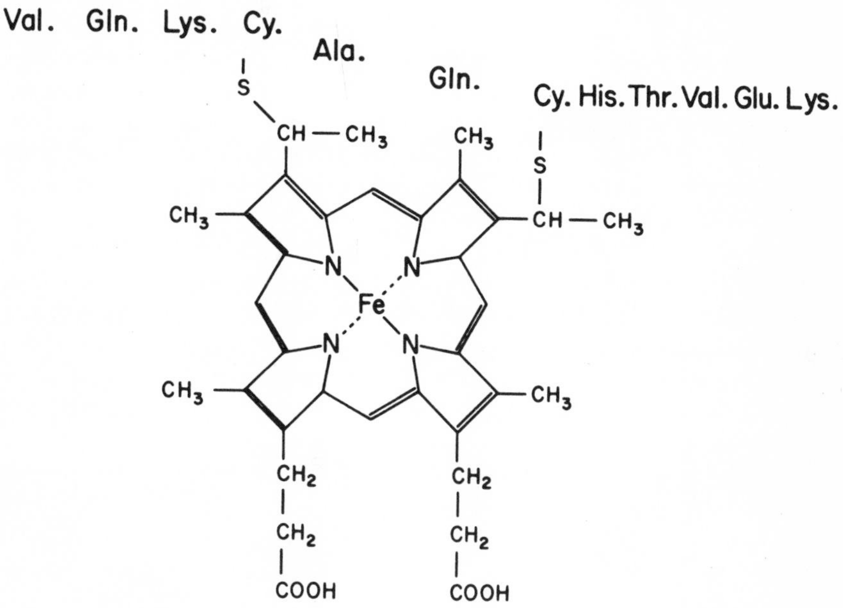

Val. Gln. Lys. Cy.

Ala.

Gln.

Cy. His. Thr. Val. Glu. Lys.

Fig. 1.   Peptic haem peptide of horse-heart cytochrome *c* (*31, 89*).

(1) The protein is basic, with isoelectric point near pH 10.5.

(2) Its molecular weight is close to 12,000.

(3) The heme prosthetic group is bound covalently by thio–ether linkage, the result of additive condensation between the vinyl side chains at positions 2 and 4 of the porphyrin structure and cysteinyl residues of the single polypeptide chain of the protein (see Fig. 1). The linkage, as shown, to the α-carbon of the side chain appears to be established (see Section V).

(4) The oxidation potential is around +0.25 V and is essentially constant over the physiological pH range.

(5) The characteristic amino-acid sequence at the site of haem binding is Lys-Cys-A-Gln-Cys-His, where A is either alanine or serine. The N-terminal residue is acetylated glycine and the COOH-terminal is glutamic acid.

(6) The total number of residues is 104. When sources such as fish, birds,

TABLE 1
ALGAL AND PLANT CYTOCHROMES c

Organism	c-Type cytochrome	Spectral characteristic (absorption maxima, mμ) Ferro	Ferri	Molecular weight	$E_{m,7}$ (mV)	pI	Remarks and additional information	References
Anacystis nidulans	c-552	552 523.5 420 — —	— 526 410 352 279	—	—	Basic		(57, 58)
Anacystis nidulans	c-554	554 522.5 416.5 — —	— 525–30 411 360 275	23,000	+350	Acidic	M.w. determined from haem content.	(57, 58)
Anacystis nidulans	c-549	549 521.5 417.5 — —	— 525–30 408 351 278	26,000* <20,000†	−260	Acidic	Reacts with CO. Autoxidizable. * Haem content. † Sedimentation studies.	(57, 58)
Porphyra tenera	c-553	553 521 (2) 417 (6) 317 272	525 409 272 — —	13,600	+335	3.5	$A_\gamma/A_\alpha = 7.0$, crystallized.	(48, 49, 56, 59–61)
Porphyra yezoensis	c-553	553 522 416	— — —	—	+342	—	$A_\gamma/A_\alpha = 7.0$, crystallized.	(56)

						Notes	Ref.
Porphyra pseudolinearis	c-553	553 522 416	— — —	+340	—	— crystallized.	(56)
Glolophloea okamurai	c-553	554 521 416	—	+340	—	$A_\gamma/A_\alpha = 5.7$	(56)
Endarachne binghamiae	c-553	553 521.5 415.5	—	+361	—	$A_\gamma/A_\alpha = 6.6$, crystallized.	(56)
Petalonia fascia	c-553	553 521.5 415.5	—	+360	—	$A_\gamma/A_\alpha = 6.9$, crystallized.	(56)
Nemalion multifidum	c-553	553 522 416	—	+363	—	—	(56)
Enteromorpha prolifera	c-553	553 522 416	—	+364	—	$A_\gamma/A_\alpha = 6.7$, crystallized.	(56)
Chaetomorpha spiralis	c-553	553 521.5 416.3	—	+382	—	$A_\gamma/A_\alpha = 5.9$ (similar c-553 from *C. crassa*).	(56)
Caulerpa brachypus	c-553	552.5 523 416	—	+385	—	$A_\gamma/A_\alpha = 6.6$	(56)
C...ium ...tum	c-555	554.5 522.8 417	—	+390	—	$A_\gamma/A_\alpha = 7.0$	(56)

TABLE 1 (Continued)

Organism	c-Type cytochrome	Spectral characteristic (absorption maxima, mμ)		Molecular weight	$E_{m,7}$ (mV)	pI	Remarks and additional information	References
		Ferro	Ferri					
Euglena gracilis	c-552	552 522 416 317 —	— 525 410 355–58 275	17,400* 13,500† 11,000‡	+370* +380†	5.5	* M.w. from haem content. † M.w. sedimentation studies. ‡ M.w. sedimentation studies. * and † Ref. (46), ‡ Ref. (102).	(50–53)
Euglena gracilis	c-556	558 554.5 sh 525 421 317	— 530 412 365 275	18,000* 27,000† 12,000‡	+307* +300‡	8	* M.w. from sedimentation. † M.w. from haem content. ‡ Ref. (102). "sh" = shoulder	(50–53)
Grateloupia sp.	c-552	552 520 415	— — —	—	+300	—		(49)
Gelidium amansii	c-553	553 521 416	— — —	—		—		(49)
Undaria pinnatifida	c-553	553 521 415	— — —	—	+340	—		(49)
Ulva sp.	c-552	552 521 415	— — —	—	+300	—		(49)
Monostroma nitidum	c-552	552 521 416	— —	—	+310	—		(49)

Organism	Type	Reduced (nm)	Oxidized (nm)	M.W.	Redox (mV)	pH	Remarks	Ref.
Tolopthrix tenuis	c-553	553, 521, 416	—, —, —	—	+300	—		(49)
Grateloupia filicina	c-552	552.5, 521.8, 416	—, —, —	—	—	—	$A_\gamma/A_\alpha = 6.6$, crystallized.	(56)
Anabaena variabiles	c-554	554, 522, 416, 317, 275	525–30, 410, 358, 275	—	—	—		(56)
Chlamydomonas reinhardi	c-553	552.5, 522.5, 416.5	—, 411	12,000	+370	—		(63) (64)
Prototheca zopfii	c-type	—	—	—	—	—	Observed.	(63)
Polytomella agilis	c-type	—	—	—	—	—	Observed.	(63)
Polytoma uvella	c-type	—	—	—	—	—	Observed.	(63)
Saprospira grandis	c-type	—	—	—	—	—	Observed.	(64)
Leucothrix mucor	c-type	—	—	—	—	—	Observed.	(64)
Vitreoscilla sp.	c-type	—	—	—	—	—	Observed.	(64)
Chlorella vulgaris var. *viridis*	c-554	554, 524, 417, 352, 319, 277	530, 412, 360, 322, 275	—	+350	Acidic		(65)

TABLE 1 (Continued)

Organism	c-Type cytochrome	Spectral characteristic (absorption maxima, mμ)		Molecular weight	$E_{m,7}$ (mV)	pI	Remarks and additional information	References
		Ferro	Ferri					
Parsley	c-555	554.5 524.5 422 330	— 530 410 —	245,000	+365	4.7	4 haems per molecule.	(46, 54, 55)
Wheat germ [a]	c-550	550	—	12,130	—	Basic	Crystallized, sequenced.	(66–68)
Navicula pelliculosa [b] (diatom)	c-550	550 522 418	525 407 353	min. 34,000	—	—	Reacts with CO and CN⁻.	(70–72a)
Navicula pelliculosa [b]	c-554	554 523 418 —	— 527 408 356	min. 13,000	+340	—		(70–72a)

[a] Samples derived from corn pollen (2) and from soybean oil (69) have been reported. The spectrochemical characteristics and other physical properties show no significant deviations from those characteristic of wheat-germ cytochrome c.

[b] Cytochromes similar in properties to those in *Navicula* sp. cited have been observed in *Chlorella pyrenoidosa* and two Chrysomonads (F. Perini, private communication, 1967). Thus, absorbance maxima (mμ) for a cytochrome c-(549, 550) in *Chlorella pyrenoidosa* and two Chrysomonads (F. Perini, private communication, 1967). Thus, absorbance maxima (mμ) for a cytochrome c-(549, 550) in the ferro form are 549–550, 520, 416, 317 and in the ferri form 523 and 408. It is autoxidizable with a haematoheme as a prosthetic group. The $E_{m,7}$ is given as +260 to 270 mV. The Soret/α ratio is 7.63. Amounts of a cytochrome c-554 relative to c-550 vary with extraction procedures.

TABLE 2

AMPHIBIAN AND REPTILIAN CYTOCHROMES *c*

Organism	*c*-Type cytochrome	Spectral characteristic (absorption maxima, mμ) Ferro	Ferri	Molecular weight	$E_{m,7}$ (mV)	pI	Remarks and additional information	Reference
Crotalus adamanteus (rattlesnake)	*c*-550	550	—	—	—	—	Sequenced.	(73)
Bufo vulgaris (toad)	*c*-550	550 520 416	— 529 410	—	—	—		(74)
Turtle	*c*-550	550 521 417	— — —	—	—	—		(75)
Chelydra serpentina (snapping turtle)	*c*-cyt.	—	—	12,200	—	—	Sequenced, crystallized.	(76)
Bullfrog	*c*-cyt.	—	—	—	—	—	Isolated.	(2)
Guanaco	*c*-cyt.	—	—	—	—	—	Isolated.	(2)

TABLE 3
INSECT CYTOCHROMES c

Organism	c-Type cytochrome	Spectral characteristic (absorption maxima, mμ)		Molecular weight	$E_{m,7}$ (mV)	pI	Remarks and additional information	References
		Ferro	Ferri					
Musca domestica L (housefly)	c-550	550 520 415 — —	— 530 410 360 280	—	—	Basic	4 extra residues at *N*-terminus (not acetylated).	(77–79)
Samia cynthia (moth)	c-550	550 521 414.5	— 530 408	12,250	—	—	Sequenced, crystallized. 4 extra residues at *N*-terminus (not acetylated).	(80)
Cecropia (silkworm moth)	c-550	—	—	—	—	—	Observed.	(81)
Bombyx mori (silkworm moth)	c-550	550 520 410 — —	— 530 400 355 280	—	—	—	Crystallized.	(82–84)
Schistocerca gregaria (desert locust)	c-type	—	—	—	—	—	Content measured.	(85)
Drosophila melanogaster (fruit fly)	c-551	—	—	—	—	—	Observed at liquid N_2 temperature.	(86)
Screw-worm fly	c-550	550 520 415	— 530 410	—	—	—		(87)

TABLE 4
FISH CYTOCHROMES *c*

Organism	*c*-Type cytochrome	Spectral characteristic (absorption maxima, mμ)		Molecular weight	$E_{m,7}$ (mV)	pI	Remarks and additional information	References
		Ferro	Ferri					
Salmon	*c*-550	550 520 415	— — 407	—	—	—	Crystallized.	(88–90)
Tunafish (*Thynnus alalunga*)	*c*-550	550 520	— —	—	—	—	Crystallized.	(91–94)
Tunafish (*Thynnus thynnus*)	*c*-550	550	—	—	—	—	X-ray studies.	(92)
Tunafish (*Neothynnus macropterus*)	*c*-550	550	—	—	—	10.2	Crystallized.	(93)
Katsuwonus vagans, Lesson (bonito)	*c*-550	550 520	— —	—	—	—	Crystallized.	(87)

TABLE 5
BIRD CYTOCHROMES c

Organism	c-Type cytochrome	Spectral characteristic (absorption maxima, mμ)		Molecular weight	$E_{m,7}$ (mV)	pI	Remarks and additional information	References
		Ferro	Ferri					
Chicken	c-550	550 520	—	12,200	—	—	Sequenced.	(88–90, 95–97)
Pigeon	c-550	550	—	—	—	—	Crystallized.	(66, 98)
Turkey	c-550	—	—	—	—	—		(2)
Peking duck	c-550	—	—	—	—	—		(2)
King penguin	c-550	—	—	—	—	—	Crystallized.	(99)

snakes, invertebrates, and fungi are used as starting materials, deviations from statements (5) and (6) occur (see Tables 3–7).

These data on soluble cytochromes *c* are considered relevant to cytochrome *c* in its functional bound form in mitochondria because no positive evidence has been found to indicate otherwise (2). Thus, unlike cytochrome b_1 from *E. coli* (27), which changes in oxidation potential by some 340 mV when solubilized—concomitant with polymerization in the soluble state—no marked change is noted, nor is polymerization marked, when cytochrome *c* is released from mitochondria (see Section V). When mitochondria are denuded of cytochrome *c* by careful extensive washing, they can be reconstituted readily by addition of the protein in its pure soluble form (see Section V). Cytochrome *c* is relatively stable, although its heat and acid stability in the earlier studies were somewhat exaggerated, owing to relatively imprecise assays for extent of denaturation. Methods of isolation have evolved from those based on relatively drastic procedures, including use of detergents, heat, and acids, to cold extractions under mild conditions of acidity and temperature [cf. Butt and Keilin (28), who have reevaluated earlier work on the chemistry of cytochrome *c*, and also the review of Margoliash and Schejter (2)].

Recently, there has been a marked upsurge in the study of cytochrome *c* structure as deduced from measurements of optical rotatory dispersion properties and circular dichroism.* As these studies are still in their infancy, we consider it premature to discuss them in this chapter. In addition, topics such as the more extended and developed researches on magnetic and potentiometric studies of cytochrome *c* and the quantum chemistry of haem proteins are outside the scope of this discussion [see Refs. (2) and (29)].

The main reason for our reluctance to include material on physicochemical analyses of structure is that there is strong likelihood tertiary structures, as determined by X-ray methods, for a number of forms of cytochromes *c* will be provided soon. [Preliminary results are now available in articles by Dickerson *et al.* and Kraut *et al.* in Ref. (7).] Hence, speculations based on the less direct evidence provided by magnetochemical, spectrochemical, potentiometric, and polarimetric methodology would appear inappropriate at this time. It is certain that the status of knowledge about the molecular architecture of cytochrome *c*, as deduced from the various approaches mentioned above, has not changed from that discussed at length in the references cited (2, 6, 30).

---

* See the articles by Urry, Flatmark, and Robinson, by Hamaguchi *et al.*, and by Van Gelder *et al.*, cited in Ref. (7).

TABLE 6[a]

YEAST, FUNGAL, AND MOLD CYTOCHROMES c

Organism	c-Type cytochrome	Spectral characteristic (absorption maxima, mμ)		Molecular weight	$E_{m,7}$ (mV)	pI	Remarks and additional information	References
		Ferro	Ferri					
Candida krusei	c-549	549 520 415	— — 410	12,000	—	Basic	Sequenced. 6 extra residues at N-terminus (not acetylated).	(70, 100–102)
Debaryomyces kloeckeri	c-550	550 520 415	— — 410	13,200	—	10.9	Partially sequenced.	(70, 101, 102)
Saccharomyces oviformis	c-type	550 520 415	— — 410	12,700	—	10.7	Sequenced. 5 extra residues at N-terminus (not acetylated).	(70, 102, 103)
Saccharomyces cerevisiae	c-550	550 520 415	— — 410	—	+281	—	Sequenced. 5 extra residues at N-terminus (not acetylated).	(101, 102, 104–106)
Aspergillus niger	c-550	550 521 412	— — —	—	—	—		(107)
Physarum polycephalum (slime mold)	c-550	550 520 417 —	530 410 358	—	—	Basic		(108)

Organism	Type	Reduced (nm)	Oxidized (nm)	Mol. wt.			Reference
Rhizoctonia solani	c-type	—	—	—	—	—	(109)
Neurospora crassa	c-550	550	—	13,500*	—	—	(110, 111)
		520	—	12,307†			
		416	409				
Aspergillus oryzae	c-550	550	—	—	—	—	(112)
		520	—				
		415	410				
Pichia polymorpha	c-550	549.5	—	—	+293	10.5	(113)
		520	530				
		418	411				
Ustilago sphaerogema	c-550	550	—	18,000	—	7.0	(114)
		520	—				
		415	—				
Protozoa: *Chaos chaos* and others	c-553	553	—	—	—	—	(115, 116)

Observed.

* From sedimentation studies.
† From amino acid composition sequenced.
4 extra residues at *N*-terminus (not acetylated).

a As noted in Section V.F, c-type cytochromes in addition to those listed here have been isolated as highly purified samples from 104 strains of the following genera of yeasts: *Schizosaccharomyces, Saccharomyces, Pichia, Lipomyces, Sporobolomyces, Candida, Rhodotorula, Endomyces, Debaryomyces, Hansenospora, Torulopsis, Kloeckera,* and *Hansenula* (117). In addition, c-type cytochromes are present in several species of *Streptomyces* (118).

TABLE 7
INVERTEBRATE CYTOCHROMES c

Organism	c-Type cytochrome	Spectral characteristic (absorption maxima, mμ)		Molecular weight	$E_{m,7}$ (mV)	pI	Remarks and additional information	References
		Ferro	Ferri					
Squid	c-550	550 520 415 —	— 530 409 278	—	—	Basic		(119)
Prawn	c-550	550 520 415 —	— 530 409 278	—	—	Basic		(119)
Oyster	c-550	550 520 415 —	— 530 409 278	—	—	Basic		(119, 120)
Dendrostomun zostericolum (marine worn)	c-550	550 520 415	— 525 407	—	—	Basic		(121)
Styela plicata (protochordate)	c-550	550 520 415 — —	— 530 410 360 280	—	—	Basic		(122)
Octopus vulgaris	c-550	—	—	—	—	—	Observed.	(123)

TABLE 8[a]

MAMMALIAN CYTOCHROMES *c*

Organism	*c*-Type cytochrome	Spectral characteristic (absorption maxima, mμ) Ferro	Ferri	Molecular weight	$E_{m,7}$ (mV)	pI	Remarks and additional information	References
Horseheart	*c*-550	550 520.5 416 315.5 280	— 528 410 360 280	12,500	+270	10.5	Sequenced, crystallized.	(2, 28, 97, 104)
Beef Heart	*c*-550	—	—	13,000	+250–260	—	Sequenced, crystallized.	(90, 104, 124–128)
Beef Kidney	*c*-cyt.	—	—	—	—	—	Crystallized.	(129)
Beef Heart	cyt. c_1	553 523 418	— 523 414 278	37,000 (minimum)	+223	—	Polymerizes on isolation, minimum m.w. based on Fe content.	(41, 43, 130)
Rabbit	cyt. *c*	—	—	—	—	—	Sequenced.	(131, 132)
Dog	cyt. *c*	—	—	—	—	—	Sequenced.	(90, 133)
Human	*c*-550	550 520 416 280	529 410 361	—	—	—	Sequenced.	(97, 134, 135)

[a] In addition to the cytochromes listed, there may be noted that those from a number of other species have been isolated and sequenced. These include sheep (136), monkey (137), impala (138), whale (138), hog (139), and kangaroo (140); *vide* Ref. (2).

TABLE 9ᵃ
BACTERIAL CYTOCHROMES c (NONPHOTOSYNTHETIC)

Organism	c-Type cytochrome	Spectral characteristic (absorption maxima, mμ)		Molecular weight	$E_{m,7}$ (mV)	pI	Remarks and additional information	References
		Ferro	Ferri					
Desulfovibrio desulfuricans	c-552 (c_3)	552 522 418 —	— 530 409 349	12,200	−215	10.5	2–3 haems per molecule, no tryptophane, crystallized. sequenced	(11–13, 141–145) (309)
Desulfovibrio desulfuricans	c-553	553 525 419	— — —	14,000* 9,100†	(−0) (−100)	8.6	* M.w. by Fe analysis. † M.w. by Sephadex filtration (monohaem), crystallized	(146)
Desulfovibrio gigas c-553	c-553	553 525 419 350	— — — —	13,000	−216	5.2	2 haems per molecule, crystallized.	(147–149)
Thiobacillus X	c-550	550	—	—	+200	—		(150)
Thiobacillus X	c-553.5	553.5	—	—	+210	—	Autoxidizable.	(150)
Thiobacillus X	c-557	557	—	—	+155	—		(150)
Thiobacillus thiooxidans	c-550	550 520 412	410	—	—	—		(151)
Thiobacillus thioparus	c-551	551 521 416	—	—	−140	—		(152, 153)
Thiobacillus denitrificans	c-552	552 522 416 317	410 —	—	+270	10.2		(154, 155)

Organism	Cytochrome	Absorption maxima (reduced)	Absorption maxima (oxidized)	MW	E_0	pI	Remarks	Ref.
Azotobacter vinelandii	c-550 (c_4)	550-51, 522, 414, 314	—, —, 409, —	11,200 (minimum)	—	Acidic		(156–158)
Azotobacter vinelandii	c-555 (c_5)	554-55, 524, 418, 318	—, —, 414, —	11,600 (minimum)	—	Acidic		(156–158)
Pseudomonas denitrificans	cyto-cc' (RHP-type)	565, 550, 435, 426	630, 500, 390–400, —	26,400	+90	Basic		(159)
Pseudomonas fluorescens	c-551	551, 520, 416, 316	—, —, 408, —	9,000	—	4.7	Sequenced.	(160)
Pseudomonas aeruginosa	c-551	551, 521, 416, 313, 290–280	530, 409, 360, 290–280	8,100	+286	4.7	Crystallized.	(161, 162)
Mycobacterium phlei	c-cyt.	—	—	—	—	—	Observed.	(163)
Hemophilus parainfluenza	c-553	553	—	—	—	—	Isolated.	(164)
Halotolerant bact. Sw	c-552	552, 525, 418	—, —	—	+279	7.2	Interconvertible to cytochrome b-574.	(165–167)
Pseudomonas saccharophila (hydrogen utilizing bact)	c-550	550, 523, 417	—, —, —	—	+80	—	Slowly autoxidable.	(168)

[a] An interesting example of a strict anaerobe, other than a sulfate reducer, that produces *c*-type cytochrome is *Vibrio cholinicus*, which ferments choline (188).

TABLE 9 (Continued)

Organism	c-Type cytochrome	Spectral characteristic (absorption maxima, mμ)		Molecular weight	$E_{m,7}$ (mV)	pI	Remarks and additional information	References
		Ferro	Ferri					
Pseudomonas saccharophila (= P. stutzeri IAM 1504)	c-552	552 523 418 —	— 530 410 280	—	+237	—		(78)
Thiobacillus ferrooxidans	c-552	552 523 417	— — —	—	+310	—		(169)
Halotolerant micrococcus	c-551	551 521 416	— 521 411	$S_{20,w}$ = 1.3S	+249	Acidic		(166, 167)
Micrococcus sp.	c-554	554 521 418	— — 414	18,000	+100	3		(166, 167)
Micrococcus sp.	c-548–554	548 554 521 418	— — 414	18,000	+113	3.2	2 hemes per molecule	(166, 167)
Hydrogenomonas ruhlandii	c-types	—	—	—	—	—	Observed.	(170)
Bordetella sp.: B. pertussis B. parapertussis B. bronchiseptica	c-550	550 520 418	520–30 408	—	+259	—	Present 6–7 times as much as c-553 of same strains.	(171)

Organism	Cytochrome						Remarks	Reference
Bordetella sp.: *B. pertussis*, *B. parapertussis*, *B. bronchiseptica*	*c*-553	553 522 416	530 409	—	+192	—	Slowly autoxidizable.	(171)
Rhizobium	*c*-551	—	—	—	—	—		(172)
Micrococcus lysodeikticus	*c*-550	550 520	—	—	—	—	Observed.	(173)
Achromobacter fisheri	*c*-550	550 520	—	—	—	—		(174)
Bacillus megaterium KM	*c*-550	419 550 520 415	530 410 278	12,000	+250	—		(175–177)
Bacillus subtilis	*c*-550	550 520 415	530 410 278	—	—	Acidic		(175, 176)
Bacillus subtilis	*c*-552	552 521 416	411	—	—	Acidic		(178)
Micrococcus sp	*c*-551, *c*-554, *c*-548/554	—	—	—	+240 +180 +110	—	Spectral properties appear normal. 2 haems/molecule.	(18, 184, 185)
Salmonella typhinurium	*c*-551	551 525 416	409	—	—	—		(183)
Escherichia coli	*c*-550	550 521 418 —	530 407 350	—	—	—		(179–182)

TABLE 9 (Continued)

Organism	c-Type cytochrome	Spectral characteristic (absorption maxima, mμ) Ferro	Ferri	Molecular weight	$E_{m,7}$ (mV)	pI	Remarks and additional information	References
Escherichia coli	c-552	552 523 420 — —	— 532 409 355 280	136,000	−250	Acidic	1 haem/12,000.	(179–182)
Serratia marcescens	c-cyt.	—	—	—	—	—	Present.	(179)
E. coli K₁₂	c-cyt.	—	—	—	—	—	Distribution studied.	(180)
E. coli B	c-cyt.	—	—	—	—	—	Distribution studied.	(180)
E. aurescens	c-cyt.	—	—	—	—	—	Distribution studied.	(180)
E. freundii	c-cyt.	—	—	—	—	—	Distribution studied.	(180)
E. intermedia	c-cyt.	—	—	—	—	—	Distribution studied.	(180)
Aerobacter cloacae	c-cyt.	—	—	—	—	—	Distribution studied.	(180)
Paracolobactrum aerogenoides	c-cyt.	—	—	—	—	—	Distribution studied.	(180)
Pseudomonad (unidentified)	c-550	550 520 416	— — —	—	+260	—	Isolated.	(186)
Pseudomonad (unidentified)	c-553	553 523 419					Isolated.	(187)

B. Primary Structure

Margoliash and Schejter (2) have given the history of developments in this area of research on cytochrome c, beginning with the first precise determinations of amino-acid composition by Theorell and Åkesson (30, 31), the early studies of Tuppy and his collaborators and others (32–35), and extending to the publications of the first complete sequence in 1961 by Margoliash et al. (36). Following rapidly, because of the relative ease of preparation of various cytochromes c from mammalian, vertebrate, and invertebrate sources, there have appeared a large number of primary-structure determinations. These have revealed a remarkable homology in amino-acid sequence. [See Ref. (2) for a complete discussion of these results.] The sequence for horse-heart cytochome c—the first such sequence published as noted above—is shown in Table 14.

For our present purposes, it is only necessary to note (1) that in tuna a single residue (lysyl) appears to be missing; (2) that in nonvertebrate species more than 104 residues are present, and that by aligning the cysteinyl residues at positions 14 and 17 of the vertebrate samples with those of the invertebrate specimens, it is seen that the extra residues are added to the N-terminal glycine; (3) that deletions appear to occur only close to the carboxyl end; and (4) that the sequence His-Thr-Val-Glu following the second cysteinyl residue is invariably present, as is the sequence extending from residues 70–80. Comparison of these results with those from the few bacterial cytochromes c available is deferred to Section III.

C. Cytochrome c_1

This protein was described first by Yakushiji and Okunuki (37), who recognized it as a component of the mammalian mitochondrial system with an oxidation potential between those of cytochrome b and cytochrome c. It has since been established as present in mitochondria from animals, plants, and yeast (38–40). The most purified preparations obtained by detergent treatment have been reported by Okunuki and his co-workers (41). Absorption maxima in the ferro form are at 553 mμ, 523 mμ, and 418 mμ and at 523 mμ, 411 mμ, and 278 mμ in the oxidized (ferri) form. From the iron content of 0.15%, the minimal molecular weight is ~37,000, but much higher molecular weights appear in physicochemical analyses, indicating polymerization during solubilization. Cytochrome c_1 appears to be rapidly autoxidized in the presence of air and cytochrome a only if some cytochrome c is present. A haem protein described as cytochrome c_1 has been isolated and partially purified from bacterial sources (42).

Further, Wada *et al.* (*43*) has announced the isolation of a 20-residue haem peptide from purified cytochrome c_1. There are marked similarities to the analogous haem peptide from cytochrome c, as suggested by comparison of the residues held in common. It appears that the sequence Lys-Cys-X-Y-Cys-His-Thr-Val, which is so characteristic of cytochrome c, is also present at the haem-binding site in cytochrome c_1.

D. Cytochrome "f"

The photosynthetic apparatus of green plants and algae contains c-type cytochromes, usually called "cytochrome f," a term derived from the Latin "folium" (leaf). The existence of such pigments in chloroplasts was pre-figured by R. Hill in his classic monograph on the chloroplast ("Hill") reaction (*44*) when he suggested that compounds analogous to respiratory pigments might be present to account for a back-oxidation process, acting to suppress photoevolution as ambient oxygen pressures were increased (see Section V). In 1951, Hill and Scarisbrick (*45*) reported the successful isolation in soluble form of a compound with spectroscopic and chemical properties that resembled cytochrome c. Davenport and Hill characterized this compound further in 1952, elaborating methods for extraction and assay in photosynthetic tissues, and demonstrated that it could be obtained from leaves of elder, parsley, and other green plants (*46*). In addition, they found it to be present in four representative species of algae: *Ulva lactuca*, *Vaucheria*, *Euglena gracilis*, and *Fucus serratus*. The haem protein, as obtained from parsley, contained two haem prosthetic groups per molecule of weight 110,000, thus being the first multiple-haem protein of c-type to be reported. With an isoelectric point at pH 4.7, notably greater heat lability than cytochrome c, and an oxidation potential (E_0', or $E_{m,7}$) of $\sim$365 mV (some 100 mV more oxidizing than that of cytochrome c), it presented differences from cytochrome c that were to be found again in the many representative cytochromes c isolated a little later by Kamen and Vernon from photosynthetic bacteria (*14*). (We will return to this aspect of comparative biochemistry of cytochromes c in Sections III and V.)

In the meantime, work in Japan was demonstrating the ubiquitous occurrence of cytochrome f in algae. Yakushiji (*47*), in fact, had described haemochromelike compounds (of the c-type spectroscopically) as early as 1935. Okunuki and his collaborators, in particular S. Katoh (*48, 49*), isolated and characterized in crystalline form c-type cytochromes from all the major groups of algae: *Rhodophyceae*, *Phaeophyceae*, *Chlorophyceae* and *Cyanophyceae*. In the particular case of *Euglena gracilis*, Gross, Wolken, and Colmano (*50–52*) demonstrated that two c-type cytochromes could be detected and isolated in soluble form; one of these, with an α-band absorption maximum in the reduced form at 552 mμ and $E_{m,7}$ of +0.38 V was cytochrome f, the other apparently a modified cytochrome c with a rather high $E_{m,7}$ (0.32 V) and α-band maximum at 556 mμ. Perini, Kamen, and Schiff (53) characterized highly purified forms of these two cytochromes and showed that the f-type occurred exclusively in chloroplasts of the *Euglena* cell, whereas the c-type was found in mitochondria-like bodies.

Later, extensive researches by Forti and his collaborators (54, 55) led to an improved method for extraction and characterization of the parsley cytochrome f, the resultant preparation showing some differences in properties from those of the earlier preparations by Davenport and Hill (46). Thus, Forti *et al.* quoted a molecular weight twice that reported previously (245,000 as compared to 110,000). Data on all cytochromes f available to the present are given in Table 1.

Cytochromes f show a considerable range in oxidation potentials, number of haem groups per molecule, molecular weights, ease of isolation, and spectrochemical properties. The criteria most often used to distinguish them as a subgroup of cytochromes c are: (1) relatively high $E_{m,7}$ (>300 mV)*; (2) high absorbance ratio of Soret to α-band (~7 as compared to ~5 for typical cytochrome c); and (3) relatively low reactivity with cytochrome c oxidase (see Section V). It has been recommended (8) that the cytochrome f appellation be changed to "cytochrome c_6."

Table 1 gives a representative sampling of algal and plant cytochromes, including specimens from all groups of algae. In addition to those shown, a number of others have been reported (56) that are not included: e.g., cytochromes c-553 from *Bangia fusco-purpurea* (crystalline), *Scinaia japonica*, *Pterocladea tenuis*, *Pachymeniopsis lanceolata*, *Gloiopeltis complanata*, *Gracillaria verricosa*, *Glacilaria textorii*, *Chondrus giganteus*, *Rhodoglossum pulcherum*, *Polysiphonia urceolata* (crystalline), *Chondria crassicaulis*, *Ishigi okamurai*, *Scytosiphon lonentaria*, *Ulva pertusa*, *Bryopsis maxima*, and *Cladophora sp.*

E. Some Bacterial Cytochromes c

We have commented (Section I) on the existence of cytochrome c in various families of bacteria. C-type cytochromes have been found in the photosynthetic apparatus of all representative species of photosynthetic bacteria, as well as in chemosynthetic sulfur-oxidizers, sulfate reducers, nitrate and

---

* A c-type cytochrome, thought to be a chloroplast component, has been isolated from the blue-green alga, *Anacystis nidulans* (57, 58) with a strongly negative oxidation potential ($E_{m,7}$ = −0.26 V).

nitrate reducers (denitrifiers), ammonia oxidizers, nitrogen fixers, in addition to the aerobic bacteria and facultative species mentioned as investigated in the early surveys (Section I). As we will note later (Section III), most of our knowledge of bacterial c-type cytochromes is derived from studies of the forms present in photosynthetic bacteria. These require separate treatment, which is accorded them in Section III. In Table 9, we have included c-type cytochromes from all the bacteria, with the exception of those from the photosynthetic bacteria, which are recorded in Tables 11 and 12.

Some generalizations may be included in this section, based on the findings given in Table 9. Thus: (1) Bacterial cytochromes c usually show acidic isoelectric points, in contrast to mammalian and aerobic cytochromes c. (2) Spectroscopic characteristics mimic those of cytochrome c but show a somewhat wider range in characteristic absorption maxima, particularly in the alpha bands of the ferro- forms. (3) There is no invariable complementarity between oxidizing potential and ease of autooxidizability; that is, many bacterial cytochromes c with a relatively high oxidizing potential $(E_{m,7} > 200 \text{ mV})$ exhibit ready auto-oxidizability, as do practically all cytochromes with negative, or low, oxidizing potentials $(E_{m,7} < 200 \text{ mV})$. (4) Multihaem cytochromes c occur with some frequency, whereas cytochrome c of mitochondrial origin is invariably a monohaem protein. (5) Oxidizing potentials exhibit values ranging from as low as -250 mV to as high as $+400 \text{ mV}$, appropriate usually to those of the terminal electron acceptors involved. Thus, sulfate-reducer cytochromes ("c_3") have $E_{m,7}$ values more reducing than those of the sulfate–sulfide system, whereas cytochromes from facultative or strictly aerobic bacteria show more positive (oxidizing) potentials, which accord well with standard potentials for the oxygen–water, nitrate–nitrite systems, etc. (6) Cytochromes c can occur either with or without cytochrome a (the mitochondrial cytochrome c oxidase, or its analogues), but no wild type species show cytochrome a without cytochromes c (for an exception, see Section V).

These findings as relevant to structural features and their relation to function are discussed further in Section V.

III. Cytochromes c of Photosynthetic Bacteria

A. Introduction

Until recently, not much attention has been given to the structure of haem proteins obtained from bacterial species because of scarcity of source material. Moreover, on the basis of their physiological role, bacterial cytochromes appeared to have little, if any, relationship to their mammalian counterparts. In fact, the first bacterial cytochrome to be investigated with regard to its amino-acid sequence, namely cytochrome c from *Pseudomonas fluorescens* (*160*) (Table 10), showed only general resemblance to mammalian cytochrome c around the haem attachment site while being at least 90% different for the rest of the molecule. However, similarity at the active site alone could be the result of convergent evolution. Hence, the possibility

TABLE 10
Amino Acid Sequence of Cytochrome c-551 of
Pseudomonas fluorescens[a]

1
H_2N-Glu-Asp-Pro-Glu-Val-Leu-Phe-Lys-Asn-Lys-Gly-Cy-Val-Ala-Cy
                                     | |
                                 S—Haem—S
      20   30
His-Ala-Ilu-Asp-Thr-Lys-Met-Val-Gly-Pro-Ala-Tyr-Lys-Asp-Val
                        40
Ala-Ala-Lys-Phe-Ala-Gly-Gln-Ala-Gly-Ala-Glu-Ala-Glu-Leu-Ala
      50   60
Gln-Arg-Ilu-Lys-Asn-Gly-Ser-Gln-Gly-Val-Trp-Gly-Pro-Ilu-Pro
                        70
Met-Pro-Pro-Asn-Ala-Val-Ser-Asp-Asp-Glu-Ala-Gln-Thr-Leu-Ala
      80
Lys-Trp-Val-Leu-Ser-Gln-Lys-COOH

---

[a] Reference (*160*).

that bacterial and mammalian cytochromes emerged from different ancestor molecules could not be ruled out.

During the last few years, structural investigations have been pursued on many c-type cytochromes from photosynthetic bacteria (*144, 191, 192*). These reveal not only sequence homology between mammalian cytochromes c and the c_2-type cytochromes found mainly in photosynthetic bacteria, but also internal homology for cytochrome c_2 from *R. rubrum*, and in addition a structural relationship to the double-haem protein, cytochrome cc'.

B. Characterization of General Types and Their Distribution

Considering the metabolic versatility of photosynthetic bacteria and the variety of cytochromes c that each individual species produces, it is obvious that any attempt to rigidly categorize all these haem proteins may be ill-advised. Nevertheless, close examination of their structural features as known at present surprisingly reveals the existence of correlations that indicate that at least four different classes of soluble haem proteins exist

(Tables 11, 12). Several of these classes exhibit vestiges of evolutionary relationship.

Admittedly, in most of the organisms listed in Table 12, one can find small quantities of additional cytochromes that do not seem to belong to any of the four classes. Our information as to whether these haem proteins represent

TABLE 11

SOLUBLE c-TYPE CYTOCHROMES OF PHOTOSYNTHETIC BACTERIA: CYTOCHROME TYPES

c_2-Type	RHP-type[a]
Single haem (probably protohaem IX), $A_{max.red.}$ at 550–552 mμ, no reaction with CO, $E_{m,7}$ = +0.28–0.33 V, single polypeptide chain, longer than mammalian cyt. c (m.w. 12.500–13.500) but showing extensive sequence homology to mammalian cyt. c; small but significant differences in light absorption and CD-spectra, as compared to mammalian cyt. c.	c' = single haem; cc' = double haem; nature of haem as yet undetermined; unique spectrum independent of number of haem groups: split Soret band in reduced state and a double peak in alpha region, additional peak at 630–640 mμ in oxidized state; autoxidizable, CO binding, $E_{m,7}$ = 0.0 to +0.1 V, m.w. 15–30,000, single polypeptide chain. Those portions of sequence known indicate homology to corresponding c_2-type.
Small c-type	Flavin c-type
Single haem (probably protohaem IX), 85–95 amino acids only (similar in size to cyt. c from *Ps. fluorescens* and *Ps. aeruginosa*), usually only a minor component of the soluble c-type fraction except in *Chl. thiosulfatophilum* and *R. molischianum*. Sequence homology to c_2-type is indicated.	1–2 haem groups (probably protohaem IX); also contain 1 flavin of yet undetermined nature, possibly bound to the proteins in form of flavin peptides. M.w. 50–72,000, $E_{m,7}$ = 0.0 to +0.14 V.

[a] See Table 12.

trace amounts of normally particle-bound cell constituents or are the result of gene suppression or possibly even result from contamination of the cultures is unfortunately incomplete. However, in some cases, it is possible to demonstrate a significant increase in amount of trace components upon extended incubation time with demonstrably pure cultures. In these cases, it appears most likely that increased production of haem proteins has rendered a normally particle-bound constituent partially soluble. In this connection, it is important to realize that the present techniques of extraction and purification of haem proteins from bacterial cells usually yield no more than

TABLE 12
SOLUBLE c-TYPE CYTOCHROMES OF PHOTOSYNTHETIC BACTERIA

Species	c_2-type	RHP[b]-type
1. Green sulfur bacteria:		
Chlorobium thiosulfatophilum (anaerobe) strain L + NCIB #8346	—	—
Chloropseudomonas ethylicum (anaerobe) strain 2K	—	—
2. Purple sulfur bacteria:		
Chromatium strain D (anaerobe)	c-551: $E_{m,7}$ = +0.32 V; acidic; 4%	cc', dihaem (5 CyS): $P_I \sim 5$; $E_{m,7}$ = 0.0 V; m.w. = 27,900; 60%
3. Purple non-sulfur bacteria:		
Rhodospirillum rubrum (facultative anaerobe) (van Niel 1.1.1)	c-550: $E_{m,7}$ = +0.32 V; P_I = 5.9; m.w. = 12,840; 40%	cc', dihaem (4 CyS): P_I = 5.4; $E_{m,7}$ = 0.0 V; m.w. = 28,840; 60%
Rhodospirillum molischianum (anaerobe) (#14031 ATCC)	c-550: $E_{m,7}$ = +0.29 V; $P_I \sim$ 10.0; m.w. = 10,180; 30%	c', monohaem (3 CyS): P_I = 7.2; m.w. = 29,000; 20%
Rhodopseudomonas capsulatus (facultative anaerobe) (van Niel 2.3.1)	c-550: $E_{m,7}$ = +0.32 V; P_I = 7.1; m.w. = 13,100; 50%	cc', dihaem (4 CyS), P_I = 4.7; $E_{m,7}$ = 0.0 V; m.w. = 27,700; 50%
Rhodopseudomonas gelatinosa ((facultative anaerobe) (van Niel 2.2.1)	c-552: P_I = 9.8; m.w. $\sim$ 12,500; 20%	c', monohaem (2 CyS): P_I = 9.6; m.w. $\sim$ 15,000; 60%
Rhodopseudomonas palustris (facultative anaerobe) (van Niel 2.1.37)	c-552: $E_{m,7}$ = +0.30 V; P_I = 9.7; m.w. = 12,500; 60%	c', monohaem (2 CyS): P_I = 9.4; $E_{m,7}$ = +0.11 V; m.w. = 14,800; 30%
Rhodopseudomonas spheroides (facultative anaerobe) (van Niel 2.4.1)	c-550: $E_{m,7}$ = +0.33 V; P_I = 6.3; m.w. = 13,800; 40%	Observed; P_I = 4.6; 20%
Rhodomicrobium vannielli (anaerobe) (van Niel 3.1.1)	c-550: $Z_{m,7}$ = +0.30 V; P_I = 7.9; m.w. $\sim$ 15,000; 20%	Observed; $P_I \sim$ 8.8; 20%

Flavin-c-type	Small c-type	Other cytochromes c	References
c-553: $E_{m,7}$ = +0.10 V; P_I = 6.7; 1 haem + 1 flavin[c]; m.w. = 50,000; 30%	c-555: $E_{m,7}$ = +0.14 V; $P_I \sim$ 10.5; m.w. $\sim$ 12,000; 50%	c-551: $E_{m,7}$ = +0.135 V; P_I = 6.0; 2 haems; m.w. = 47,000; 20%	(189, 190, 193, 194)
	c-555: acidic; 20%	c-552: acidic; 80%	(189)
c-552: $E_{m,7}$ = +0.01 V; P_I = 5.1; 2 haems + 1 flavin[c]; m.w. = 72,000; 30%	c-550: basic; 2%	c-550: M.w. $\sim$ 30,000; acidic; 4%	(193, 195, 196)
—	—	—	(142, 144, 192, 197, 202)
—	c-550: $E_{m,7}$ = +0.38 V; $P_I \sim$ 10.5; m.w. = 10,200; 50%	—	(192)
—	—	—	(199)
—	c-555: acidic 20%	—	(200)
—	c-555: $E_{m,7}$ +0.23 V; basic; m.w. = 11,000; 3%	c-552: $E_{m,7}$ −0.15 V; acidic; m.w. $\sim$ 13.000; 3% c-554: $E_{m,7}$ = 0.0 V; basic, m.w. $\sim$ 40,000; 3%	(201, 203–205)
—	c-551: P_I = 5.1; m.w. $\sim$ 10,000; 20%	c-553: $E_{m,7}$ = +0.12 V; acidic; m.w. $\sim$ 22.000; 20%	(206, 207, 207a)
—	c-550: P_I 8.2; m.w. $\sim$ 9,000; 60%	—	(208, 208a)

[a] Characteristic α-band maxima for the ferro forms are indicated as usual. Other band maxima are similar to those of c type cytochromes. · (Census to August 1967.)
[b] RHP = Rhodospirillum haem protein ≡ c', or cc'.
[c] Nature of flavin undetermined.

50% of any given protein. Therefore, it is premature to attempt to balance accounts for bacterial cytochromes.

While it is generally assumed that all these haem proteins found in anaerobically grown photosynthetic bacteria are part of the photosynthetic electron-transfer chain, they are also available for respiration as part of a common system for both anaerobic photometabolism and aerobic non-photochemical metabolism in the facultative bacteria. However, most cytochromes listed in Table 12, if found at all in dark-grown cells of facultative organisms, are present in significantly smaller amounts.

C. Primary Structure and Evolutionary Relationships

Up to now, most efforts have been spent on the structural investigation and characterization of the c_2-type, because, among the cytochromes c found in photosynthetic bacteria, this haem protein usually is the most abundant component (Table 12). As demonstrated by the amino-acid composition of four representative members of this group, the c_2-cytochromes of different bacterial species are closely related to each other (Table 13). In contrast to mammalian cytochromes, these haem proteins have a longer polypeptide chain with a freely available amino terminus (144, 198, 199). Their midpoint potentials are higher (more oxidizing than mammalian cytochromes). Small, but significant, differences in light absorption and CD spectra are also noted (210). Of great interest is the occurrence of a single histidyl residue in the cytochrome c_2 from Rps. capsulatus (199). It may be inferred from this observation that histidine need not provide invariably the group chelating to the haem-iron at position 6 in the central iron coordination sphere. Previously similar conclusions had been reached for a mammalian cytochrome c, that from horse heart, which has more than one histidine available for chelation. In this case, the methionyl residue in position 80 has been implicated as the chelating group (211). The second histidine (replaced by Gln in Neurospora crassa) and the third histidine (replaced by Trp in tuna and by Asn in kangaroo) have been ruled out as possible chelators in position 6 on the grounds that contrary to expectations they are not found among the invariant amino-acid residues in proteins belonging to this group (212).

Since c_2-type cytochromes are part of the photosynthetic electron-transfer chain, their physiological activity is different from that of mammalian cytochromes (21). They do not react appreciably with mammalian cyto-chrome oxidase (21, 70, 178), and may, at best, be listed as functionally related to mammalian cytochrome c. Nevertheless, as indicated by the extensive sequence homology seen in Table 14 (209), cytochrome c_2 and mammalian cytochromes c could have evolved from the same prototype.

This finding is of great importance because it clearly demonstrates for the first time the existence of an intimate genetic relationship between proteins from eucaryotic and procaryotic organisms.

In the comparison of these homologous sequences, certain gaps or deletions are invoked to permit attainment of substantially improved fits. Thus, 20

TABLE 13
CYTOCHROMES c_2 OF PHOTOSYNTHETIC BACTERIA:
COMPARISON OF AMINO ACID COMPOSITIONS

Amino acids	R. rubrum	R. molischianum	Rps. capsulatus	Rps. spheroides
Cysteic acid	2	2	2	2
Aspartic acid	13	12	10	12
Threonine	8	8	8	8
Serine	5	6	5	2
Glutamic acid	10	6	9	16
Proline	3	8	5	4
Glycine	8	13	15	15
Alanine	15	15	17	16
Valine	6	8	6	6
Methionine	2	1	1	2
Isoleucine	2	4	3	2
Leucine	8	9	5	5
Tyrosine	5	4	5	5
Phenylalanine	5	4	5	5
Histidine	2	2	1	3
Lysine	17	14	17	12
Tryptophan	1	1	1	1
Arginine	0	2	1	3
Total	112	119	116	119
Protohaem IX	1	1	1	1
M.w.	12,840	13,100	12,760	13,440

residues of those found to be invariant in cytochrome *c* of over twenty different species coincide with identical residues in cytochrome c_2. The most significant gap in cytochrome *c* corresponds to residues 81 through 88 of cytochrome c_2. The presence of this additional segment, which interrupts the longest sequence of invariant residues in cytochrome *c*, could account for the difference in reactivity toward the oxidase.

Clustering of basic residues is limited in cytochrome c_2 to a single lysyl–lysine sequence (positions 12 and 13), and clustering of hydrophobic residues is less pronounced than in the mammalian *c*-type. This fact could provide the basis for the suggestion that these prominent features of mammalian cytochromes *c* evolved comparatively late.

TABLE 14

SEQUENCE HOMOLOGY BETWEEN MAMMALIAN CYTOCHROME c AND *Rhodospirillum rubrum* CYTOCHROME c_2 [a]

R. rubrum cytochrome c_2
[1]
NH₂-Glu-Gly-Asp-Ala-Ala-Ala-Gly-Glu-Lys-D-D-Val-Ser-Lys-Lys- Cys -Leu-Ala- Cys - His -Thr-Phe-   [10] [20]

Cytochrome c of horse-heart muscle
[1]
Acetyl-Gly-Asp-Val-Glu-Lys-Gly-Lys-Lys-Ile-Phe-Val-Gln-Lys-D- Cys -Ala-Gln- Cys - His -Thr-Val-   [10] [20]

R. rubrum cytochrome c_2
Asp-Gln-Gly-Gly-Ala-Asn-Lys-Val-Gly-Pro-Asn-Leu-Phe-Gly-Val-Phe-Glu-Asn-Thr-Ala-Ala-His-Lys-   [30] [40]

Cytochrome c of horse-heart muscle
Glu-Lys-Gly-Gly-Lys-His-Lys-Thr-Gly-Pro-Asn-Leu-His-Gly-Leu-Phe-Gly-Arg-Lys-Thr-Gly-Gln-Ala-   [30] [40]

R. rubrum cytochrome c_2
Asp-Asn-Tyr-Ala-Tyr-Ser-Glu-Ser-Tyr-Thr-Glu-Met-Lys-Ala-Lys-Gly-Leu-Thr- Trp -Thr-Glu-Ala-Asn-   [50] [60]

Cytochrome c of horse-heart muscle
Pro-Gly-Phe-Thr-Tyr-Thr-Asp-Ala-Asn-Lys-Asn-D-D-Lys-Gly-Ile-Thr- Trp -Lys-Glu-Glu-Thr-   [50] [60]

R. rubrum cytochrome c_2
Leu-Ala-Ala-Tyr-Val-Lys-Asn-Pro-Lys-Ala-Phe-Val-Leu-Glu-Lys-Ser-Gly-Asp-Pro-Lys-Ala-Lys-Ser-   [70] [80]

Cytochrome c of horse-heart muscle
Leu-Met-Glu-Tyr-Leu-Glu-Asn-Pro-Lys-Lys-Tyr-Ile-Pro-Gly-D-D-D-D-D-D-D-D-Thr-   [70]

R. rubrum cytochrome c_2
Lys- Met -Thr-Phe-D-Lys-Leu-Thr-Lys-Asp-Asp-Glu-Ile-Glu-Asn-Val-Ile-Ala-Tyr-Leu-Lys-D-D-   [100]

Cytochrome c of horse-heart muscle
Lys- Met -Ile-Phe-Ala-Gly-Ile-Lys-Lys-Lys-Thr-Glu-Arg-Glu-Asp-Leu-Ile-Ala-Tyr-Leu-Lys-Lys-Ala-   [90] [100]

R. rubrum cytochrome c_2
[11] [2] *
Thr-Leu-Lys-COOH

Cytochrome c of horse-heart muscle
[104]
Thr-Asn-Glu-COOH

[a] Areas of invariance in mammalian species are underlined. D = gap or deletion.
Invariant residues in all c-type cytochromes encircled.

Recent studies on the CD spectra of bovine-heart cytochrome *c* and cytochrome c_2 from *R. rubrum* (*210*) indicate that there are no gross differences in the conformation of the polypeptide chain. However, subtle differences are noted in the haem environment, as well as a slightly higher content of alpha helix in cytochrome c_2. A preliminary electron-density map of cytochrome c_2 from *R. rubrum* at 5 Å (*213*) indicates that the haem group may be deeply embedded within the molecule. In comparison to data on mammalian cytochrome *c* (*214*), there also appears to be an additional segment of polypeptide chain in the middle of this protein.

TABLE 15

INDICATIONS OF INTERNAL HOMOLOGY IN THE AMINO ACID SEQUENCE OF
Rhodospirillum rubrum CYTOCHROME c_2 [a]

1   12
Glu-Gly-Asp-Ala-Ala-Ala-Gly-Glu-Lys-Val-Ser-Lys
64   85
Glu-Ala-Asn-Leu-Ala-Ala-Tyr-Val-Lys-Asn-Pro-Lys
102   112
Glu-D-Asn-Val-Ile-Ala-Tyr-Leu-Lys-Thr-Leu-Lys

1   13
Glu-Gly-Asp-Ala-Ala-Ala-Gly-Glu-Lys-Val-Ser-Lys-Lys
80   94
Glu-Lys-Ser-Gly-Asp-Pro-Lys-Ala-Lys-Ser-Lys-Met-Thr-Phe-Lys
50   61
Glu-Ser-Tyr-Thr-Glu-Met-Lys-Ala-Lys-Gly-D-Leu-Thr

20   35
Phe-Asp-Gln-Gly-Gly-Ala-Asn-Lys-Val-Gly-Pro-Asn-Leu-Phe-Gly-Val
36   50
Phe-Glu-Asn-Thr-Ala-Ala-His-Lys-Asp-Asn-D-Tyr-Ala-Tyr-Ser-Glu

[a] D = gap or deletion.

Table 15 shows indications of internal homology in this haem protein, which are consistent with a suggestion that cytochrome c_2 may have evolved from a small ancestral peptide fragment (about 13–20 residues long) by the process of gene duplication. The areas of internal homology comprise about 83% of the total molecule. Again, this finding implies that the c_2-type probably represents a very old form of *c*-type cytochrome.

The best evidence, so far, for homology between the c_2-type and other cytochromes *c* from photosynthetic bacteria has been established for the cytochromes *cc'* from *Chromatium* (*144, 192, 215*) and from *R. rubrum* (*144, 192, 215*), as seen in Table 16. However, apart from the amino terminus and the haem attachment site, not much is known about the primary structure of these dihaem proteins. They are characterized by a unique spectrum (Fig. 2),

M. D. Kamen, K. M. Dus, T. Flatmark, and H. de Klerk

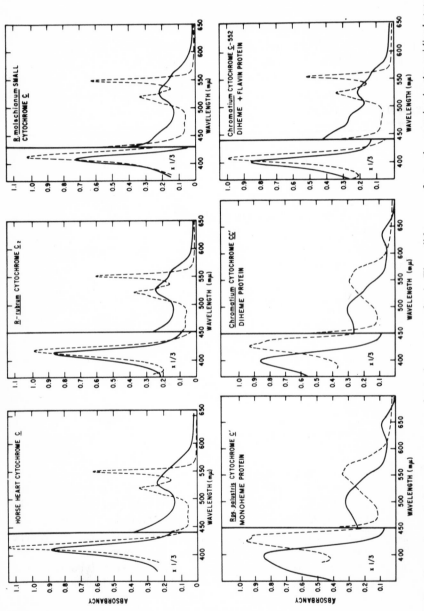

Fig. 2. Comparison of absorption spectra of cytochromes c and cc'. The solid traces refer to the proteins in the oxidized state [$K_3Fe(CN)_6$], the dashed traces to the reduced forms ($Na_2S_2O_4$). All proteins were dissolved in 0.05 M potassium phosphate buffer, pH 7.0.

TABLE 16

COMPARISONS OF SEQUENCES AROUND THE HAEM ATTACHMENT SITE OF MAMMALIAN CYTOCHROME *c*, CYTOCHROME *c₂* FROM *R. rubrum*, AND THE CYTOCHROMES *cc′* FROM *R. rubrum* AND FROM *Chromatium*

R. rubrum cytochrome c_2:   H₂N-Glu-Gly-Asp-Ala-Ala-Gly-Glu-Lys-Val-D-D-Ser-Lys-Lys-CyS-Leu-Ala-CyS

R. rubrum cytochrome *cc′*:   · · · Ala-Gln-Ala-D-D-Leu-D-Gly-Ser-Lys-D-CyS-Leu,Ala,CyS

Chromatium cytochrome *cc′*:   · · · Phe-Ala-Gly-Lys-D-CyS-Ser-Gln-CyS

                                          10

Horse-heart cytochrome *c*:   Acetyl-Gly-Asp-Val-Glu-Lys-Gly-Lys-Lys-Ilu-Phe-Val-Gln-Lys-D-CyS-Ala-Gln-CyS

                                          20

R. rubrum cytochrome c_2:   His-Thr-Phe-D-D-Asp-Gln-Gly-Gly-Ala-Asn-Lys-Val-Gly-Pro-Asn-Leu-Phe · · ·

R. rubrum cytochrome *cc′*:   His-Thr · · ·

Chromatium cytochrome *cc′*:   His-Thr-Leu-Val-Ala-Asp-Glu-Gly-Ser-Ala-Lys-D-D-D-CyS-His-Thr-Phe · · ·

Horse-heart cytochrome *c*:   His-Thr-Val-D-D-Glu-Lys-Gly-Gly-Lys-His-Lys-Thr-Gly-Pro-Asn-Leu-His · · ·

a D = gap or deletion.

which most notably includes a split Soret band and a double peak in the alpha region for the reduced state of the proteins, while the oxidized form has an additional peak at 630–640 mμ. As in cytochrome c, the haem groups are attached to a single polypeptide chain through thioether linkages between the cysteine residues of the protein and the vinyl side chains of the haem. A case of special interest is provided by the cytochrome cc' from *Chromatium*, which bears two haem groups attached by only three cysteines (*191*). The nature of the haem groups has been repeatedly investigated (*216, 217*) but is not yet completely established. In all probability, these prosthetic groups are similar to protohaem IX. They are attached to the polypeptide chain near the amino terminus, which usually is not acetylated (*144*). The two haem groups in cytochromes cc' seem located close to each other with less than 20 amino-acid residues in between (only 11 residues in the case of *Chromatium cc'*).

Several species of photosynthetic bacteria have been found to contain a member of this group bearing only a single haem (*200, 203, 204, 210*). Nevertheless, the spectral characteristics of the monohaem proteins are the same as the ones previously established for the proteins containing two haem groups (Fig. 2), thus invalidating the interpretation that the splitting of absorption bands arises from different contributions of each of the two haem groups.

All cytochromes c' and cc' are autoxidizable, bind CO, and exhibit a redox potential at, or close to, 0.00 V. The only exception is cytochrome c' from *Rps. palustris*, which has a potential $E_{m,7} = +0.11$ V (*201*). The ligand-binding characteristics of these variant cytochromes—anomalous in that they bind only CO and NO despite the unsaturated type of spectrum they exhibit—have been studied intensively only for the *R. rubrum* protein (*202*).

A comparison of the amino-acid composition of representative members of this group is shown in Table 17. Each dihaem protein is roughly twice as large as the corresponding cytochrome c_2 from the same organism, and this—together with the evidence for sequence homology around the haem-attachment site—is taken as an indication that cytochromes cc' may have evolved from the c_2-type through gene duplication and subsequent divergent evolution (*209*). The analogy to the myoglobin–haemoglobin pair is very intriguing. However, it is not easy to reconcile this concept with the un-usually high content of alanine that seems to be a characteristic feature of cytochromes c' and cc'.

Only two members of the flavin–c-type have been characterized so far. The nature of the flavin is still in question, and it has been suggested that flavin may be bound to these proteins in form of flavin peptides that can be dissociated (*193*). The two members of this group are distinguished from each other, in that *Chromatium* cytochrome *c-552* contains two haem groups and binds CO, while the protein from *Chlorobium thiosulfatophilum* has only one haem and does not bind CO. Not much is known about the structure of these proteins except that they are unusually large for the *c*-type cytochromes.

Small *c*-type cytochromes (below 100 amino-acid residues) are present in many species of photosynthetic bacteria. They are frequently overlooked

TABLE 17

CYTOCHROMES *c′* AND *cc′* FROM PHOTOSYNTHETIC BACTERIA: COMPARISON OF AMINO ACID COMPOSITIONS

	Cytochromes *cc′*			Cytochromes *c′*	
	Chromatium	*R. rubrum*	*Rps. capsulatus*	*R. molischianum*	*Rps. palustris*
Cysteic acid	5	4	4	3	2
Aspartic acid	23	19	27	22	15
Threonine	12	15	15	14	7
Serine	7	20	12	11	7
Glutamic acid	34	25	24	31	11
Proline	6	8	8	15	5
Glycine	28	15	23	21	9
Alanine	47	60	60	45	29
Valine	20	9	9	13	4
Methionine	7	4	8	7	2
Isoleucine	10	14	5	7	7
Leucine	8	16	15	24	11
Tyrosine	8	7	5	3	1
Phenylalanine	10	8	9	9	5
Histidine	2	4	4	3	1
Lysine	17	32	24	25	19
Arginine	8	3	4	9	2
Tryptophan	2	2	2	4	0
Total	254	265	258	266	137
Haem	2	2	2	1	1
M.w.	27,900	28,840	27,720	29,020	14,820

because they usually amount only to a small fraction of total soluble *c*-type cytochromes. However, they appear to be the most prominent *c*-type component in *Chl. thiosulfatophilum* and *R. molischianum*. Again, the structure of these proteins has not yet been investigated. However, some preliminary experiments (*198*) with cytochrome c_2 and the small *c*-type cytochrome from *R. molischianum* (see also Section IV. C.4) have yielded fingerprints of striking similarity. Investigation of their carboxyl termini with carboxypeptidases also have indicated homology between those two proteins, as well as between them and cytochrome c_2 from *R. rubrum*. This is of interest because cytochrome *c-551* of *Pseudomonas fluorescens*, which is of a similarly small size and does not seem to be related to any other type of cytochrome *c*, contains a COOH-terminal region that strongly resembles the corresponding area in cytochrome c_2 from *R. rubrum*. In this connection, it is of importance to recall a recent investigation by Cantor and Jukes (*218, 219*). With the help of a suitable computer program, these authors were able to demonstrate a certain degree of homology between sequences contained in *Ps. fluorescens* cytochrome *c-551* and portions of the amino-acid sequence of cytochrome *c* from *Neurospora crassa*.

Thus, we may conclude that, despite the great variety of bacterial *c*-type cytochromes, it appears quite probable that all of them are structurally interrelated. Furthermore, taking into account the extensive sequence homology pointed out for *c*-type cytochromes of organisms as widely separated on the evolutionary scale as *R. rubrum* and mammals, it is tempting to suppose that all *c*-type cytochromes, however dissimilar with regard to their physiological function, may have evolved from the same ancestral molecule. Unquestionably, much more information on the amino-acid sequences and the three-dimensional structures of these proteins will have to be accumulated before convincing conclusions can be reached. Nonetheless, it is important to realize the accessibility of such biochemical complexities to exploration by comparative amino-acid sequence analysis of the essential protein components in different species.

IV. Multiple Molecular Forms

A. Introduction

One of the most significant contributions in the past few years to the studies of yeast genetics in general and of cytochrome *c* in particular has been the demonstration and the further characterization of the multiple molecular forms of this haem protein in aerobically grown baker's yeast (*Saccharomyces cerevisiae*). Thus, this yeast has provided the first example of the simultaneous occurrence in the same cell of at least two cytochromes *c* of different primary structures that are under separate genetic control (*220*). Furthermore, recent studies on mammalian cytochromes *c* have proven the physiological occurrence of multiple molecular forms (*221*) that have been found to be catabolic products (deamidated forms) of the synthesized, native protein which, however, are all formed *in vivo*. These two types of multiple molecular forms have both a biological origin and a functional significance.

The high solubility, the ease of isolation, the low molecular weight, and the characteristic colour of *c*-type cytochromes have made these ideal proteins for study of multiple molecular forms. The techniques most successfully used for their detection and isolation have been chromatography on cation-exchange resins, such as Amberlite CG-50 (*220*) and Duolite CS-101 (*221*), disk electrophoresis on polyacrylamide gel (*221, 222*) and electrofocusing in a combined density and natural pH gradient (*223*).

B. Heterogeneity Observed As Artifacts

As we have mentioned above (Section II.A), *c*-type cytochromes have generally been considered particularly stable proteins. Thus, isolation procedures applied have often been rather drastic, and different types of modified proteins have been reported to result from extraction and purification. It may, therefore, be appropriate first to comment on the various artifacts that have been observed for cytochrome *c* obtained from different biological materials. It is not intended to give an exhaustive description of these artifacts. We shall be concerned with them only to the extent that they may represent a potential source of misinterpretation as far as the occurrence of multiple molecular forms *in vivo* is concerned. For a more detailed review on this subject, see Ref. (*2*).

1. MAMMALIAN CYTOCHROME *c*

a. Formation of Polymeric Species

If cytochrome *c* is extracted from bovine- or horse-heart muscle by trichloroacetic acid (*224*) and chromatographed on the cation-exchange resin Amberlite CG-50, three main fractions are usually observed (*225–229*); these are a small fraction of reduced cytochrome *c* (I), and two fractions (II and III) of oxidized cytochrome *c*. The minor component (III) is eluted at a higher cation concentration than fractions I and II, and it has been shown to represent a series of polymers of cytochrome *c* (*230, 231*). These modified forms have been clearly shown to be artifacts resulting from the purification procedure, i.e., an effect of trichloroacetic acid (*230, 231*). Treatment with organic solvents (*231*), as well as heating (*232*) of cytochrome *c*, also causes the formation of polymers. The polymers are distinguished from the monomer in their spectral properties (*233, 234*) and in their oxidation–reduction potentials (*235–237*), as well as in their reduced ability to transfer electrons in various systems derived from the terminal electron transport chain (*231, 234*). Furthermore, the polymers show increased reactivities with carbon monoxide and oxygen (*231*).

b. Aggregates of Cytochrome c with Other Basic Proteins

A second type of artifact that has occasionally been observed (*2*) appears to consist of aggregates of cytochrome *c* with other basic proteins, possibly histones, which occur in acid extracts of tissues; in cation-exchange chromatography they tend to move more slowly than the pure protein. These chromatographic fractions have a relatively low iron content and a low ratio $A_{550 \text{ m}\mu}$(reduced)/$A_{280 \text{ m}\mu}$(oxidized), but exhibit haemochrome properties identical to those of the pure protein. The importance of these observations lies in the fact that basic proteins inhibit the cytochrome *c* oxidase (EC 1.9.3.1) reaction, and such contamination may affect the activity of the cytochrome *c* preparation in respiratory-chain enzyme systems.

2. YEAST CYTOCHROME *c*

It has been pointed out that yeast cytochrome *c* is a less stable protein than mammalian cytochrome *c* (*238–240*). Thus, Nozaki *et al.* (*238*) found that even if the yeast cytochrome *c* was prepared by a mild procedure, using neither trichloroacetic acid nor ethyl acetate, the cytochrome *c* showed five chromatographic species when subjected to chromatography on Amberlite CG-50 in the fully oxidized form (ferricyanide). This heterogeneity has been explained partly as due to the formation of a dimer (*241*), but of another type than described above for mammalian cytochrome *c*. Thus, the main component of baker's-yeast cytochrome *c* has a single cysteinyl residue near the COOH-terminal end, which exists as a free sulfhydryl group (*103*). This group can be oxidized to a disulfide bond with the formation of a dimeric type of cytochrome *c* via an intermolecular linkage (*241*), and the dimer can easily be transformed back to monomer by suitable reducing agents (*241*). This type of dimer (disulfide type), however, has essentially the same absorption spectrum and enzymic activity as the monomer (sulfhydryl) type (*241*).

C. Native Multiple Molecular Forms (Isocytochromes)

1. TERMINOLOGY

At the present time, there is no official system of nomenclature for multiple molecular forms of cytochromes. In this review the term "isocytochromes" is used to designate multiple molecular forms of cytochrome *c* isolated from a single organism or a single tissue and which occur *in vivo*, i.e., they cannot be attributed either to the extraction or to the purification procedure (see Section IV.B.1 and -2), no matter whether there is or is not a genetic basis for their difference in physicochemical properties and whether or not differences in primary structure exist. This definition is analogous to that introduced by Markert and Möller (*242*) for multiple molecular forms of enzymes, i.e., "isoenzymes" or "isozymes," which has been accepted previously by the International Commission on Enzyme Nomenclature. There is, however, strong opposition to this extended use of the prefix "iso-" for forms that show no differences in primary structure; the matter requires resolution by official bodies.

2. AEROBICALLY GROWN BAKER'S YEAST

Several components are usually found when cytochrome *c* from aerobically grown baker's yeast (wild type) is subjected to either electrophoresis (*220, 222*) or column chromatography (see Section IV.B.2). Although some of the minor components may be artifacts, it is now well established from the studies of Slonimski and co-workers (*220, 243*) and of Sherman, *et al.* (*222, 244, 245*) that the same haploid cell of baker's yeast (*Saccharomyces cerevisiae*) synthesizes at least two molecular forms of the cytochrome-*c* monomer (Fig. 3), which are designated "iso-1-cytochrome *c*" and "iso-2-cytochrome *c*" respectively (*220, 222*).

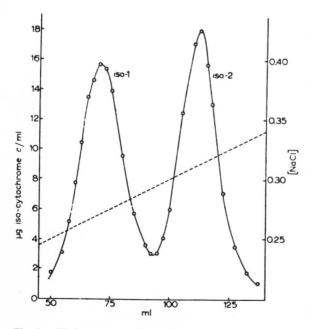

Fig. 3. Elution pattern of yeast isocytochromes.

TABLE 18

AMINO ACID SEQUENCE OF THE CHYMOTRYPTIC HAEM PEPTIDE OF BAKER'S
YEAST ISO-1-CYTOCHROME c^a AND ISO-2-CYTOCHROME c^b

$$
\begin{array}{ccccc}
 & Gln^c & Ile & Glu & AsN \\
\end{array}
$$

$$\cdots \text{Lys-Thr-Arg-Cy-Glu-Leu-Cy-His-Thr-Val-Glu-Lys-Gly-Gly-Pro-His} \cdots$$

$$\text{S——Haem——S}$$

20

[a] This sequence was determined on the main component of cytochrome c isolated from S. oviformis M_2 (103), which in total amino acid composition and peptide map is identical with iso-1 from S. cerevisiae (105).

[b] Isolation from S. cerevisiae (244).

[c] Amino acids which, in iso-2, differ in structure and position from those in iso-1 are in italics.

a. Physicochemical Differences

The first comparative study of iso-1- and iso-2-cytochrome c (220) revealed a difference in their amino-acid composition, notably in the tryptic haem peptides; in Table 18 are compared the primary structures of the chymo-tryptic haem peptides (244). This discovery has been followed by the determination of the complete amino-acid sequence of iso-1 (102) as well as

TABLE 19

SOME PHYSICOCHEMICAL PROPERTIES OF ISOCYTOCHROMES c OF S. cerevisiae (WILD TYPE)

Property	Iso-1	Iso-2	Reference
Percentage distribution[a]	95	5	(246)
pI	Basic	Basic	(220)
Amino acid residues[b]	107	112	
Amide-N/molecule[b]	10	10	
Molecular weight[c]	12,510	12,960	
$E_{m,7}$ (volts)	+0.247	+0.247	(220)
Spectral properties (Fe^{2+})			(220)
α-band (mμ)	549.4	548.8	
β-band (mμ)	520	520	
γ-band (mμ)	415	415	
Ratio $\varepsilon_\beta/\varepsilon_\alpha$	0.54	0.53	
Ratio $\varepsilon_\gamma/\varepsilon_\alpha$	4.50	4.33	(243)
ε_α (cm^{-1} mM^{-1})	29.5	29.5	(243)

[a] This percentage distribution is usually found in aerobically grown laboratory strains, but the relative amounts of the two isocytochromes c vary in different normal strains and with different conditions of growth (e.g., see Fig. 5).

[b] From the determinations of the amino acid sequences (Table 18).

[c] Minimum molecular weight calculated on the basis of the known amino acid sequences (Table 18), together with the contribution of the haem group.

of iso-2 (*245, 246*) except for four of the residue positions. Thus, it has been found that iso-2 differs from iso-1 in 20 residue positions (*245, 246*).

No information is as yet available on the tertiary structure of iso-1 and iso-2, but the small differences in the light-absorption spectra of these two cytochromes *c* may indicate small differences also in their overall conformation as shown for the multiple forms of bovine-heart cytochrome *c* (see Section IV.C.3.a). Otherwise, the two cytochromes *c* have essentially the same physicochemical properties (Table 19).

Recent chromatographic studies on baker's-yeast cytochrome *c* indicate, however, that the heterogeneity problem may be more complex than shown in Fig. 3. Thus, Fukuhara and co-workers (*247, 248*) have presented evidence in support of the presence of two distinct molecular species of iso-1 (a major and a minor form) as well as of iso-2, termed "iso-2R" and "iso-2S-cytochrome *c*" (see Section IV.C.2.c) (*3*). Iso-2R was eluted from a column of Amberlite CG-50 slightly after iso-2S, when studied by a special technique of cochromatography of one labeled (with ^{59}Fe) and one non-labelled fraction (*247*). However, the significance of these observations is uncertain at the present time because no complete separation of the subfractions has been obtained, and their physicochemical properties (notably their amino-acid compositions and primary sequences) are not known.

b. Functional Significance

One of the most interesting questions about the isocytochromes in yeast is whether or not the two components are equivalent, as far as their function in the respiratory chain is concerned. As a result of the extensive studies by Slonimski *et al.* (*220*), we know that the minor form iso-2 can function in place of the predominant iso-1. Thus, the two forms are oxidized at equal rates by yeast cytochrome *c* oxidase (EC 1.9.3.1). Furthermore, Mattoon and Sherman (*249*) have recently found that both electron transport and oxidative phosphorylation in mitochondria isolated from a cytochrome-*c*–deficient (cy_1) mutant (see Section IV.C.2.c) (*2*) may be equally efficiently reconstituted by addition of purified iso-1 and iso-2, when phosphorylation is coupled to oxidation of citric-acid-cycle intermediates, e.g., succinate. However, some important problems still concern the functional differentiation of the two cytochromes *c* observed *in vitro* with respect to lactate utilization. Thus, Slonimski *et al.* (*220*) find that, in nonphosphorylating respiratory particles from *S. cerevisiae* (wild type), iso-2 is only about 25% as efficient as iso-1 in stimulating *D*-lactate oxidation, whereas little or no difference is observed in *L*-lactate oxidation. On the other hand, Mattoon and Sherman (*249*) have shown that under more favorable test conditions (phosphorylating, cytochrome-*c*–depleted mitochondria isolated from a cy_1 mutant supplemented with malate to avoid limitations of supply of citric-acid-cycle intermediates), iso-2 is as effective as iso-1 in restoring D-lactate oxidation. Phosphorylation efficiency is, however, considerably less with iso-2. Finally, iso-2 is even less effective than iso-1 in restoring phosphorylation coupled to L-lactate oxidation. Further studies by Pradines (250) on mutants have provided important information concerning this problem *in vivo*. Thus, it has been possible to isolate a series of mutants starting from the $cy_{1\text{-}1}$ strain containing various intracellular concentrations of iso-2, but no iso-1. A linear relationship is obtained between Q_{O_2} and the concentration of iso-2, and when this concentration is the same as the total cytochrome c (i.e., mainly iso-1) concentration of the wild-type yeast, the two lactates are oxidized with the same Q_{O_2} as in the wild type. This fact has been interpreted to indicate that the maximal respiratory capacity on D- and L-lactate is the same *in vivo* whether the respiration is catalyzed by iso-1 or by iso-2 (250, 251).

c. Biological Significance

The most intriguing question about isocytochromes in yeast concerns their biological significance. The difference in primary structure of iso-1- and iso-2-cytochrome c (Table 18) means that they are synthesized independently by the yeast cell. It is known from work in Sherman's laboratory (246), as well as in Slonimski's (220), that the structural gene for iso-1-cytochrome c is the chromosomal Cy_1. However, there remain for the most part the questions as to where the iso-cytochromes are synthesized, where their structural genes are located (chromosomal or nonchromosomal), and how their biosynthesis is controlled. While definitive answers to all of these questions cannot be given at the present time, it is evident that knowledge on these matters has increased rapidly in the last few years, and that much activity in this field can be expected.

(1) *Site of Cytochrome c Biosynthesis.* The existence of two protein synthesizing systems in *S. cerevisiae*, one mediated by 80 s-type ribosomes in the cytoplasm and the other by 70 s-type ribosomes in the mitochondria, has recently been demonstrated (252–256). Thus, the ability of isolated mitochondria to carry out energy-dependent incorporation of labeled amino acids into protein is well established, and this synthesis appears to occur via the "classical" mechanisms, such as have been observed in extramitochondrial ribosomes (257). As far as cytochrome c is concerned, the significance of this fact is uncertain because, so far, net synthesis of this haem protein in isolated rat-liver mitochondria has not been successfully demonstrated (258, 259). Furthermore, direct evidence indicating that the microsomes represent the site of synthesis of rat-liver cytochrome c has been obtained

recently by Campbell and Cadavid (*260*), and Kadenbach (*261*) has demonstrated a transfer of labeled cytochrome c (as well as of other soluble mitochondrial proteins) from microsomes to mitochondria *in vitro*. Finally, as remarked above, genetic analyses of cytochrome-c–deficient mutants of S. *cerevisiae* have shown that the structural gene for iso-1-cytochrome c is located on a chromosome (see next section).

(*2*) *Genetic Control and Regulation of the Biosynthesis of Isocytochromes c.* At the present time, there is no universal agreement as to the nature of the regulatory mechanism that operates to control the biosynthesis of the isocytochromes c in yeast, but all studies so far indicate that, genetically, control is complex.

Studies of the biosynthesis of cytochrome c in different mutants have provided important information concerning genetic control. Particularly, cytochrome-c–deficient mutants ("*cy* mutants") can be easily induced, e.g., by nitrous acid or ultraviolet light* (*220, 262–264a* and -*b*), and since they are characterized by their ability to utilize nonfermentable carbon sources, they do not have complete blocks in the electron transport system from substrate to oxygen. Genetic studies of mutants deficient in cytochrome c have distinguished six unlinked genes, $CY_1 \ldots CY_6$, that effect the synthesis of cytochrome c (*264b*), independent of cytochromes $a + a_3$, b, and c_1. This difference in genetic control has been suggested to be due to the fact that the latter cytochromes are strongly particle bound, while cytochrome c is not (*254, 264b*).

Mutants of one of these genes, CY_1, completely lack iso-1-cytochrome c while containing approximately the amount of iso-2 found in normal strains† (*220, 262–264a* and -*b*). The mutants cy_1 are unique in that they exhibit approximately normal amounts of cytochromes $a + a_3$, b, and c_1 (*262–264a* and -*b*) and that electron transport and oxidative phosphorylation in mitochondria isolated from them may be efficiently reconstituted by addition of purified yeast cytochrome c (*249*) (see preceding Section IV.C.2.b). Furthermore, these mutants do not grow when *DL*-lactate is used as the sole carbon source (*251, 264b*), making it possible to select revertants and

---

* *Genetic Nomenclature:* The symbol *cy* is used to denote single-gene mutants that have below-normal amounts of cytochrome c. *CY* denotes the wild type, and if no subscript is utilized, it will imply the presence of all wild-type genes controlling cytochrome c. Nonallelic mutants are designated by different subscripts, i.e., cy_1, cy_2, etc., and independent mutants at the same locus are denoted by a second number, i.e., cy_{1-1}, cy_{1-2}, etc. Independent intragenic revertants of a particular cy_1 mutant are distinguished by letters following the original mutant number. Thus, CY_{1-2-A} and CY_{1-2-B} specify two independent revertants obtained from cy_{1-2} in which the reversions occurred at the cy_1 gene.

† In fact, it was the investigations of cytochrome c from cy_1 mutants that led to the finding of two types of cytochrome c in normal yeast (*220*).

recombinants (246).   In contrast, the other cy mutants ($cy_2 \ldots cy_6$) have reduced amounts of both isocytochromes, in approximately the same proportion as found in the wild type (264b).

Other studies of cy mutants in Slonimski's laboratory have also shown that there are numbers of unlinked genes involved in the biosynthesis of cytochrome c (248, 265–268).

(a) The Structural Gene for Iso-1-cytochrome c.   Genetic studies of mutants at the cy_1 locus have shown that the CY_1 gene is strictly required for the formation of iso-1-cytochrome c (220, 246, 262–269), and that the rate of synthesis of this monomer is proportional to the dosage of this gene in strains of different ploidy (220).   Furthermore, a revertant, $CY_{1\text{-}2\text{-}A}$, has recently been isolated from a $cy_{1\text{-}2}$ mutant strain (269), and the sites of the original mutation and of the reversion were found to be at, or very close to,

TABLE 20

THE ORIGINAL AND REVERTANT cy_1 STRAINS OF *S. cerevisiae*[a]

Original cy_1 strains			Revertant strains			
					Alteration in iso-1-cytochrome c	
Genotype	Isolation methods	Mutagen[b]	Genotype	Mutagen[b]	Peptide map	Amino acid composition
$cy_{1\text{-}1}$	See text		See text			
$cy_{1\text{-}2}$	Spectroscope	NA	$CY_{1\text{-}2\text{-}A}$	UV	C-3, T-5	Glu → Tyr
$cy_{1\text{-}3}$	Benzidine	NA	$CY_{1\text{-}3\text{-}A}$	UV	None	None
$cy_{1\text{-}4}$	Benzidine	NA	$CY_{1\text{-}4\text{-}A}$	UV	None	None
$cy_{1\text{-}5}$	Benzidine	NA	$CY_{1\text{-}5\text{-}A}$	UV	None	None
$cy_{1\text{-}6}$	Benzidine	NA	$CY_{1\text{-}6\text{-}B}$	UV	C-2, T-3	Ala → Ser
$cy_{1\text{-}7}$	Benzidine	UV	$CY_{1\text{-}7\text{-}C}$	UV	None	None
$cy_{1\text{-}9}$	Benzidine	UV	$CY_{1\text{-}8\text{-}B}$	UV	C-3, T-6	His → neutral residue
$cy_{1\text{-}9}$	Benzidine	UV	$CY_{1\text{-}9\text{-}A}$	UV	C-1, T-1	Not tested
$cy_{1\text{-}10}$	Benzidine	UV	$CY_{1\text{-}10\text{-}A}$	UV	C-2, T-3	Thr → Asn
$cy_{1\text{-}11}$	Benzidine	UV	$CY_{1\text{-}11\text{-}A}$	UV	None	None
$cy_{1\text{-}12}$	Benzidine	NIL	$CY_{1\text{-}12\text{-}B}$	UV	None	None
$cy_{1\text{-}13}$	Benzidine	NIL	$CY_{1\text{-}13\text{-}A}$	UV	C-1, T-1	Not tested
$cy_{1\text{-}14}$	Benzidine	NIL	$CY_{1\text{-}14\text{-}C}$	Spont	None	None
$cy_{1\text{-}15}$	Benzidine	ICR-170	$CY_{1\text{-}15\text{-}A}$	UV	None	None
$cy_{1\text{-}16}$	Benzidine	ICR-170	$CY_{1\text{-}16\text{-}A}$	Spont	Not tested	Not tested

[a] Reprinted from Ref. (246), p. 120, by courtesy of University of Tokyo Press.

[b] Mutagen abbreviations: Spont, none: NA, nitrous acid; UV, ultraviolet light; NIL, 1-nitrosoimidazolidone-2;   ICR-70, 2-methoxy-6-chloro-9-(3-[ethyl-2-chloroethyl] aminopropylamino) acridine dihydrochloride.

the same genetic locus. The revertant gene segregated from the wild type in the 2–2 Mendelian manner expected for a chromosomal gene. Thus far, a total of 15 revertant cy_1 strains have been tested in this manner, and they all give similar results (*246*). The proximity of the sites of reversion of cy_1 mutations has thus been confirmed. Furthermore, six of the revertants (including $CY_{1\text{-}2\text{-}A}$) synthesized functionally acceptable iso-1-cytochrome *c* that had altered primary structures (Table 20). The residue substitutions detected so far in the revertant proteins have been restricted to the portion of the sequence amino-terminal to, or within, the haem peptide (*246*). Further studies of these cy_1 mutational events may therefore provide information concerning the peptide-chain requirements for the biosynthetic attachment of the prosthetic group.

(*b*) *The Structural Gene for Iso-2-cytochrome c.* The structural gene for iso-2-cytochrome *c* is less well studied and characterized. These studies are complicated by a very recent observation (*247, 248*) that iso-2 appears not to be a single component (see Section IV.C.2.a). Studies in Slonimski's laboratory, however, have shown that there are probably four different unlinked genes (*248, 265–268*) denoted CY_{2A}, CY_{2B}, CY_{2C}, and CY_{2D} (*248, 267, 268*) that all concern the synthesis of iso-2. Thus, mutants at the $cy_{2A\text{-}2D}$ loci have in common the ability to synthesize increased amounts of iso-2 (*248, 265–268*). In addition, the iso-2 synthesized by the cy_{2A} mutants is chromatographically different from iso-2 of the wild-type strain, which has led to the suggestion that the mutant and wild-type iso-2 have different primary structures (*248, 267, 268*). This observation, as well as genetic studies of mutants at the cy_{2A} locus, have been interpreted as supporting the concept that the CY_{2A} gene is the structural gene for iso-2 (*267*), in particular, iso-2S-cytochrome *c* (*248, 268*). This interpretation is, however, complicated by another observation, i.e., that the synthesis of iso-1-cytochrome *c* is depressed in the cy_{2A} mutants (*248, 267, 268*) and that it is chromatographically different from iso-1 of the wild-type strain (*248, 268*). Again, conclusions as to the significance of these observations must await isolation and primary sequence analyses.

(*c*) *Regulation of the Biosynthesis of Isocytochromes c.* There are quite striking differences in the content of respiratory enzymes and in the respiratory activity of normal laboratory strains of baker's yeast dependent upon whether they are grown in the presence or absence of oxygen [for review, see Ref. (*270*)]. Whereas yeast grown aerobically possesses an active respiratory system similar to that of mammalian mitochondria (including cytochromes *b*, *c*, c_1, and $a + a_3$), the same yeast under anaerobic conditions shows only little respiratory activity and no evidence of the $a + a_3$ and *b* cytochromes.

However, it still synthesizes small amounts of cytochrome *c* (see Fig. 4), and several other hematin compounds (*270*). In fact, even the production of morphologically normal mitochondria [for review, see Ref. (*271*)], as well as of mitochondrial DNA (*272*), are extremely sensitive to oxygen. Thus, yeast grown anaerobically on glucose media contains no mitochondria;

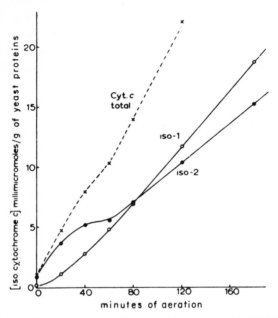

Fig. 4. Kinetics of the biosynthesis of iso-1- and iso-2-cytochrome *c* from *S. cerevisiae* during adaption to oxygen in glucose–succinate buffer. Reprinted from Ref. (*243*), p. 499, by courtesy of *Biochim. Biophys. Acta.*

but during aeration, normal mitochondria, which exhibit the complete cytochrome pattern, appear.

There seems to be a nutritional basis for this "respiratory adaption" in yeast. Thus, e.g., *glucose* has been shown to be a potent repressor of the synthesis of respiratory and citric-acid-cycle enzymes under anaerobic and aerobic conditions, as well as of the development of true mitochondrial structures [for reviews, see Refs. (*264b*) and (*273*)]. On the basis of kinetic studies of the "aerobic adaption" on glucose, it has been proposed (*273, 274*) that the effect is due to a "derepression" upon consumption of glucose, rather than to an induction by oxygen. As far as the mechanism responsible for this glucose effect is concerned, very recent studies (*275*) have demonstrated that glucose represses the synthesis of catalase in *S. cerevisiae* by creating a deficiency of inorganic phosphate, which is analogous to the Crabtree effect observed with tumor cells (*276, 277*).

A characteristic feature of the "respiratory adaptation" in yeast is that the different functionally related enzymes are corepressed and coinduced. Thus, spectroscopic examination of yeast grown under various degrees of catabolic repression have revealed that the concentrations of all the cytochromes change proportionally; only cytochromes $a + a_3$ appear to be slightly more sensitive to catabolite repression (*264b*). The hypothesis has therefore been put forward that the differentiation from promitochondria to mitochondria is controlled by regulator-type genes of an operon unit (*278*). In yeast, it may be the genes within this operon that easily undergo mutation (respiratory-deficient "petit" mutants), thus blocking the normal differentiation of the organelle [for review, see Ref. (*279*)]. Furthermore, the fact that the flavoproteins and cytochromes of the mitochondrion appear to exist in simple ratios to each other (*280*) and that certain enzymes of the Krebs cycle also appear to occur in constant molar proportions to each other and to the cytochromes (*281*) may suggest that the biosynthesis of mitochondrial enzymes is under operon control (*282*). It is not yet known, however, whether the "respiratory adaption" in yeast occurs by way of a direct effect on the genes of the organelle or whether the effect is mediated by activation of nuclear genes.

One of the most noticeable features of the "aerobic adaption" in yeast is that the kinetics of the synthesis is different for iso-1- and iso-2-cytochrome *c* (Fig. 4) (*220, 243*). It has therefore been postulated (*220, 243*) that the apoprotein of iso-2 exists in the repressed cell and that this apoprotein acts as a specific genetic repressor of the biosynthesis of the iso-1 polypeptide chain. Though further circumstantial evidence in support of this proposal has been given (*283, 284*), the situation is complicated, as remarked above (Section IV.C.2.a), by results of recent chromatographic studies on iso-2, as well as of the kinetics of its synthesis during "aerobic adaption" in the wild-type strain and in certain mutants, in that two distinct molecular species of iso-2, termed "iso-2R" and "iso-2S" cytochrome *c* (*247, 248*) occur. The former presumably represents molecules formed during the primary phase of the adaptation period by transformation of a "precursor" polypeptide chain (apoprotein)—a process resistant to cycloheximide—and the latter component represents molecules synthesized *de novo* from free, exchangeable amino acids—a process sensitive to cycloheximide. However, this "precursor iso-2R" does not satisfy all the criteria of a repressor protein (*268*) as proposed by Jacob and Monod (*285*) for the regulation of protein synthesis in bacteria.

Whatever final conclusions may be reached on the genetic control of the biosynthesis of the isocytochromes c in yeast, it seems already clear that the mechanism is far more complex than initially suggested (220), and that it is premature to discuss in details the different hypothetical models that have been proposed so far (268).

3. MAMMALIAN SPECIES

It has long been known that mammalian cytochromes c reveal heterogeneity as a consequence of different extraction and purification procedures (see Section IV.B.1), but it was not until lately that conclusive proof was presented for the existence of more than one molecular form of the monomer *in vivo* (221). In considering multiple molecular forms of mammalian cytochromes c, emphasis will be placed on cytochrome c isolated from bovine-heart muscle before discussing more briefly the other mammalian cytochromes c. This selection reflects largely the fact that more information is available about this cytochrome c than about those of other mammalian species. In general, however, the principles, and even most of the details, that derive from studies on bovine-heart cytochrome c appear to be broadly applicable to cytochrome c of other mammalian species.

a. Bovine-heart Cytochrome c

The occurrence of more than one molecular form of monomeric bovine-heart cytochrome c is now well documented (221, 223, 286–289). Thus, four subfractions have been quantitatively separated from one another by disk electrophoresis on polyacrylamide gel (Fig. 5), and termed, in order of their decreasing mobility toward the cathode, Cy I, Cy II, Cy III, and Cy IV (221). The relative percentages of the three main fractions (Cy I to Cy III) are approximately 90:9:1 in terms of absorbacy at λ_{max} in the Soret region. The same pattern of heterogeneity is found by chromatography on Duolite CS-101 (221), by moving-boundary electrophoresis (286) and by electrofocusing in a combined density and natural pH gradient (223). All results obtained support the conclusion that this multiplicity represents the intracellular state of the haem protein. Thus, the pattern of heterogeneity is almost independent of the extraction procedure, and each of the four fractions is quite stable when subjected to the whole purification procedure (221). Finally, tracer studies with ^{59}Fe in rats have shown a difference in, and a change of, the specific activity of the iron in the different subfractions that is compatible only with a conversion of Cy I to the other fractions *in vivo* (290). Since all forms can be isolated from a single organ (221), the term "isocytochromes" has been used (288) tentatively, *vide* the terminology

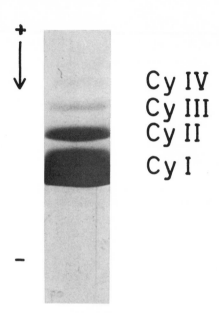

Fig. 5. Photograph of disk electrophoresis pattern of fully reduced cytochrome *c* purified from bovine heart muscle. Aniline Blue Black staining revealed four zones: Cy I, Cy II, Cy III, and Cy IV; they were all visible as pink zones in the unstained gel and demonstrated a positive peroxidase staining. Approximately 0.1 mg cytochrome *c* was applied. Reprinted from Ref. (*221*), p. 1659, by courtesy of *Acta. Chem. Scand.*

of multiple molecular forms of enzymes (see Section IV.C.1). The different forms can be distinguished from one another by their chemical and physical as well as their functional properties.

(*1*) *Physicochemical Differences.* The multiple forms are separated from one another by electrophoresis and chromatography on the basis of a difference in net charge, the main component (Cy I) being the most positively charged (Table 21). Thus, Cy II and Cy III differ from Cy I by having one and two more carboxyl groups respectively; Cy I contains one amide group more than Cy II and two amide groups more than Cy III (*286, 287*). These isocytochromes differ, however, from each other in a variety of respects other than in net charge. Thus, the differences in light absorption spectra at 25° (Fig. 6) (*286*) and at −190° (*289*), in optical rotatory dispersion (*289*), as well as in circular dichroic absorption spectra (*289*), indicate that the various forms differ in conformation and that Cy I represents the native form. This, coupled with the fact that their oxidation–reduction potentials are different (Table 21), indicates that the environment of the haem group, i.e.,

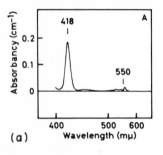
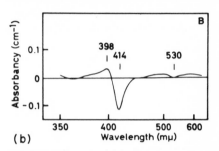

Fig. 6. Difference absorption spectra between Cy III and Cy I (Cy III minus Cy I) in 40 mM phosphate buffer, pH 6.9, 25°. (a). Ferrous form (dithionite). The concentration of Cy I is 14.3 μM and that of Cy III is adjusted to give the same absorbancy at 550 mμ. (b) Ferric form (ferricyanide). The concentration of Cy I is 10.5 μM and that of Cy III is adjusted to give the same absorbancy at 530 mμ. This corresponds to an absorbancy at 550 mμ (reduced) of 0.306 (Cy I) and 0.263 (Cy III). Reprinted from Ref. (286), p. 1482, by courtesy of Acta Chem. Scand.

TABLE 21

SOME PHYSICOCHEMICAL PROPERTIES OF THE MULTIPLE FORMS
OF BOVINE HEART CYTOCHROME c

Property	Cy I	Cy II	Cy III	Cy IV	Reference
Percentage distribution[a]	89.0	9.7	1.1	0.2	(221)
pI (ferrous form)					
Moving-boundary					
electrophoresis (0°)	10.80	10.60	10.32	—	(286)
Electrofocusing (4°)	10.80	10.58	10.38	—	(223)
Difference in net charge[b]		1	1		(286)
Amino acid analysis		No difference			(286)
Amide-N/molecule	8.40	7.51	—	—	(286)
$E_{m,7}$ (volts)[c]	+0.247	+0.212	+0.187	—	(289)
Binding of CO (pH 7)	None	None	None	—	(286)
Autoxidability (pH 7)[d]	3.0	6.1	8.3	—	(289)
[$(k_1 \sec^{-1}) \times 10^5$]					

[a] Cytochrome c was extracted by means of dilute sulphuric acid (pH 4.0; 0°) and the percentage distribution was determined by disk electrophoresis on polyacrylamide gel.

[b] Calculated from the moving-boundary electrophoresis and the acid–base titration curve.

[c] Oxidative titration.

[d] Oxidation by molecular oxygen at a concentration of 278 μM.

the haem–protein interaction, has been affected in Cy II and Cy III. However, neither of these forms combines with carbon monoxide at neutral pH, whereas they are slowly oxidized by molecular oxygen (Table 21).

(2) *Functional Significance.* As expected from the differences in net charge, in conformation and in oxidation–reduction potentials between the various forms, significant differences are observed in catalytic properties in different systems derived from the terminal electron-transfer chain. Thus, the various forms are all able to restore succinate oxidation in mitochondria depleted of endogeneous cytochromes c, but they are not isokinetic (289). Though the V'_{max} values are essentially the same, the K_m' values for cytochrome c increase in the order Cy I < Cy II < Cy III and numerically by about 5.5 times (from 2.9 to 15.9 μM). Furthermore, the various forms are reduced at a decreasing rate by ascorbate (289) and by $NADH_2$ cytochrome c reductase (EC 1.6.2.1) (288) in the order Cy I > Cy II > Cy III, which indicates that the effect seen in the succinate oxidase system is due mainly to the lower potential of Cy II and Cy III.

b. Cytochrome c of Other Mammalian Species and Tissues

All of the cytochromes c of mammalian species so far studied by disk electrophoresis on polyacrylamide gel have revealed four subfractions of the

TABLE 22

PERCENTAGE DISTRIBUTION OF THE MULTIPLE FORMS OF CYTOCHROME c IN VARIOUS TISSUES OF THE RAT[a]

Tissue	Percentage of total[b]			
	Cy I	Cy II	Cy III	Cy IV
Heart	92.0	6.1	1.5	0.4
Liver	88.9	7.6	2.9	0.6
Kidney	86.2	8.0	4.2	1.6

[a] From Ref. (290).
[b] The values are the average result for 29 animals.

monomeric form with almost the same percentage distribution as originally found for bovine-heart cytochrome c (291). It is therefore reasonable to apply the same terminology (i.e., Cy I to Cy IV) to the multiple molecular forms of all mammalian species. In one and the same species, however, small differences are observed in the percentage distribution of these forms in various tissues, e.g., of the rat (Table 22).

c. Biological Significance

The data presented above support the view that the minor forms (Cy II to Cy IV) are not preparative artifacts, but that they all exist *in vivo*. Although their percentage distribution can be changed artificially (change in the quantity of Cy I relative to Cy II–Cy IV) by certain environmental conditions, these are not usually present during extraction and purification of the haem protein (see below), but they exist *in vivo*.

It is well known from the study of several proteins, notably hormones, that the amide group of certain asparagine and/or glutamine residues of the

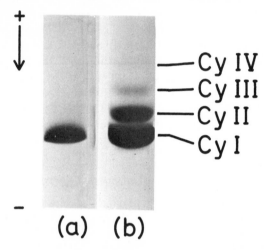

Fig. 7. Photograph of disk electrophoresis patterns showing the conversion of sub-fraction Cy I into Cy II–Cy IV at 37°. Borate buffer, pH (37°) = 10.4 and $\mu \approx 0.1$. The electrophorograms were obtained after an incubation period of 0 hr (a) and 54.6 hr (b). Reprinted from Ref. (*287*), p. 1490, by courtesy of *Acta Chem. Scand.*

polypeptide chain are very susceptible to hydrolysis. Mammalian cytochromes *c* represent no exception in this behavior. Thus, recent studies on the kinetics of the hydrolysis of labile amide groups in native bovine-heart cytochrome *c* (Cy I) have revealed that this form is rather easily converted to Cy II to Cy IV (Fig. 7), even when it is exposed to a physiological environmental condition as far as buffer composition, pH, and temperature are concerned (*287*). The conversions proceed as consecutive reactions in the following sequence:

$$\text{Cy I} \xrightarrow{k_1} \text{Cy II} \xrightarrow{k_2} \text{Cy III} \xrightarrow{k_3} \text{Cy IV} \qquad (1)$$

At 37°, the pseudo first-order rate constants k_1 and k_2 reveal minimum values at about pH 5 (Fig. 8), and they increase markedly with increasing concentrations of protons and hydroxyl ions. However, at 4° no measurable rates are observed at $3 < \text{pH} \leqslant 9.3$ (formate and borate buffers, $\mu \approx 0.1$) over a period of at least one week (Fig. 8). These studies are compatible with the concept that the minor components Cy II to Cy IV arise from Cy I by simple

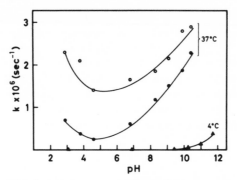

Fig. 8.   Effect of pH and temperature on the first-order rate constant for the conversion of subfraction Cy I into Cy II (▲ and ●) and of Cy II into Cy III (○). Formate and borate buffers of ionic strength 0.1 were used. Reprinted from Ref. (*287*), p. 1491, by courtesy of *Acta Chem. Scand.*

deamidation *in vivo*, and that they all represent the intracellular state of the haem protein.   Recently, isotope experiments in rats (referred to in Section IV.C.3.a) have provided conclusive evidence for this concept (*290*).   Thus, a distinct heterogeneity of ^{59}Fe labeling of Cy I to Cy III is observed in the kidney (Fig. 9), and the change in the degree of labeling as a function of time reveals the characteristics of a precursor–product relationship.   The experimental values are consistent with $T/2 \approx 80$ days for Cy I *in vivo*, which is in good agreement with the rate of conversion *in vitro* ($T/2 \approx 95$ days) under conditions that are close to those that exist *in vivo*, as far as buffer composition, pH, and temperature are concerned (*290*).   These results also indicate that no enzymic mechanism is involved in the conversion *in vivo*.

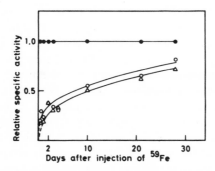

Fig. 9.   Changes as a function of time in the specific radioactivity of ^{59}Fe in Cy I (●), Cy II (○), and Cy III (△) isolated from the kidneys of normal rats given a single standard dose of ^{59}Fe.   The ordinate gives the relative specific activities.   The specific activity of Cy I was taken as unity at each time interval after injection of the isotope, and the values for Cy II and Cy III are expressed as ratios. [Reference (*290*).]

4. Photosynthetic Bacteria (*Rhodospirillum molischianum*)

The existence of *c*-type cytochromes in most photosynthetic bacteria is now well established (see Section III), but it was not until recently that more than one molecular form of cytochrome *c* was demonstrated in these photo-heterotrophes (as already mentioned in Section III.C). A single cell of *R. molischianum* synthesizes at least two distinct molecular forms of cytochrome *c* monomer, which can be resolved completely by disk electrophoresis on

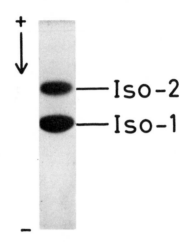

Fig. 10. Photograph of disk electrophoresis pattern of purified "iso-1-" and "iso-2-" cytochrome *c* from *R. molischianum* (wild type) in the fully reduced form. Aniline blue-black staining. [Reference (*198*).]

polyacrylamide gel (*198*) (Fig. 10). One of these, the more basic fraction, is a small molecule containing only 92 amino-acid residues, and is the only component observed in one day-old light-grown cells. The other fraction, which appears in increasing amounts as the cell suspensions age (3–5 days), has a slightly less basic isoelectric point and contains 27 more amino-acid residues. This component resembles, both in size and amino-acid composition, the c_2-type of cytochromes *c* usually found in photosynthetic bacteria (Table 12). No information is available, as yet, on the primary structures of these two "isocytochromes," but on the basis of their known amino-acid compositions (Table 23), major differences in their primary structures may be anticipated. Unlike the situation for the yeast isocytochromes, they may also perform diverse functions because their oxidation potentials are different by at least 93 mV (Table 24). The functional and biological significance of these isocytochromes *c* is, however, not yet understood. It is interesting to note that in spite of the fact that they are strongly basic proteins and have

TABLE 23

AMINO ACID COMPOSITION OF THE ISOCYTOCHROMES OF *R. molischianum* (WILD TYPE)[a]

Residue	Iso-1	Iso-2
Cysteic acid	2	2
Asp + Asn	10	12
Thr	4	8
Ser	4	6
Glu + Gln	2	6
Pro	8	8
Gly	10	13
Ala	15	15
Val	5	8
Met	2	1
Ile	4	4
Leu	7	9
Tyr	5	4
Phe	2	4
His	2	2
Lys	9	14
Trp	0	1
Arg	1	2
Amide-N	5	5
Total Number	92	119

[a] Reference (*198*).

TABLE 24

SOME PHYSICOCHEMICAL PROPERTIES OF THE ISOCYTOCHROMES OF *R. molischianum* (WILD TYPE)

Property	Iso-1	Iso-2	Reference
Percentage distribution	The ratio Iso-1/Iso-2 decreases during the growth of the cells		(*198*)
pI (21°)	9.78	9.44	(*291*)
Amino acid residues	92	119	(*198*)
Haem/molecule	1	1	(*198*)
Amide-N/molecule	5	5	(*198*)
Molecular weight[a]	10,180	13,100	(*198*)
$E_{m,7}$ (volts)[b]	+0.381	+0.288	(*291*)
Spectral properties (Fe^{2+}):			
α-band(mμ)	550	550	(*198*)
ε_α (cm^{-1} mM^{-1})	29.15	29.76	(*198*)
Autoreduction (pH 7)	Fast	Slow	(*291*)

[a] Minimum molecular weight calculated on the basis of the known amino acid composition (Table 23), together with the contribution of the haem group.

[b] Oxidative titration; $n = 1$ for both isocytochromes.

spectral properties that are very similar to those of mammalian cytochromes *c* (in contrast to other *c*-type cytochromes of photosynthetic bacteria), they are both completely inactive in the mammalian succinate oxidase system (*291*).

V. Structure–Function Relations

A. Introduction

Much has been written about special structural features as related to function for cytochrome *c*, based for the most part on data that, while biochemical, are essentially indirect in nature. It is unlikely that another such discussion placed in this chapter would do more than burden the literature. Indeed, it is difficult to imagine a poorer time for an attempt to fashion such a discussion out of the great bulk of suggestive, but not definitive, data that exist. For one thing, a sufficient effort to do so has already been made (*2*). For another, definitive data based on high-resolution X-ray-diffraction analyses of at least two cytochromes *c* should appear shortly (*213, 214*).* It appears wise to defer inferences from extensive chemical studies available until these tertiary structure data become available. For the sake of completeness, it may be stated that the inferences made in the past appear on the whole to be consistent with what is already known about tertiary structure. Thus, in horse-heart ferricytochrome *c*, the prosthetic haem group is attached to the peptide chain, as indicated in Fig. 1, and it is essentially buried in a hydrophobic region of the protein (*214*).

We will confine our discussion in this concluding section to questions that remain for the class of cytochromes *c* as a whole, and that are unlikely to be clarified fully by precise knowledge of tertiary structure for one or two cytochromes *c*.

B. The "Native" Protein

As already remarked (Section IV), uncertainties created first by effects of extraction procedures, then by artifacts produced during purification, raise the question of identification of the "native" form and, further, the relation of this protein specimen to the functional protein in its bound form. For cytochrome *c*, it has been traditional to assume that, because of its low molecular weight and its apparent stability, these uncertainties can be

* The complete tertiary structures, at 2.8 Å resolution, of horse heart and bonito ferricytochromes *c*, have appeared; *vide* R. E. Dickerson, T. Takano, D. Eisenberg, O. B. Kallai, L. Samson, A. Cooper, and E. Margoliash, *J. Biol. Chem.*, in press (1971).

minimized, as compared to those for more complicated proteins and enzymes. However, as sophistication in methods of detection of heterogeneity in proteins has increased, it has become apparent, even for cytochrome c (let alone for cytochromes c in general), that no single simple isolation method is universally applicable or likely to give complete satisfaction. Each tissue presents its own problems.

To minimize heterogeneity, it is obvious that the mildest conditions compatible with efficient extraction should be used. Margoliash and Schejter (2), for example, recommend extraction with aluminum sulfate for the various mammalian forms of cytochrome c they have characterized. On the other hand, Flatmark, investigating the significance of his "isocytochromes c" (Section IV), has used extraction with dilute sulfuric acid at pH 4 in the cold. Various other procedures, which include sonication, use of detergents, organic solvents, lyophilization followed by extraction with buffer, etc., will be found in the references cited in Tables 1–9.

Most modern preparative procedures depend upon judicious applications of absorption chromatography. Molecular sieve filtration involving the use of Sephadex G-75 gels, together with disk gel electrophoresis or the natural pH-gradient column electrophoresis, introduced by Svenssen and Vesterberg (292, 293), appear as the most recent methods of choice.

Criteria for isolation of the native protein in the case of cytochrome c can be specified on the basis of long experience. Thus, homogeneity determined by successive chromatographic essays on weak cationic ion exchangers, or other columns of the type mentioned above, is the first requirement. Second, the preparation should exhibit no reactivity with carbon monoxide, very low rates of autoxidation, and a rapid rate of reaction in the cytochrome c oxidase system. As an example, we may refer the reader to the investigations of Flatmark (223) on the properties of native bovine-heart cytochrome c, compared to its isocytochromes (Section IV). Many other good examples will be found described in the literature (2).

The criteria for specification of the "native" form in cytochromes c derived from sources other than mammalian or yeast mitochondria are less certain. One may define the native form as the major component, that is, the least deamidated form, by disk gel electrophoresis or by the Svenssen–Vesterberg procedure. Such a preparation should exhibit satisfactory reactivity in the corresponding electron-transfer system prepared from the homologous tissue. Criteria based on reactivity with carbon monoxide, oxygen, or other reagents cannot be recommended in general, because of the wide range of such reactivities exhibited by bacterial cytochromes c. Thus, the flavocytochrome c-552 from Chromatium (Section III) persists in showing partial reactivity towards carbon monoxide, as well as a relatively rapid rate of autoxidation (195). The cytochromes c from Thiobacillus species all show lack of reactivity with carbon monoxide, but varying rates of autoxidation (*150*).

An interesting study relevant to the nature of ligand reactions in cyto-chromes *c* is that of Gibson and Kamen (*294*) on the kinetics of cytochrome *cc'* reactions with carbon monoxide, oxygen, and redox reagents. Using stopped-flow techniques, together with flash spectrophotometry, it was found that in this variant dihaem protein, combination with carbon monoxide took place in two steps, the first involving a remote association of the ligand with the central iron atoms, and the second the final close combination. Kinetic characteristics of the two partial equilibria varied with the particular cytochrome studied. While carbon monoxide appeared merely to be hindered in its approach to the ligand binding site, oxygen apparently reacted only at the periphery of the protein. More remarkable results were obtained with ferricyanide, which was found to oxidize the ferrocytochrome *cc'* at rates corresponding to second-order constants of 10^9, values consistent only with diffusion-limited processes. Thus, it appeared that movement of electrons from the interior of the protein to the periphery could be effected practically instantaneously, and direct contact of the central iron atoms with the redox reagents was not essential. These results are suggestive as to enzyme-catalyzed mechanisms operative in the cytochrome-*c*–cytochrome-oxidase system.

Once a preparation of the protein is obtained that is adequate for purposes of identification of the prosthetic haem group, the problem of fission of the prosthetic group without modification arises. Numerous procedures for cleavage have been used since the original work of Hill and Keilin (*295*), the most popular being that originated by Paul (*296*)—use of heavy-metal (Ag, Hg) salts in dilute acid—with modifications introduced later by others. However, the most convincing evidence for the linkage of the prosthetic group, as given in Fig. 1, comes from studies on synthesis of porphyrin model systems, rather than from studies based on degradation of the native protein. Sano and his associates (*297*) have synthesized various dimethyl esters, prepared as authentic samples of iso-haematoporphyrin (where the hydroxyl group is on the β-carbon) and haematoporphyrin. These have been compared with dimethyl esters prepared from haematoporphyrin derived from porphyrin *c*. The results are consistent with linkage of the peptide chain of cytochrome *c* to the α-carbon of the side-chain group of the haem. From the same laboratory, reconstitution of cytochrome *c*, in yields up to 10%, beginning with apoprotein and the prosthetic group as protoporphyrinogen, followed by iron insertion, has been reported (*298*). This demonstration leads to the speculation that, in biosynthesis, addition of the prosthetic group may proceed first by incorporation of the iron-free porphyrinogen followed later by enzymic insertion of the iron. Again, the final demonstration of the actual linkage should come from the X-ray analyses of the tertiary structure at resolutions of a few angstroms.

There remains the question whether all cytochromes c have the same haem group(s), linked in the same manner found for vertebrate cytochrome c. It is certain this is not the case for *Chromatium* cytochrome cc', because only three cysteines are available in the molecule for attachment to the four vinyl groups of the two haems (*191*). The most recent evidence (*217*) on the nature of the porphyrins liberated from various cytochromes cc' by a modified Paul procedure indicates that α-haematoporphyrin is the only product, and is produced essentially quantitatively if care is exercised in the cleavage procedures.

It is to be hoped that revelations of the X-ray analyses will indicate a structural basis for the interaction between cytochrome c and cytochrome oxidase. In addition, an explanation may be provided for the large variation of reactivity of cytochromes c with various oxidases, since the tertiary structure for at least one bacterial cytochrome—cytochrome c_2 of *Rhodospirillum rubrum* is expected to appear in the near future (213).

The final elaboration of the cytochrome c structure in the horse-heart preparation and in that from *R. rubrum* will settle the nature of the ligand in the sixth position of the haem iron, but it will leave open the question of the ligands bound in other cytochromes c. At present, studies of bacterial cytochromes c indicate that no cytochrome c exists without at least one histidine (see Section III.C). This histidine always appears adjacent to one of the two cysteines involved in covalent bonding to the haem. Again, *Chromatium* cytochrome cc' [cf. Ref. (*191*)] affords an extreme example in that the second haem-binding site contains a single histidine even though the necessary quota of two cysteines is not filled, only one cysteine being present (cf. Section III.C).

C. IN VIVO *versus* IN VITRO **Cytochrome c**

It is agreed generally that mitochondrial cytochrome c is functional only when bound to its site in the respiratory chain [see Ref. (*2*) for a complete discussion]. Little is known about the extent to which binding influences the structure of soluble cytochrome c. It is certain that when cytochrome c is extracted from heart muscle, purified, then incubated with cytochrome-freed muscle mince, a particulate preparation from the resultant reconstituted system shows full oxidase activity (*299*). Thus the exogenous cytochrome c acts like the endogenous protein. Measurements of the affinity (Michaelis) constant for exogenous cytochrome c in cytochrome-c–deficient heart-muscle preparations (*300–303*) show practically the same efficiency for exogenous as for endogenous cytochrome *c*, provided salt concentrations are kept low. Smith and co-workers (*304, 305*) have clarified salt effects in the phenomenon of reactivation by exogenous cytochrome *c*, by demonstrating that cytochrome *c* added to heart-muscle particulates reacts at different sites (with differential responses to salt concentrations) on the pyridine-linked cytochrome *c* reductase and cytochrome oxidase systems. Furthermore, they find that added cytochrome *c* can react directly with the particle oxidase, increasing oxygen uptake by just the amount to be expected from the oxidation of the added cytochrome, reduced by the reductase. From these observations, it appears unnecessary to suppose the existence of an activated endogenous form of the protein formed after addition of the soluble protein. On the other hand, mitochondria are a more closed system in that cytochrome *c*, added from without, becomes incorporated to some extent, and shows the customary lowering of reactivity characteristic of endogenous cytochrome relative to added ligand reagents (*304, 305*). The question remains as to the modification in structure, if any, produced by binding.

Insofar as redox potential reflects structural change, there is some evidence that no drastic modifications occur on binding. As an example, we may cite the experience of Taniguchi and Kamen (*306*) with bound cytochrome c_2 in membrane fragments containing the active oxidase system of *R. rubrum*, grown as an aerobic hetereotrophe. In this bacterium (see below), cytochrome c_2 is not a substrate for oxidase activity. Nevertheless, it persists as a bound component in the structure of the oxidase-active membrane fragments, where it remains reduced even under strong aeration. Its reduction state is maintained by a strong pyridine-linked reductase present in the membrane system. If exogenous soluble pure cytochrome c_2 extracted from the same system is added, it is found that aeration causes a rapid oxidation of the added cytochrome c_2 without effecting the steady state of reduction of the endogenous bound form. Thus, the exogenous protein does little more than would any oxidizable dye capable of coupling with the membrane oxidase system. When the redox potential of the endogenous cytochrome c_2 is measured by spectrochemical potentiometric titration against the standard ferri–ferro cyanide system, the midpoint potential is essentially identical with that of the soluble form.

D. Structural Variance

The present definition of cytochromes *c* leaves room for much structural variation. A few variants have already been found among the bacteria. The most well known are the cytochromes of the *cc'* class (see Section III). Another, discovered in the green sulfur photosynthetic bacteria (Section

III), includes multihaem forms in which bound flavin occurs, analogous to the flavin–haem protein in the b class of cytochromes (cytochrome b_2). The multihaem cytochrome c_3 of the sulfate reducers is another interesting variant, exhibiting two or three haem groups per molecular weight of 12,000 (Table 9). The present status of the chemistry of cytochrome c_3 is uncertain. Earlier results (162) gave the following amino-acid composition: $Gly_{10}Ala_{11}$ $Val_9Leu_3Ser_5Thr_5Asp_{13}Glu_6Tyr_3Phe_2Cys_8Met_3Pro_4Lys_{22}His_9Arg_1$. Later (307), analyses on protein from which haem had been removed prior to digestion gave the composition: $Gly_8Ala_9Val_8Leu_2Ser_5Thr_4Asp_{10}Glu_4Tyr_3$-$Phe_2Cys_5Met_2Pro_4Lys_{16}His_7Arg_1$. Both determinations were based on the content of arginine. No quantitative data on tryptophane content are yet available, although present experience (145, 215) indicates qualitatively that none is present. Difficulties such as unusual resistance to proteolytic digestion, limitations in amounts available, and anomalous behavior of haem peptides on various adsorbents usually effective in isolation of haem peptides have impeded work on primary sequence. It is obvious that cytochrome c_3 presents unusual opportunities for exploitation in structure research. The high iron content together with the low molecular weight and the low redox potential must reflect remarkable alterations in the cytochrome-c structure.

E. Functional Variations

In extension of comments in Section I, we shall discuss briefly the variations in cytochrome-c distribution and function that arise in response to changes in environment, in particular in bacterial systems. A few examples will be given.

In *E. coli* the usual complement of cytochromes includes only insoluble b-type cytochromes and cytochrome "o." When cultured under strictly anaerobic conditions, a labile form of low-potential cytochrome c appears, in addition to soluble b-type cytochromes (179, 180). In *Chromatium*, photosynthetic growth under strictly autotrophic conditions produces an acidic c-type cytochrome that is absent from the same bacterium when grown photosynthetically under heterotrophic conditions (196). In *Ps. fluorescens*, high oxygen tension in the absence of iron causes cytochrome c to drop from 3.5% of soluble cell protein to less than 0.1% (181). In *R. rubrum*, aerobic heterotrophic growth conditions suppress completely appearance of soluble cytochrome cc' and markedly diminish synthesis of soluble cytochrome c_2 (306, 308).

The obvious analogy between sulfate reduction in *Desufovibrio sp.*, employing cytochrome c_3 as an intermediate (309, 310) and nitrate reduction in anaerobic *E. coli*, has prompted suggestions that the *E. coli* low-potential cytochrome c may be utilized as an electron donor in nitrate or nitrite reduction (*311*) and also possibly as an intermediate in the formic hydrogenlyase reaction (*311*), which also occurs in *Desulfovibrio* (*312*). Studies with *E. coli* mutants lacking nitrate reductase and incapable of releasing H_2 and CO_2 from formate show large amounts of the low-potential cytochrome c, but these are reduced by growth in the presence of nitrate. Conclusive data are still to be obtained (*313*).

These examples illustrate two points: First, the cytochrome composition can change both qualitatively and quantitatively as nutrition conditions vary. Second, aerobic metabolism can be associated with the absence, rather than the presence, of cytochromes c.

The fact that cytochrome c can function as an oxidation catalyst characteristic of anaerobic systems has already been remarked (Section I). Typical anaerobic oxidative systems are the facultative nitrate reduction (*310*) sulfate-reduction (*309*), and bacterial photosynthetic systems (*21, 314*). We can illustrate in the latter case a typical example in which cytochromes c exist side by side with aerobic systems, yet function wholly as catalysts in anaerobic oxidation. This appears to be the general condition among the nonsulfur purple bacteria. These organisms, of which *R. rubrum* appears to be typical, grow either anaerobically in the light, or aerobically in the dark; both processes are competitive with regard to utilization of substrate hydrogen donor (*314*). Cytochrome c_2, cytochrome cc', and cytochrome b are found, but no cytochrome a (*306*). The oxidase system has been characterized for the dark-grown aerobic organism as cytochrome "o"—a protohaem enzyme (*306, 315*).

As remarked above, aeration fails to effect cytochrome-c oxidation. Cytochrome c_2 is oxidized only by the photochemical apparatus in which bacteriochlorophyll acts as the photooxidase (*21, 306*). In fact, as a great bulk of literature attests (e.g., *316–321*), the photooxidation of c-type cytochromes in bacteria, and of its analog—cytochrome f—in plants, is an early event in photochemical phases of photosynthesis. In plants, the chloroplast structure likewise contains no cytochrome a, but large amounts of both cytochrome b and f are present. There is also evidence for existence of cytochrome-f-specific reductases in chloroplasts (*322*). In this connection, we may recall some interesting observations by Davenport and Hill (*46*), who reported in their original paper in 1951 that while cytochrome f with its high midpoint potential ($\sim$365 mV) failed to react with cytochrome a, it could exhibit slow autoxidation when incubated with the oxidase in the presence of traces of mammalian cytochrome c, the natural substrate for the oxidase. The case for functional independence of the two types of cytochrome c is completed by the demonstration that cytochrome f is located invariably in the chloroplast, whereas cytochrome c is found only in plant mitochondria (*53*).

Another striking example of variance in oxidase systems is afforded by the anaerobic *Staphylcocci*. In *S. aureus*, no cytochrome c occurs; only b-type cytochromes are found, together with cytochrome a and "o," and it is the cytochrome "o" that is functional as the oxidase, not the cytochrome a (*323*).

Finally, we may note that structural modifications and species differences in cytochrome c appear detectable by enzyme assay and immunochemical procedures, as well as by the present highly sensitive electrophoretic methods. Thus, the *D*-lactate-cytochrome c reductase of yeast distinguishes cytochrome c prepared from different sources, as well as heterogeneities in preparations of a given cytochrome c (*324*). In recent years, there have been reports of success in production of antibodies of various cytochromes c. If rabbits are subjected to a prolonged series of inoculations of horse-heart cytochrome c, or tuna cytochrome c conjugated with acetylated bovine γ-globulin, good, antibody titres are produced and active sera obtained (*325, 326*). These antibodies, although present in large amounts in the circulatory system of the inoculated animals, do not produce deleterious immune reactions, presumably because they do not reach the cytochrome c locked in the mitochondria. Bacterial cytochromes c, coming from tissues wholly unrelated serologically to mammalian blood plasma, can be expected to evoke good antibody production. In fact, they have been used in the earlier successful production of antisera in rabbits inoculated with cytochromes cc' and cytochrome c_2 from various photosynthetic bacteria (*327, 328*). These antisera have been exploited in a number of investigations on structural modifications of bacterial cytochrome c (*328*) and mechanisms of membrane synthesis in chromatophores (*329*).

Most recently, the cross reactivities of 26 different mammalian-type cytochromes c with antisera evoked by tuna, chicken, turkey, kangaroo, horse and human proteins have been examined (*330*). In general, the antisera show the greatest specificity when challenged by the particular protein used to elicit the antibody response. Competition for binding sites is the better, the more nearly the cytochromes c resemble each other in primary sequence. It is also found that the differences in conformation between ferri- and ferro- forms of horse-heart cytochrome c can be detected readily by complement-fixation procedures.

F. Molecular Archaeology

Among the fundamental questions in biology, none are more fascinating than those that arise concerning structure–function relations that underlie evolution and development of species at the molecular level. While it is not certain that evolution of the many complex traits of organisms can be related to evolutionary development of single proteins, nevertheless attempts to find correlations are indicated. In particular, the availability of a large number of amino-acid sequences of cytochromes c for different aerobic species encourages comparisons of these sequences of homologous proteins. These have led to some proposals concerning the relation of protein structure to the evolution of species. [For reviews, see Refs. (2) and (331).]

By fixing the N-terminal point at the glycine residue, usually the N-terminal acid in acylated form for the homologous series of mammalian-type cytochrome c, one can overlay all the sequences so as to reveal a strikingly large number of residues that remain constant at given positions in the sequence, despite the wide range in species. From such observations, it appears reasonable to conclude that a common origin for all mitochondrial cytochromes c existed in the past. There are 35 residues that remain unsubstituted: the glycines at positions 1, 6, 29, 34, 41, 45, 77, and 84; phenylalanines at positions 10 and 82; cysteines at positions 14 and 17; histidine at position 18; lysines at positions 27, 72, 73, 79, and 87; prolines at positions 30, 71, and 76; leucines at positions 32 and 68; isoleucine at position 75, arginines at positions 38 and 91; tyrosines at positions 48, 67, and 74; alanine at position 51; asparagines at positions 52 and 70; tryptophan at position 59; threonine at position 78, and methionine at position 80. The only long constant sequence strictly preserved is that of residues 70 through 80, consisting of Asn·Pro·Lys·Lys·Try·Ile·Pro·Gly·Thr·Lys·Met.

Comparisons of this type shed some light on specific questions, such as the nature of the ligand-binding groups, alpha-helix content essential for function, the hydrophobic clusters critical for structural integrity, etc. Thus, only one invariant histidine is found—namely, the histidine immediately adjacent to cysteine at the haem binding site. It follows that this histidine occupies one of the extraplanar positions of the haem as a requisite ligand in the cytochrome-c haemochrome structure. As noted previously (Section III.C), the variation in position of the extra histidines in many of the mammalian cytochromes (212), the constant appearance of methionine at position 80 in the invariant long sequence (212), and demonstrations that methionyl residues can coordinate to haem giving typical haemochrome spectra (211) reinforce the suggestion that the second ligand in horse-heart cytochrome c could be methionine, which binds by virtue of its sulfur function.

Likewise, the invariant appearance of the two cysteines near the N-terminal end, always separated by two amino-acid residues (which themselves are variable), implies that the two thioether bonds in the protein–haem complex remain the distinguishing characteristic of cytochrome c. This generalization, incidentally, holds for the two cytochromes c of bacterial origin mentioned above, although not for the variant cytochromes cc' (Section III.C). These deductions should be placed on firm ground when the tertiary structure determinations become available.

On the basis of the supposition that an increase of number of substitutions is an index of length of time for appearance in evolution of one species compared to another, phylogenetic trees can be constructed (332). These have permitted some interesting inferences about ancestral origins that appear remarkably consistent with deductions based on archaeological findings. Some deviations are noted, such as that kangaroo (nonplacental) is more nearly related to rabbit (placental) than rabbit to pig or dog (placental). Likewise, penguin seems closer to chicken than chicken to pigeon. However, these are rather few deviations to find when one considers that only a small portion of a single gene has been examined.

An effort to obtain phylogenetic relations from primary structures has been reported with some success for the case of fibrinopeptides from several ungulates (333).

Results obtained with bacterial cytochromes c are not sufficient at present for applications to molecular archaeology. However, the data available are sufficient to encourage some brief speculations, such as those presented in Section III.C.

Nevertheless, as has been emphasized by Margoliash and Smith (331), preoccupation with primary structure is unlikely to shed light on the manner in which selective pressures or functional demands operate to alter protein structure. It is not even certain that protein structure uniquely reflects evolutionary circumstances. The approach based on analyses of amino-acid sequence has a major disadvantage in practical terms. It requires isolation of relatively large amounts of pure protein. As an example, the original sequence determination of horse-heart cytochrome c required the dissipation of many grams of starting material. It is granted that, after the initial breakthrough occurred, Margoliash, Smith, and co-workers were able to determine any sequences from related source materials in a comparatively short time, using reasonably small amounts of protein.

The difficulties that existed in the early work on cytochrome c are multiplied manyfold in researches on primary structure in bacterial cytochromes c. It is not possible in most cases to obtain pure bulk cultures of bacteria of the types desired by a procedure as simple as that of going to the slaughterhouse and buying several kilos of muscle. The expense and labor incidental to maintenance and production of source materials from a large variety of bacterial species militate against development of a program of primary-structure analyses for bacterial cytochromes c like that which has supported the work on cytochrome c. It is possible, however, that the large amount of material needed may be reduced considerably by availability of automated installations, such as the stepwise sequence analysis machine developed by Edman (334).

We may mention other approaches to the comparative biochemistry of cytochrome *c* based on functionality rather than structure. Thus, there are numerous examples of the use of enzyme–substrate kinetic studies to determine structural variations. These have been applied with remarkable success in at least one series of studies on the evolution of species. Kaplan *et al.* showed that by use of coenzyme I analogues the lactate dehydrogenases isolated from heart and other organs of a great variety of species could be ordered on an evolutionary scale, according to characteristic rates of reactions with the coenzyme analogues (*335*). Similar results could be obtained by study of the cross reactivities of sera to purified enzyme (*336*).

In the case of cytochromes *c*, it has been shown that rates of oxidation of these proteins vary greatly when incubated with various oxidases. In particular, there are available studies on cross reactions of various cytochromes *c* with two extreme examples of oxidase enzymes. Yamanaka (*70, 178*) has exploited these phenomena to develop an approach to analyses of evolutionary development of cytochromes *c* that requires only trace amounts of pure material. He has used two oxidases as test systems for a large variety of purified cytochromes *c*. One oxidase is the soluble cytochrome *cd* of *Ps. aeruginosa*, which functions as a kind of primitive oxidase as well as a nitrite reductase (*70, 161*). It consists of a dihaem protein of molecular weight 80,000, in which the two prosthetic groups are a haem *c* and a haem *d*. It can be obtained in a crystalline state. The second oxidase is the conventional cytochrome *c* preparation of beef heart. The reactivities of the various ferro-cytochromes *c*, when incubated in air with either of these oxidases, have been examined. Studies have included a wide range of cytochromes *c*—44 in number at the last count—isolated from microorganisms, plants, and animals. The bacterial cytochromes *c* include specimens from *Pseudomonas* species, *Micrococcus* species, *B. subtilis*, and all groups of photosynthetic bacteria. Algal forms include representatives from the red algae, diatoms and *Euglena*. Five kinds of fungi are represented, i.e., yeast (*Saccharomyces*, *Kloeckera*, *Candida*) and molds (*Aspergillis* and *Physarum*). In addition, various sipunculids, molluscs, arthropods, fish, amphibia, birds, and mammals have been included.

When the relative velocities of oxidation of all these cytochromes *c* were examined in the presence of the *Pseudomonas* preparation, it was found that they could be classified in a series, based on reactivity, ranging over 5 orders of magnitude. The reference system was taken as the reactivity of the natural substrate, *Pseudomonas aeruginosa* cytochrome *c-551*, incubated with the *Pseudomonas* oxidase. Assuming that the more primitive the source material the faster it would react with this "primitive oxidase," it was found that denitrifying bacteria were the most primitive (most reactive), followed closely by the green sulfur photosynthetic bacteria. These all preceded the alga, which in turn preceded the aerobic organisms in the list. This finding was agreeable, as it accorded with the notion that oxygen production preceded development of aerobic organisms. The aerobe *B. subtilis* occupied a position close to the algae. The purple photosynthetic bacteria appeared as the most recent. (However, it should be recalled [Section III.C] that in some of these bacteria the commonly occurring cytochromes c_2 are accompanied by smaller *c*-type cytochromes that appear to resemble the *Pseudomonas* cytochrome *c*.) A number of other interesting conclusions have been presented and may be examined by reference to the original articles (*70, 178*).

Recently, data for a considerable extension of this approach have been supplied by Haneishi and Shirasaka (*117*), who examined cytochrome *c* isolated and purified from 14 genera of yeasts, including 104 strains tested for affinity to Amberlite CG-50, relative reactivities with two antisera prepared from *Saccharomyces* and *Candida* strains, quantitative precipitin reactions using antiserum against *Candida* cytochrome *c*, and the relative activities in two oxidase systems—one from *Saccharomyces oviformis S.* and the other the beef-heart cytochrome *c* oxidase. Four groups could be distinguished—one in particular, which reacted very weakly with both oxidases, included cytochromes *c* from eight strains of the genera *Schizosaccharomyces, Saccharomyces, Pichia, Lipomyces, Sporobolomyces, Candida,* and *Rhodotorula.* These data appear, on the basis of Yamanaka's criterion, to place some yeast strains with the purple photosynthetic bacteria, while others are placed either as originally by Yamanaka, or at other points intermediate between algae and the photosynthetic bacteria.

REFERENCES

The list of references that follows is not claimed to be complete. It is offered as a base for compilation of a complete bibliography to all who may be interested in pursuing further researches into the many areas of interest concerned directly or peripherally with the cytochromes *c*. Every effort has been made to check the accuracy of the references cited; many in our initial lists have been deleted because of inaccessibility. It would be fatuous to suppose, however, that no errors have crept into the text. For these, we apologize in advance.

1. D. Keilin, *The History of Cell Respiration and Cytochrome*, Cambridge Univ. Press, London and New York, 1966.

2. E. Margoliash and A. Schejter, in *Advances in Protein Chemistry* (C. B. Anfinsen *et al.*, eds.), Vol. 21, Academic, New York, 1966, pp. 114 *et seq.*

3. R. Lemberg and J. W. Legge, *Haematin Compounds and Bile Pigments*, Wiley-Interscience, New York, 1949.

4. R. K. Morton, *Rev. Pure Appl. Chem.*, **8,** 161 (1958).

5. J. E. Falk, R. Lemberg, and R. K. Morton, eds., *Haematin Enzymes*, Pergamon, Oxford, 1961.

6. B. Chance, R. W. Estabrook, and T. Yonetani, eds., *Hemes and Heme Proteins*, Academic, New York, 1966.

7. K. Okunuki, M. D. Kamen, and I. K. Sekuzu, eds. *Symposium on Structure and Function of Cytochromes, Osaka, 1967*, University of Tokyo Press, 1968.
8. M. Florkin and E. H. Stotz, eds., *Comprehensive Biochemistry*, Vol. 13, 2nd ed., Elsevier, Amsterdam, 1965, Chap. 5.
9. H. Tamiya and K. Tawaka, *Biochem. Z.*, **266**, 1 (1933).
10. A. Fujita and T. Kodama, *Biochem. Z.*, **273**, 186 (1934).
11. J. Postgate, *Biochem. J.*, **56**, xi (1954).
12. J. Postgate, *J. Gen. Microbiol.*, **15**, 186 (1956).
13. M. Ishimoto, J. Koyama, T. Omura, and Y. Nagai, *J. Biochem. Japan*, **41**, 537 (1954).
14. M. D. Kamen and L. P. Vernon, *Biochem. Biophys. Acta*, **17**, 10 (1958).
15. M. D. Kamen and L. P. Vernon, *J. Bacteriol.*, **67**, 617 (1954).
16. M. D. Kamen, in *Proc. Intern. Symp. Enzyme Chem., 1957*, Pergamon, London, 1958, pp. 245 *et seq.*
17. H. D. Peck, *Proc. Natl. Acad. Sci. (U.S.)*, **45**, 701 (1959); *Bacteriol Rev.*, **62**, 67 (1962).
18. S. Taniguchi, R. Sato, and F. Egami, in *Inorganic Nitrogen Metabolism* (W. D. McElroy and B. Glass, eds.), Johns Hopkins Press, Baltimore, Md., 1956, pp. 87 *et seq.* and pp. 163 *et seq.*
19. A. Nason, *Bacteriol. Rev.*, **26**, 16 (1962); S. Taniguchi, *Allgem. Mikrobiol.*, **1**, 341 (1961).
20. L. P. Vernon, *J. Biol. Chem.*, **222**, 1035 (1956).
21. M. D. Kamen, in *Biological Structure and Function* (T. W. Goodwin and O. Lindberg, eds.), Vol. II, Academic, New York, 1961, pp. 276 *et seq.*
22. R. Hill and F. Bendall, *Nature*, **186**, 136 (1960).
23. L. M. N. Duysens, *Nature*, **173**, 692 (1954).
24. J. M. Olson and B. Chance, *Arch. Biochem. Biophys.*, **88**, 26 (1960).
25. B. Chance and M. Nishumura, *Proc. Nat. Acad. Sci. (U.S.)*, **46**, 19 (1960).
26. M. D. Kamen, in *Bacterial Photosynthesis* (H. Gest, L. P. Vernon, and A. San Pietro, eds.), Antioch Press, Antioch, Ohio, 1963, pp. 61 *et seq.*
27. S. S. Deeb and L. P. Hager, *J. Biol. Chem.*, **239**, 1024 (1964).
28. W. D. Butt and D. Keilin, *Proc. Roy. Sci. (London)*, Ser. B, **156**, 429 (1962).
29. J. Duchesne, ed., *Structure and Properties of Biomolecules and Biological Systems*, Wiley-Interscience, New York, 1964, Chaps. 1 (J. I. Fernandez-Alonso), 3 (M. Kotani), 4 (G. Schoffa), 10 (P. S. Braterman, R. C. Davies, and R. J. P. Williams), and 16 (A. Ehrenberg).
30. H. Theorell and A. Åkesson, *J. Am. Chem. Soc.*, **63**, 1804 (1941).
31. H. Theorell, *Enzymol.*, **6**, 88 (1939).
32. H. Tuppy and G. Bodo, *Monatsh. Chem.*, **85**, 897, 1024, 1182 (1954).
33. H. Tuppy and K. Dus, *ibid.*, **89**, 407 (1959).
34. G. Leaf and N. E. Gillies, *Biochem. J.*, **61**, vii (1955).
35. E. Margoliash, N. Frohwirt, and E. Wiener, *Biochem. J.*, **71**, 559 (1959).
36. E. Margoliash, E. L. Smith, G. Kreil, and H. Tuppy, *Nature*, **192**, 1125 (1961).
37. E. Yakushiji and K. Okunuki, *Proc. Imp. Acad. (Tokyo)*, **16**, 299 (1940).
38. D. Keilin and E. F. Hartree, *Nature*, **164**, 254 (1940).
39. D. Keilin and E. F. Hartree, *ibid.*, **176**, 200 (1955).
40. R. W. Estabrook, *J. Biol. Chem.*, **227**, 1093 (1957).
41. K. Okunuki, I. Sekuzu, Y. Orii, M. Higuchi, S. Takemori, and T. Yonetani, *Proc. Jap. Acad.*, **34**, 379 (1958); see also K. Okunuki, Ref. (*8*), Vol. XIV, p. 240.
42. L. P. Vernon, *J. Biol. Chem.*, **222**, 1035, 1045 (1956).
43. K. Wada, H. Matsubara, and K. Okunuki, *vide*, Ref. (*7*), pp. 309 *et seq.*

44. R. Hill, *Proc. Roy. Soc. (London), Ser. B*, **120**, 472 (1939).
45. R. Hill and R. Scarisbrick, *New Phytol.*, **50**, 98 (1951).
46. H. E. Davenport and R. Hill, *Proc. Roy. Soc. (London), Ser. B*, **139**, 327 (1952).
47. E. Yakushiji, *Acta Phytochem.* **8**, 325 (1935).
48. S. Katoh, *Nature*, **186**, 138 (1960).
49. S. Katoh, *J. Biochem. (Japan)*, **46**, 629 (1959).
50. J. A. Gross and J. J. Wolken, *Science*, **132**, 357 (1960).
51. J. J. Wolken and J. A. Gross, *J. Protozool.*, **10**, 189 (1963).
52. G. Colmano and J. J. Wolken, *Nature*, **198**, 784 (1963).
53. F. Perini, M. D. Kamen, and J. A. Schiff, *Biochim. Biophys. Acta*, **88**, 74 (1964).
54. G. Forti, M. L. Bertolé, and B. Parisi, in *Photosynthetic Mechanism of Green Plants*, Publ. No. 1145, Natl. Acad. Sci., Nat. Res. Council, Washington, 1963, p. 284.
55. G. Forti, M. L. Bertolé, and G. Zanetti, *Biochim. Biophys. Acta*, **109**, 33 (1965).
56. E. Yakushiji, Y. Sugimura, F. Toda, and T. Furukawa, in Ref. (7).
57. R. W. Holton and J. Myers, *Science*, **142**, 234 (1963).
58. R. W. Holton and J. Myers, *Biochim. Biophys. Acta*, **131**, 375 (1967).
59. S. Katoh, *Plant Cell Physiol. (Tokyo)*, **1**, 91 (1960).
60. S. Katoh, *ibid.*, **1**, 29 (1959).
61. E. Yakushiji, Y. Sugimura, I. Sekuzu, I. Morikawa, and K. Okunuki, *Nature*, **185**, 105 (1960).
62. W. A. Susor and D. W. Krogmann, *Biochim. Biophys. Acta*, **120**, 65 (1966).
63. D. S. Gorman and P. Levine, *Plant Physiol.*, **41**, 1643 (1966).
64. A. Webster and D. P. Hackett, *Plant Physiol.*, **41**, 599 (1966).
65. S. I. Honda, *Plant Cell Physiol. (Tokyo)*, **2**, 151 (1961).
66. B. Hagihara, I. Sekuzu, K. Tagawa, M. Yoneda, and K. Okunuki, *Nature*, **181**, 1588 (1958).
67. B. Hagihara, K. Tagawa, I. Morikawa, M. Shin, and K. Okunuki, *ibid.*, **181**, 1656 (1958).
68. F. C. Stevens, A. N. Glaser, and E. L. Smith, *J. Biol. Chem.*, **242**, 2764 (1967).
69. C. Fridman, H. Lis, N. Sharon, and E. Katchalski, in Ref. (7), in press (1968).
70. T. Yamanaka, *Nature*, **213**, 1183 (1963).
71. T. Yamanaka and K. Okunuki, in Ref. (7), pp. 390 *et seq.* (1968).
72. T. Yamanaka and M. D. Kamen, *Biochim. Biophys. Acta*, **112**, 436 (1966).
72a. T. Yamanaka and M. D. Kamen, *Biochim. Biophys. Acta*, **143**, 416 (1967).
73. O. P. Bahl and E. L. Smith, *J. Biol. Chem.*, **240**, 3585 (1965).
74. T. Yamanaka, K. Okunuki, and T. Horio, *Biochim. Biophys. Acta*, **73**, 167 (1963).
75. G. Biorck and S. Paléus, *Nature*, **191**, 712 (1961).
76. S. K. Chan, I. Tulloss, and E. Margoliash, *Biochem.*, **5**, 2586 (1966).
77. T. Yamanaka, K. Miki, and K. Okunuki, *Biochim. Biophys. Acta*, **77**, 654 (1963).
78. T. Yamanaka, S. Tokuyama, and K. Okunuki, *ibid.*, **77**, 592 (1963).
79. R. W. Estabrook and B. Sacktor, *Arch. Biochem. Biophys.*, **76**, 532 (1958).
80. S. K. Chan and E. Margoliash, *J. Biol. Chem.*, **241**, 335 (1966).
81. D. G. Shappiric, *Proc. Roy. Soc. (London), Ser. B.*, **147**, 218 (1957).
82. H. Tuppy, *Z. Naturforsch.*, **126**, 784 (1958).
83. H. Tuppy and S. Paléus, *Acta. Chem. Scand.*, **9**, 353 (1955).
84. K. Ueda, *Compt. Rend. Soc. Biol.*, **153**, 1666 (1959).
85. R. Hess and A. G. F. Pearse, *Enzymol. Biol. Chem.*, **1**, 15 (1961).
86. H. H. Goldin and M. H. Farnsworth, *J. Biol. Chem.*, **241**, 3590 (1966).
87. E. Margoliash and O. Walasek, in *Methods in Enzymology* (S. P. Colowick and N. O. Kaplan, eds.), Vol. 10, 1967, 339.
88. S. Paléus, *Svensk. Kem. Tridskr.*, **67**, 6 (1955).

89. A. Ehrenberg and H. Theorell, *Acta. Chem. Scand.*, **9**, 1193 (1955).

90. S. Paléus, *ibid.*, **4**, 1024 (1950).

91. B. Hagihara, K. Tagawa, I. Morikawa, M. Shin, and K. Okunuki, *J. Biochem. (Japan)*, **45**, 725 (1958).

92. D. M. Blow, G. Bodo, M. G. Rossman, and C. P. S. Taylor, *J. Mol. Biol.*, **8**, 606 (1964).

93. F. Matsuura and K. Hashimoto, *Nippon Suisangaku Kaishi*, **24**, 216 (1958).

94. G. Kreil and H. Tuppy, *Nature*, **192**, 1123 (1961).

95. S. Vinogradov and H. A. Harbury, *Biochim. Biophys. Acta*, **115**, 494 (1966).

96. S. K. Chan and E. Margoliash, *J. Biol. Chem.*, **241**, 507 (1966).

97. E. Margoliash, *Acta. Chem. Scand.*, **17**, 5250 (1963).

98. S. Paléus, *ibid.*, **6**, 969 (1952).

99. G. Bodo, *Nature*, **176**, 829 (1955).

100. K. Narita and K. Titani, *Proc. Japan Acad.*, **41**, 831 (1965).

101. K. Narita and K. Sugeno, in Ref. (7), pp. 304 *et seq.* (1968).

102. T. Yamanaka, H. Mizushima, H. Katano, and K. Okunuki, *Biochim. Biophys. Acta*, **85**, 11 (1964).

103. K. Narita, K. Titani, Y. Yaoi, and H. Murakami, *ibid.*, **77**, 688 (1963).

104. K. Takahashi, K. Titani, and S. Minakawa, *J. Biochem. (Japan)*, **46**, 1323 (1959).

105. Y. Yaoi, *J. Biochem. (Japan)*, **61**, 54 (1967).

106. M. Nozaki, H. Mizushima, T. Horio, and K. Okunuki, *ibid.*, **45**, 815 (1958).

107. J. Lavollay, *Bull. Soc. Physiol. Veg.*, **12**, 81 (1966).

108. T. Yamanaka, H. Nakajima, and K. Okunuki, *Biochim. Biophys. Acta*, **63**, 510 (1962).

109. J. L. Van Etten, H. P. Molitoris, and D. Gottlieb, *J. Bacteriol.*, **91**, 169 (1966).

110. J. Heller and E. L. Smith, *J. Biol. Chem.*, **241**, 3158, 3165 (1966).

111. J. Heller and E. L. Smith, *Proc. Natl. Acad. Sci. U.S.*, **54**, 1621 (1965).

112. T. Yamanaka, T. Nishimura, and K. Okunuki, *J. Biochem. (Japan)*, **54**, 61 (1963).

113. T. P. Levchuk, L. V. Kolesnik, and A. B. Losinov, *Mikrobiol.*, **36**, 233 (1967).

114. J. B. Neilands, *J. Biol. Chem.*, **197**, 701 (1952).

115. K. M. Moller and D. M. Prescott, *Exptl. Cell. Res.*, **9**, 373 (1956).

116. I. F. Ryley, *Biochem. J.*, **52**, 483 (1952).

117. T. Haneishi and M. Shirasaka, in Ref. (7), p. 404 (1968).

118. D. I. Niederpruem, *Nature*, **184**, 1954 (1959).

119. T. Yamanaka, H. Mizushima, and K. Okunuki, *Biochim. Biophys. Acta*, **81**, 223 (1964).

120. T. Hoshi, *Sci. Repts., Tohoku Univ.*, **24**, 131 (1958).

121. T. Yamanaka and M. D. Kamen, *Biochim. Biophys. Acta*, **143**, 425 (1967).

122. T. Yamanaka, H. Mizushima, K. Miki, and K. Okunuki, *ibid.*, **81**, 386 (1964).

123. A. Ghiretti-Magaldi, A. Guiditta, and F. Ghiretti, *Biochem. J.*, **66**, 303 (1957).

124. K. T. Yasunobu, T. Nakashima, H. Higa, H. Matsubara, and H. M. Benson, *Biochim. Biophys. Acta*, **78**, 791 (1963).

125. S. Paléus and H. Theorell, *Acta. Chem. Scand.*, **11**, 905 (1957).

126. S. Paléus, *ibid.*, **14**, 1743 (1960).

127. B. Hagihara, T. Horio, J. Yamashita, M. Nozaki, and K. Okunuki, *Nature*, **178**, 629 (1956); also, see I. Marikawa and I. Sekuzu, *Nature*, **178**, 630, 631 (1956).

128. T. Nakashima, H. Higa, H. Matsubara, A. M. Benson, and K. T. Yasunobu, *J. Biol. Chem.*, **241**, 1166 (1966).

129. B. Hagihara, K. Tagawa, I. Morikawa, M. Shin, and K. Okunuki, *J. Biochem. (Japan)*, **46**, 11 (1959).

130. E. Bernstein and W. W. Wainio, *Arch. Biochem. Biophys.*, **91**, 138 (1960).

131. S. B. Needleman and E. Margoliash, *J. Biol. Chem.*, **241**, 853 (1966).

132. L. D'Alessandio, *Biochem. Appl.*, **10**, 231 (1963).

133. M. A. McDowell and E. L. Smith, *J. Biol. Chem.*, **240**, 4635 (1965).

134. H. Matsubara and E. L. Smith, *J. Biol. Chem.*, **238**, 2732 (1963).

135. S. Paléus, *Arch. Biochem. Biophys.*, **96**, 60 (1962).

136. S. K. Chan, S. B. Needleman, J. W. Stewart, and E. Margoliash, in Ref. (*2*), p. 176.

137. J. A. Rothfus and E. L. Smith, *J. Biol. Chem.*, **240**, 4277 (1965).

138. A. Goldstone and E. L. Smith, *J. Biol. Chem.*, **241**, 4480 (1966).

139. J. W. Stewart and E. Margoliash, *Can. J. Biochem.*, **43**, 1187 (1965).

140. C. Nolan and E. Margoliash, *J. Biol. Chem.*, **241**, 1049 (1966).

141. M. Ishimoto, J. Koyama, T. Yagi, and M. Shiraki, *J. Biochem.*, **44**, 233, 413 (1957).

142. T. Horio and M. D. Kamen, *Biochim. Biophys. Acta*, **48**, 266 (1961).

143. J. Miller, *J. Gen. Microbiol.*, **37**, 419 (1964).

144. K. Dus and M. D. Kamen, *Biochem. Z.*, **338**, 364 (1963).

145. H. Drucker and L. L. Campbell, private communication.

146. J. LeGall and M. Bruschi-Heriaud, in Ref. (*7*), p. 467 (1968).

147. J. LeGall, G. Mazza, and N. Dragoni, *Biochim. Biophys. Acta*, **99**, 385 (1965).

148. H. D. Peck, Jr., *Biochem. Biophys. Res. Commun.*, **22**, 112 (1966).

149. M. Bruschi-Heriaud, *Contribution de la Biochimie Comparée de Cytochromes Bactériens de Type C* (Doctoral Thesis, 3rd Cycle, Fac. Sci. Marseilles, 1969).

150. P. Trudinger, *Biochim. Biophys. Acta*, **30**, 211 (1958).

151. T. Cook and W. W. Umbreit, *Biochem.*, **2**, 194 (1963).

152. T. W. Szcepskowski and B. Skarzynski, *Bull. Acad. Polon. Sci.*, **1**, 93 (1952).

153. R. Klimek, B. Skarzynski, and T. W. Szcepskowski, *Acta. Biochim. Polon.*, **3**, 261 (1956).

154. J. P. Aubert, G. Milhaud, C. Moncel, and J. Millet, *Compt. Rend. Acad. Sci.*, **246**, 1616, 1766 (1958).

155. J. P. Aubert, J. Millet, and G. Milhaud, *Ann. Inst. Pasteur*, **96**, 559 (1959).

156. N. P. Neumann and R. H. Burris, *J. Biol. Chem.*, **234**, 3286 (1959).

157. Y. I. Shethna, P. W. Wilson, and H. Beinert, *Biochim. Biophys. Acta*, **113**, 225 (1966).

158. A. Tissieres, *Biochem. J.*, **64**, 582 (1956).

159. H. Suzuki and H. Iwasaki, *J. Biochem. (Japan)*, **52**, 193 (1962).

160. R. P. Ambler, *Biochem. J.*, **89**, 341, 349 (1963).

161. T. Horio, T. Higashi, M. Sasagawa, K. Kusai, M. Nakai, and K. Okunuki, *Biochem. J.*, **77**, 194 (1960).

162. M. Coval, T. Horio, and M. D. Kamen, *Biochim. Biophys. Acta*, **51**, 246 (1961).

163. A. Asano and A. F. Brodie, *J. Bio. Chem.*, **239**, 4280 (1964).

164. D. C. White and L. Smith, *J. Biol. Chem.*, **237**, 1332, 1337 (1962).

165. T. Mori and K. Hirai, in Ref. (*7*), p. 681 (1968).

166. K. Hori, *J. Biochem. (Japan)*, **50**, 440 (1961).

167. K. Hori, *ibid.*, **53**, 354 (1963).

168. D. H. Bone, *Nature*, **197**, 517 (1963).

169. L. P. Vernon, J. H. Mangum, J. V. Beck, and F. M. Shafia, *Arch. Biochem. Biophys.*, **88**, 227 (1960).

170. L. Packer, *ibid.*, **78**, 54 (1958).

171. I. Sutherland, *Biochim. Biophys. Acta*, **73**, 162 (1963).

172. C. Appleby and F. J. Bergersen, *Nature*, **182**, 1174 (1958).

173. N. S. Gellman, *Biokhim.*, **25**, 482 (1960).

174. I. C. Sadana and W. D. McElroy, *Arch. Biochem. Biophys.*, **67**, 16 (1957).

175. T. Yamaguchi, G. Tamura, and K. Arima, *Biochim. Biophys. Acta*, **124**, 413 (1966).

176. T. Yamanaka and K. Okunuki, *J. Biol. Chem.*, **239**, 1813 (1964).

177. L. P. Vernon and J. H. Mangum, *Arch. Biochem. Biophys.*, **90**, 103 (1960).

178. T. Yamanaka, *Ann. Rept. Sci. Works, Fac. Sci., Osaka Univ.*, **11**, 77 (1963).

179. T. Fujita and R. Sato, *Biochim. Biophys. Acta*, **77**, 693 (1963).

180. C. T. Gray, J. W. T. Wimpenny, D. E. Hughes, and M. Ramlett, *ibid.*, **67**, 157 (1963).

181. H. Lenhoff, *Nature*, **199**, 601 (1963).

182. T. Fujita, *J. Biochem. (Japan)*, **60**, 304, 329, 568, 691 (1966).

183. M. Richmond and N. O. Kjelgard, *Acta Chem. Scand.*, **15**, 226 (1961).

184. L. P. Vernon, *J. Biol. Chem.*, **222**, 1045 (1956).

185. L. P. Vernon and F. G. White, *loc. cit.*, **25**, 321 (1957).

186. M. D. Kamen and L. P. Vernon, *ibid.*, **17**, 10 (1955).

187. M. Kono and S. Taniguchi, *Biochim. Biophys. Acta*, **43**, 419 (1960).

188. H. R. Hayward and T. C. Stadtman, *J. Bacteriol.*, **78**, 557 (1959).

189. J. M. Olson and E. Stanton, in press (1967).

190. T. Meyer, R. G. Bartsch, M. A. Cusanovich, and J. H. Mathewson, *Biochem. Biophys. Acta*, **153**, 854 (1968).

191. K. Dus, R. G. Bartsch, and M. D. Kamen, *J. Biol. Chem.*, **237**, 3083 (1962).

192. K. Dus, K. Sletten, K. de Klerk, and M. D. Kamen, *J. Biol. Chem.*, in press (1968).

193. R. G. Bartsch, T. A. Meyer, and A. B. Robinson, in Ref. (7), in press (1968).

194. J. Gibson, *Biochem. J.*, **79**, 151 (1961).

195. R. G. Bartsch and M. D. Kamen, *J. Biol. Chem.*, **235**, 825 (1968).

196. M. Cusanovitch and M. D. Kamen, *Biochim. Biophys. Acta*, in press (1968).

197. K. Sletten and M. D. Kamen, in Ref. (7), p. 422 (1968).

198. K. Dus, T. Flatmark, H. de Klerk, and M. D. Kamen, in preparation (1968).

199. K. Dus, H. de Klerk, K. Sletten, and M. D. Kamen, in preparation (1968).

200. R. G. Bartsch and R. Chenn, unpublished.

201. H. de Klerk, R. G. Bartsch, and M. D. Kamen, *Biochim. Biophys. Acta*, **97**, 275 (1965).

202. S. Taniguchi and M. D. Kamen, *Biochim. Biophys. Acta*, **74**, 438 (1963).

203. K. Dus, H. de Klerk, R. G. Bartsch, T. Horio, and M. D. Kamen, *Proc. Natl. Acad. Sci. U.S.*, **57**, 367 (1967).

204. R. W. Henderson and D. K. Nankiville, *Biochem. J.*, **98**, 587 (1966).

205. R. G. Bartsch and T. Horio, unpublished.

206. J. A. Orlando, *Biochim. Biophys. Acta*, **57**, 373 (1962).

207. R. G. Bartsch and K. Dus, unpublished.

207a. Y. Motokawa and G. Kikuchi, *Biochim. Biophys. Acta*, **120**, 274 (1966).

208. S. Morita and S. F. Conti, *Arch. Biochem. Biophys.*, **100**, 302 (1963).

208a. K. Dus and H. de Klerk, unpublished.

209. K. Dus, K. Sletten, and M. D. Kamen, *J. Biol. Chem.*, **243**, 5507 (1968).

210. T. Flatmark and A. B. Robinson, in Ref. (7), p. 422 (1968).

211. H. A. Harbury, J. R. Cronin, M. W. Fanger, T. P. Hettinger, A. J. Murphy, Y. P. Myer, and S. N. Vinogradov, *Proc. Natl. Acad. Sci. U.S.*, **54**, 1658 (1965).

212. E. L. Smith, in Ref. (7), in press (1968); also E. L. Smith, *Harvey Lectures*, in press (1968).

213. J. Kraut, S. Singh, and E. A. Alden, in Ref. (7), in press (1968).

214. R. E. Dickerson, M. L. Kopka, J. Weinzierl, J. Varnum, D. Eisenberg, and E. Margoliash, in Ref. (7), p. 225 (1968).

215. K. Dus, unpublished.
216. J. Barrett and M. D. Kamen, *Biochim. Biophys. Acta,* **50,** 573 (1961).
217. M. Morrison, K. Dus, and M. D. Kamen, in preparation (1968).
218. C. R. Cantor and T. H. Jukes, *Information Exch. Group #7,* (NIH) #331.
219. C. R. Cantor and T. H. Jukes, *Proc. Natl. Acad. Sci. U.S.,* **56,** 177 (1966).
220. P. P. Slonimski, R. Acher, G. Pere, A. Sels, and M. Somlo, in *Mechanismes de Regulation des Activites Cellulaires chez les Microorganisms, Marseille, 1963,* Centre National de la Recherche Scientifique, Paris, 1965, p. 435.
221. T. Flatmark, *Acta. Chem. Scand.,* **18,** 1956 (1964).
222. H. W. Taber and F. Sherman, *Ann. N.Y. Acad. Sci.,* **121,** 600 (1964).
223. T. Flatmark and O. Vesterberg, *Acta. Chem. Scand.,* **20,** 1497 (1966).
224. D. Keilin and E. F. Hartree, *Proc. Roy. Soc. (London), Ser. B,* **122,** 298 (1937).
225. S. Paléus and J. B. Neilands, *Acta. Chem. Scand.,* **4,** 1024 (1950).
226. E. Margoliash, *Biochem. J.,* **56,** 535 (1954).
227. R. W. Henderson and W. A. Rawlinson, *Biochem. J.,* **62,** 21 (1956).
228. N. K. Boardman and S. M. Partridge, *Nature,* **171,** 208 (1953).
229. N. K. Boardman and S. M. Partridge, *Biochem. J.,* **59,** 543 (1955).
230. M. Nozaki, *J. Biochem. (Tokyo),* **47,** 592 (1960).
231. E. Margoliash and J. Lustgarten, *J. Biol. Chem.,* **237,** 3397 (1962).
232. T. Flatmark, unpublished experiments, 1966.
233. A. Schejter, S. C. Glauser, P. George, and E. Margoliash, *Biochim. Biophys. Acta,* **73,** 641 (1963).
234. R. Havez, A. Hayem-Levy, J. Mizon, and G. Biserte, *Bul. Soc. Chim. Biol.,* **48,** 117 (1966).
235. R. W. Henderson and W. A. Rawlinson, in *Haematin Enzymes* (J. E. Falk, R. Lemberg, and R. K. Morton, eds.), Part 2, Pergamon, Oxford, 1961, p. 370.
236. E. Margoliash, *ibid.,* Part 2, Pergamon, Oxford, 1961, p. 390.
237. N. Frohwirt, *The Relation between the Structure and Enzymatic Activity of Cytochrome c,* Ph.D. Thesis, Hebrew Univ., Jerusalem, 1961.
238. M. Nozaki, T. Yamanaka, T. Horio and K. Okunuki, *J. Biochem. (Tokyo),* **44,** 453 (1957).
239. M. Nozaki, H. Mizushima, T. Horio, and K. Okunuki, *J. Biochem. (Tokyo),* **45,** 815 (1958).
240. J. McD. Armstrong, J. H. Coates, and R. K. Morton, in *Haematin Enzymes* (J. E. Falk, R. Lemberg, and R. K. Morton, eds.), Part 2, Pergamon, Oxford, 1961, p. 386.
241. K. Motonaga, E. Misaka, E. Nakajima, S. Ueda, and K. Nakanishi, *J. Biochem. (Tokyo),* **57,** 22 (1965).
242. C. L. Markert and F. Möller, *Proc. Natl. Acad. Sci. U.S.,* **45,** 753 (1959).
243. A. A. Sels, H. Fukuhara, G. Pere, and P. P. Slonimski, *Biochim. Biophys. Acta,* **95,** 486 (1965).
244. J. W. Stewart, E. Margoliash, and F. Sherman, *Federation Proc.,* **25,** 647 (1966).
245. J W. Stewart, G. J. Putterman, E. Margoliash, and F. Sherman, in preparation [cit. Ref. (*246*)].
246. F. Sherman, J. W. Stewart, J. Parker, G. J. Putterman, and E. Margoliash, in Ref. (*7*), p. 257 (1968).
247. H. Fukuhara, *J. Mol. Biol.,* in press (1967).
248. H. Fukuhara, L. Clavilier, G. Pere, and P. P. Slonimski, *J. Mol. Biol.,* in press (1967).
249. J. R. Mattoon, and F. Sherman, *J. Biol. Chem.,* **241,** 4330 (1966).

250. M. Pradines, *Analyse des facteurs limitants in situ la respiration de la levure, Oxydation du D- et du L-lactate par les mutants de l'iso-2 cytochrome C*, These de 3² cycle, Paris, 1966.
251. M. Somlo, *Étude physiologique des trois lacticodeshydrogenases de la levure*, These de Doctorat, Paris, 1967.
252. M. Huang, D. R. Biggs, G. D. Clark-Walker, and A. W. Linnane, *Biochim. Biophys. Acta*, **114**, 434 (1966).
253. G. D. Clark-Walker and A. W. Linnane, *Biochem. Biophys. Res. Commun.*, **25**, 8 (1966).
254. G. D. Clark-Walker and A. W. Linnane, *J. Cell. Biol.*, **34**, 1 (1967).
255. A. W. Linnane, D. R. Biggs, M. Huang, and G. D. Clark-Walker, in *Aspects of Yeast Metabolism* (R. K. Mills, ed.), Blackwell Scientific Publ., Oxford, 1966.
256. P. J. Rogers, B. N. Preston, E. B. Titchener, and A. W. Linnane, *Biochem. Biophys. Res. Commun.*, **27**, 405 (1967).
257. L. W. Wheeldon and A. L. Lehninger, *Biochem.*, **5**, 3533 (1966).
258. M. V. Simpson, D. M. Skinner, and J. M. Lucas, *J. Biol. Chem.*, **236**, PC81 (1961).
259. D. B. Roodyn, J. W. Suttie, and T. S. Work, *Biochem. J.*, **83**, 29 (1962).
260. N. F. Gonzalez Cadavid and P. N. Campbell, *Biochem. J.*, **102**, 38P–39P (1967).
261. B. Kadenbach, *Biochim. Biophys. Acta*, **138**, 651 (1967).
262. F. Sherman, in *Mechanismes de Regulation des Activities Cellulaires chez les Microorganisms, Marseille, 1963*, Centre National de la Recherche Scientifique, Paris, 1965, p. 465.
263. F. Sherman, *Genetics*, **49**, 39 (1964).
264a. F. Sherman and P. P. Slonimski, *Biochim. Biophys. Acta*, **90**, 1 (1964).
264b. F. Sherman, H. Taber, and W. Campbell, *J. Mol. Biol.*, **13**, 21 (1965).
265. L. Clavilier, G. Pere, P. P. Slonimski, and M. Somlo, *Proc. 6th Intern. Congr. Biochem. New York, 1964*, Abstr., p. 673.
266. G. Pere, L. Clavilier and P. P. Slonimski, *Ann. Genet.*, **8**, 112 (1965).
267. L. Clavilier, H. Fukuhara, G. Pere, M. Pradines, M. Somlo, and P. P. Slonimski, *Bull. Soc. Franc. Physiol. Veg.*, **12**, 7 (1966).
268. H. Fukuhara, *Synthese des macromolecules au cours de l'adaptation respiratorie de la levure*, These de Doctorat, Paris, 1967.
269. F. Sherman, J. W. Stewart, E. Margoliash, J. Parker, and W. Campbell, *Proc. Natl. Acad. Sci. U.S.*, **55**, 1498 (1966).
270. A. Lindenmayer and L. Smith, *Biochim. Biophys. Acta*, **93**, 445 (1964).
271. A. W. Linnane, in *Oxidases and Related Redox Systems* (T. E. King, H. S. Mason, and M. Morrison, eds.), Vol. 2, Wiley, New York, 1965, p. 1102.
272. M. Rabinowitz, E. Snoble, P. Sanghavi, G. S. Getz, and J. Heywood, *Proc. 4th Meeting, Federation European Biochem. Soc., Oslo, 1967*, Abstr., p. 46.
273. E. S. Polakis, W. Bartley, and G. A. Meek, *Biochem. J.*, **90**, 369 (1964).
274. E. S. Polakis, W. Bartley, and G. A. Meek, *ibid.*, **97**, 298 (1965).
275. G. A. Sulebele and D. V. Rege, *Nature*, **215**, 420 (1967).
276. H. G. Crabtree, *Biochem. J.*, **23**, 539 (1929).
277. M. Brin and R. W. McKee, *Cancer Res.*, **16**, 364 (1956).
278. A. Gibor and S. Granick, *Science*, **145**, 890 (1964).
279. R. K. Mortimer and D. C. Hawthorne, *Ann. Rev. Microbiol.*, **20**, 151 (1966).
280. B. Chance and G. R. Williams, *Advan. Enzymol.*, **17**, 65 (1956).
281. D. Pette and T. Bücher, *Z. Physiol. Chem.*, **331**, 180 (1963).
282. A. L. Lehninger, in *Molecular Organization and Biological Function* (J. M. Allen, ed.), Harper & Row, New York, 1966, p. 107.
283. H. Fukuhara and A. Sels, *J. Mol. Biol.*, **17**, 319 (1966).

284. H. Fukuhara, *J. Mol. Biol.*, **17**, 334 (1966).
285. F. Jacob and J. Monod, in *Cytodifferentiation and Macromolecular Synthesis* (M. Locke, ed.), Academic, New York, 1963, p. 30.
286. T. Flatmark, *Acta Chem. Scand.*, **20**, 1476 (1966).
287. T. Flatmark, *Acta. Chem. Scand.*, **20**, 1487 (1966).
288. T. Flatmark, in Ref. (*6*), p. 411.
289. T. Flatmark, *J. Biol. Chem.*, **242**, 2454 (1967).
290. T. Flatmark and K. Sletten, in Ref. (*7*), p. 413 (1968).
291. T. Flatmark, unpublished experiments, 1967.
292. H. Svensson, *Arch. Biochem. Biophys.*, *Suppl. 1*, 132 (1962).
293. O. Vesterberg and H. Svensson, *Acta. Chem. Scand.*, **20**, 820 (1966).
294. Q. Gibson and M. D. Kamen, *J. Biol. Chem.*, **241**, 1969 (1966).
295. R. Hill and D. Keilin, *Proc. Roy. Soc.* (*London*), *Ser. B*, **107**, 286 (1930).
296. K. G. Paul, *Acta Chem. Scand.*, **4**, 239 (1950).
297. S. Sano, N. Nanzyo, and C. Rimington, *Biochem. J.*, **93**, 270 (1964).
298. S. Sano and K. Tanaka, *J. Biol. Chem.*, **239**, PC3109 (1964).
299. C. L. Tsou, *Biochem. J.*, **50**, 493 (1952).
300. D. Keilin and E. F. Hartree, *Nature*, **176**, 200 (1955).
301. D. Keilin and T. E. King, *ibid.* **181**, 1520 (1958).
302. R. W. Estabrook, in Ref. (*5*), p. 436.
303. P. Nicholls, *Arch. Biochem. Biophys.*, **106**, 25 (1964).
304. L. Smith and K. Minnaert, *Biochim. Biophys. Acta*, **105**, 1 (1965).
305. L. Smith and P. W. Camerino, *Biochem.*, **2**, 1432 (1963).
306. S. Taniguchi and M. D. Kamen, *Biochim. Biophys. Acta*, **96**, 395 (1965).
307. R. P. Ambler, *Biochem. J.*, **109**, 47P (1968).
308. D. M. Geller, *J. Biol. Chem.*, **237**, 2947 (1962).
309. J. Postgate, in Ref. (*5*), Vol. II, p. 407.
310. F. Egami, M. Ishimoto, and S. Taniguchi, *ibid.*, Vol. II, p. 392.
311. T. Fujita and R. Sato, *J. Biochem.* (*Japan*), **60**, 204, 691 (1966).
312. S. P. Williams, J. T. Davidson, and H. D. Peck, Jr., *Bacteriol. Proc.*, 110 (1964).
313. C. T. Gray and J. O'Hara, in Ref. (*7*), in press (1968).
314. H. Gest and M. D. Kamen, in *Handbuch d. Pflanz. Physiologie*, Vol. V, 1960, p. 559.
315. L. N. Castor and B. Chance, *J. Biol. Chem.*, **234**, 1587 (1959).
316. L. M. N. Duysens, in *Research in Photosynthesis* (H. Gaffron *et al.*, eds.), Wiley-Interscience, New York, 1957, p. 16A.
317. R. Hill and F. Bendall, *ibid.*, **186**, 136 (1960).
318. M. Nishimura and B. Chance, *Biochim. Biophys. Acta*, **66**, 1 (1963).
319. J. M. Olson and B. Chance, *Arch. Biochem.*, **88**, 26 (1960).
320. D. De Vault, in Ref. (*7*), in press (1968).
321. D. De Vault and B. Chance, *Nature*, **215**, 642 (1967).
322. G. Forti, *Brookhaven Symp. Biol.*, **19**, 195 (1966).
323. H. Taber and M. Morrison, *Arch. Biochem. Biophys.*, **105**, 367 (1964).
324. C. Gregolin and T. S. Singer, *Nature*, **193**, 659 (1962).
325. Y. Okada, S. Watanabe, and T. Yamanaka, *J. Biochem.* (*Japan*), **55**, 342 (1964).
326. M. Reichlin, S. Fogel, A. Nisonoff, and E. Margoliash, *J. Biol. Chem.*, **241**, 251 (1966).
327. J. W. Newton and L. Levine, *Arch. Biochem. Biophys.*, **83**, 456 (1959).
328. J. A. Orlando, L. Levine, and M. D. Kamen, *Biochim. Biophys. Acta*, **46**, 126 (1961).

329. J. W. Newton, *ibid.*, **58,** 474 (1962).

330. E. Margoliash, M. Reichlin, and A. Nisonoff, in Ref. (7), in press (1968).

331. E. Margoliash and E. L. Smith, in *Evolving Genes and Proteins* (V. Bryson and H. D. Vogel, eds.), Academic, New York, 1965, p. 221.

332. W. M. Fitch and E. Margoliash, *Nature,* **155,** 279 (1967).

333. R. F. Doolittle and B. Blomback, *ibid.*, **202,** 147 (1964).

334. P. Edman and G. Begg, *European J. Biochem.*, **1,** 80 (1967).

335. N. O. Kaplan, M. M. Ciotti, M. Hanolsky, and R. E. Bieber, *Science,* **131,** 392 (1960).

336. A. C. Wilson, N. O. Kaplan, L. Levine, A. Pesce, M. Reichlin, and W. S. Allison, *Federation Proc.*, **23,** 1258 (1964).